MECHANICS OF MATERIALS

CRC Series in
COMPUTATIONAL MECHANICS and APPLIED ANALYSIS
Series Editor: J.N. Reddy, *Texas A&M University*

Published Titles

ADVANCED THERMODYNAMICS ENGINEERING, Second Edition
Kalyan Annamalai, Ishwar K. Puri, and Miland Jog

APPLIED FUNCTIONAL ANALYSIS
J. Tinsley Oden and Leszek F. Demkowicz

COMBUSTION SCIENCE AND ENGINEERING
Kalyan Annamalai and Ishwar K. Puri

**COMPUTATIONAL MODELING OF POLYMER COMPOSITES:
A STUDY OF CREEP AND ENVIRONMENTAL EFFECTS**
Samit Roy and J.N. Reddy

CONTINUUM MECHANICS FOR ENGINEERS, Third Edition
Thomas Mase, Ronald Smelser, and George E. Mase

DYNAMICS IN ENGINEERING PRACTICE, Tenth Edition
Dara W. Childs

EXACT SOLUTIONS FOR BUCKLING OF STRUCTURAL MEMBERS
C.M. Wang, C.Y. Wang, and J.N. Reddy

**THE FINITE ELEMENT METHOD IN HEAT TRANSFER AND FLUID DYNAMICS,
Third Edition**
J.N. Reddy and D.K. Gartling

MECHANICS OF MATERIALS
Clarence W. de Silva

**MECHANICS OF LAMINATED COMPOSITE PLATES AND SHELLS:
THEORY AND ANALYSIS, Second Edition**
J.N. Reddy

**MICROMECHANICAL ANALYSIS AND MULTI-SCALE MODELING USING THE
VORONOI CELL FINITE ELEMENT METHOD**
Somnath Ghosh

NUMERICAL AND ANALYTICAL METHODS WITH MATLAB®
William Bober, Chi-Tay Tsai, and Oren Masory

**NUMERICAL AND ANALYTICAL METHODS WITH MATLAB®
FOR ELECTRICAL ENGINEERS**
William Bober and Andrew Stevens

PRACTICAL ANALYSIS OF COMPOSITE LAMINATES
J.N. Reddy and Antonio Miravete

**SOLVING ORDINARY AND PARTIAL BOUNDARY VALUE PROBLEMS IN SCIENCE
AND ENGINEERING**
Karel Rektorys

STRESSES IN BEAMS, PLATES, AND SHELLS, Third Edition
Ansel C. Ugural

MECHANICS OF MATERIALS

CLARENCE W. DE SILVA

CRC Press
Taylor & Francis Group
Boca Raton   London   New York

CRC Press is an imprint of the
Taylor & Francis Group, an **informa** business

CRC Press
Taylor & Francis Group
6000 Broken Sound Parkway NW, Suite 300
Boca Raton, FL 33487-2742

© 2014 by Taylor & Francis Group, LLC
CRC Press is an imprint of Taylor & Francis Group, an Informa business

No claim to original U.S. Government works

Printed on acid-free paper
Version Date: 20130531

International Standard Book Number-13: 978-1-4398-7736-4 (Hardback)

This book contains information obtained from authentic and highly regarded sources. Reasonable efforts have been made to publish reliable data and information, but the author and publisher cannot assume responsibility for the validity of all materials or the consequences of their use. The authors and publishers have attempted to trace the copyright holders of all material reproduced in this publication and apologize to copyright holders if permission to publish in this form has not been obtained. If any copyright material has not been acknowledged please write and let us know so we may rectify in any future reprint.

Except as permitted under U.S. Copyright Law, no part of this book may be reprinted, reproduced, transmitted, or utilized in any form by any electronic, mechanical, or other means, now known or hereafter invented, including photocopying, microfilming, and recording, or in any information storage or retrieval system, without written permission from the publishers.

For permission to photocopy or use material electronically from this work, please access www.copyright.com (http://www.copyright.com/) or contact the Copyright Clearance Center, Inc. (CCC), 222 Rosewood Drive, Danvers, MA 01923, 978-750-8400. CCC is a not-for-profit organization that provides licenses and registration for a variety of users. For organizations that have been granted a photocopy license by the CCC, a separate system of payment has been arranged.

Trademark Notice: Product or corporate names may be trademarks or registered trademarks, and are used only for identification and explanation without intent to infringe.

Library of Congress Cataloging-in-Publication Data

De Silva, Clarence W.
   Mechanics of materials / Clarence W. de Silva.
      pages cm. -- (Computational mechanics and applied analysis)
   Includes bibliographical references and index.
   ISBN 978-1-4398-7736-4 (hardback)
   1. Materials--Mechanical properties. I. Title.

TA404.8.D43 2014
620.1'1292--dc23   2013014811

Visit the Taylor & Francis Web site at
http://www.taylorandfrancis.com

and the CRC Press Web site at
http://www.crcpress.com

In memory of my parents for all that I have achieved in life.

*Science can amuse and fascinate us all, but it is
engineering that changes the world.*

Isaac Asimov, 1988

Contents

Preface xiii
Acknowledgments xv
Author xvii

Chapter 1 Mechanics of Materials 1

    Chapter Objectives 1
    1.1   What Is Mechanics of Materials? 1
    1.2   Subject Definition 1
    1.3   Application of the Subject 3
    1.4   Applicable Engineering Fields 4
    1.5   Useful Terms 5
    1.6   History of Mechanics of Materials 7
    1.7   Basic Problem Scenarios 8
    1.8   Problem Solution 8
            1.8.1   Problem Solution Steps 10
    1.9   Organization of the Book 10
    Problems 12

Chapter 2 Statics: A Review 15

    Chapter Objectives 15
    2.1   Statics 15
            2.1.1   Equations of Equilibrium 15
    2.2   Support Reactions 18
            2.2.1   Free-Body Diagrams 18
            2.2.2   Principle of Transmissibility 22
    2.3   Analysis of Trusses 24
            2.3.1   Method of Joints 24
            2.3.2   Method of Sections 25
            2.3.3   Two-Force Members 25
    2.4   Distributed Forces 34
    2.5   Statically Indeterminate Structures 44
    Problems 47

Chapter 3 Stress 57

    Chapter Objectives 57
    3.1   Introduction 57
    3.2   Definition of Stress 58
    3.3   Normal Stress under Axial Loading 60
            3.3.1   Solution Steps for Stress Problems 62
            3.3.2   Axial Force Diagram 66
    3.4   Bearing Stress 68

3.5 Shear Stress 71
    3.5.1 Average Shear Stress 71
    3.5.2 Single Shear and Double Shear Connectors...................................77
    3.5.3 Shear Stress in a Key...78
    3.5.4 Complementarity Property of Shear Stress.....................................80
3.6 Stress Transformation in a Bar under Axial Loading 82
Problems..89

Chapter 4   Strain 99

Chapter Objectives 99
4.1 Introduction 99
    4.1.1 Types of Strain 100
4.2 Normal Strain...100
    4.2.1 Local Normal Strain and Average Normal Strain 100
        4.2.1.1 Average Normal Strain 102
4.3 Shear Strain 108
    4.3.1 Local Shear Strain and Average Shear Strain................................108
        4.3.1.1 Average Shear Strain 109
4.4 Thermal Strain..114
    4.4.1 Coefficient of Thermal Expansion (α) 114
4.5 Measurement of Strain 115
    4.5.1 Bridge Circuit 116
    4.5.2 Bridge Constant 118
    4.5.3 Calibration Constant..119
Problems..120

Chapter 5   Mechanical Properties of Materials 133

Chapter Objectives 133
5.1 Introduction 133
    5.1.1 Problem of Mechanics of Materials 133
    5.1.2 Homogeneity and Isotropy 134
5.2 Stress–Strain Behavior...134
    5.2.1 Tensile Test 134
    5.2.2 Stress–Strain Diagram 136
5.3 Stress–Strain Characteristics..137
    5.3.1 Strain Hardening (Work Hardening) 140
    5.3.2 Necking 142
    5.3.3 True Stress–Strain Diagram 142
5.4 Hooke's Law 143
5.5 Poisson's Ratio...151
5.6 Material Types and Behavior...155
    5.6.1 Ductile Materials..155
    5.6.2 Ductility Measures 156
    5.6.3 Brittle Materials 157
    5.6.4 Hardness 157
    5.6.5 Creep 157
    5.6.6 Fatigue 158
5.7 Strain Energy..158
    5.7.1 Strain Energy in Shear 161

Contents

		5.7.2	Modulus of Resilience 161
		5.7.3	Modulus of Toughness 162
	Problems 165

Chapter 6 Axial Loading 179
    Chapter Objectives 179
    6.1   Introduction 179
          6.1.1   Basic Types of Loading 180
          6.1.2   Chapter Objectives 181
    6.2   Saint-Venant's Principle 181
    6.3   Axially Loaded Member 183
          6.3.1   Continuously Varying Nonuniform Section 184
          6.3.2   Multiple Segments of Uniform Cross Section 193
    6.4   Principle of Superposition 195
          6.4.1   Linear Elastic Systems 195
          6.4.2   Load Reversal 196
          6.4.3   Principle of Superposition 196
          6.4.4   Summary of PoS 199
    6.5   Statically Indeterminate Structures 200
          6.5.1   Solution Approach 200
    6.6   Thermal Effects 204
          6.6.1   Principle of Superposition Applied to Thermal Problems 208
    6.7   Stress Concentrations 211
          6.7.1   Nature of Stress Concentration 211
          6.7.2   Stress Concentration Factor 212
          6.7.3   Residual Stresses 215
    Problems 217

Chapter 7 Torsion in Shafts 233
    Chapter Objectives 233
    7.1   Introduction 233
          7.1.1   Chapter Objectives 234
    7.2   Analysis of Circular Shafts 234
          7.2.1   Approach of the Analysis 235
    7.3   Formulation of Strain 236
          7.3.1   Sign Convention 238
          7.3.2   Geometry of Torsional Deformation 240
          7.3.3   Formulation of Strain 241
    7.4   Formulation of Stress 242
          7.4.1   Linear Elastic Case 242
          7.4.2   Polar Moment of Area 243
          7.4.3   Internal Torque at a Cross Section 243
    7.5   Angle of Twist 244
    7.6   Statically Indeterminate Torsional Members 254
    7.7   Solid Noncircular Shafts 260
    7.8   Thin-Walled Tubes 263
          7.8.1   Shear Stress Relation 263
    7.9   Composite Shafts 266
    Problems 270

Chapter 8 Bending in Beams 279
    Chapter Objectives 279
    8.1   Introduction 279
    8.2   Shear and Moment Diagrams 279
           8.2.1   Steps of Derivation of Shear and Moment Diagrams 280
           8.2.2   Sign Convention 281
           8.2.3   Governing Relations... 281
           8.2.4   Effect of Point Load on Shear and Moment............................... 282
           8.2.5   Area Method (Graphical Method) for Shear Diagram and Moment Diagram 288
           8.2.6   Coordinate-Reversal Method 292
    8.3   Flexure Formula 295
           8.3.1   Neutral Surface and Neutral Axis.. 295
           8.3.2   Assumptions 296
           8.3.3   Flexure Analysis... 296
           8.3.4   Application of the Flexure Formula 300
           8.3.5   Parallel Axis Theorem 300
           8.3.6   Area Removal Method 300
    8.4   Composite Beams.. 306
           8.4.1   Transformed Section Method... 306
    8.5   Transverse Shear... 312
           8.5.1   Shear Formula 312
           8.5.2   Shear Flow 316
    8.6   Beam Deflection 319
           8.6.1   Bending Moment–Deflection Relation...................................... 319
           8.6.2   Slope Relation... 320
           8.6.3   Boundary Conditions 321
           8.6.4   Deflection by Integration.. 322
           8.6.5   Deflection by Superposition 330
    8.7   Statically Indeterminate Beams 332
    Problems... 337

Chapter 9 Stress and Strain Transformations 349
    Chapter Objectives 349
    9.1   Introduction 349
           9.1.1   Stress Transformation... 349
           9.1.2   Strain Transformation 350
           9.1.3   Coordinate System 350
    9.2   Stress Transformation... 351
           9.2.1   Specification of Stress 351
           9.2.2   Sign Convention 351
           9.2.3   General State of Stress 352
           9.2.4   Plane-Stress Problem 352
           9.2.5   Plane-Stress Transformation.. 352
           9.2.6   Principal Stresses 355
           9.2.7   Maximum In-Plane Shear Stress.. 358
    9.3   Mohr's Circle of Plane Stress 361
           9.3.1   Principal Stresses 363
           9.3.2   Maximum In-Plane Shear Stress.. 364

9.4	Three-Dimensional State of Stress		366
	9.4.1	Stress Transformation in 3-D	367
	9.4.2	Principal Stresses in 3-D	367
	9.4.3	Absolute Maximum Shear Stress in Plane Stress	368
9.5	Thin-Walled Pressure Vessels		372
	9.5.1	Cylindrical Pressure Vessels	372
	9.5.2	Hoop Stress	374
	9.5.3	Longitudinal Stress	374
	9.5.4	Absolute Maximum Shear Stress	375
	9.5.5	Spherical Pressure Vessels	379
9.6	Strain Transformation		380
	9.6.1	Sign Convention	380
	9.6.2	General State of Strain	381
	9.6.3	Plane-Strain Problem	381
	9.6.4	Comparison of Plane-Stress and Plane-Strain Problems	381
	9.6.5	Plane-Strain Transformation	381
	9.6.6	Principal Strains	388
	9.6.7	Maximum In-Plane Shear Strain	390
9.7	Mohr's Circle of Plane Strain		392
	9.7.1	Principal Strains	392
	9.7.2	Maximum In-Plane Shear Strain	393
9.8	Three-Dimensional State of Strain		395
	9.8.1	Strain Transformation in 3-D	395
	9.8.2	Principal Strains in 3-D	396
	9.8.3	Absolute Maximum Shear Strain in Plane Strain	396
9.9	Strain Measurement		398
9.10	Theories of Failure		400
	9.10.1	Failure Theories for Ductile Material	401
		9.10.1.1 Maximum Shear Stress Theory	401
		9.10.1.2 Maximum Distortion Energy Theory	401
	9.10.2	Failure Theories for Brittle Material	403
		9.10.2.1 Maximum Normal Stress Theory	403
		9.10.2.2 Mohr's Failure Criterion	403
	Problems		408

Appendix A: Geometric Properties of Planar Shapes 413

Appendix B: Deflections and Slopes of Beams in Bending 415

Appendix C: Buckling of Columns 419

Appendix D: Advanced Topics 425

Index 441

Preface

This is an introductory book on the subject of mechanics of materials. It serves both as a textbook and a reference book for engineering students and practicing professionals. As a textbook, it is suitable as a first course on the subject.

Mechanics of materials deals with the internal effects (primarily stresses and strains) in a deformable solid body due to external loads acting on it. The subject is also known as *strength of materials* or *mechanics of deformable solids*, and in more advanced study, as *theory of elasticity* and *continuum mechanics*. The subject is useful in a variety of areas including mechanical, civil, mining, materials, electrical, aerospace, and biomechanical engineering. It provides theory, formulas, methods, and techniques that are directly applicable in the modeling, analysis, design, testing, and regulating of engineering devices and structures such as automobiles, airplanes, robots, machine tools, engines, bridges, elevated guideways, and buildings.

SCOPE OF THE BOOK

External loads (forces, moments, torques, etc.) applied on a body result in reaction loads at support locations (bearings, basement, anchors, suspension points, etc.) and internal loads throughout the body. Under static conditions, these loads satisfy *equations of equilibrium*, which come under the field of statics. Stresses in the body are caused by the internal loading and are a determining factor of the *strength* of the object. Strains caused by loading are directly related to the *deflection* or *deformation* or *compatibility* of the object. The stress–strain relations (or *constitutive relations*) determine the *response due to loading*, particularly the *stiffness*, of an object and are governed by the physics of the object. Typically, stress–strain relations are determined experimentally because analytical procedures for determining them using material physics can be quite complex if not impossible. In addition to strength, deformation, and stiffness, the field of mechanics of materials also concerns *stability*, which studies the possibility of deformations that can grow suddenly and continuously without limit (in theory). The underlying techniques and procedures of mechanics of materials (particularly concerning strength, deformation, and stability) are directly needed and applicable in modeling, analysis, computer simulation, design, testing and diagnosis, operation, regulation, and maintenance of a variety of engineering systems. This book systematically presents the theory, procedures, and applications of mechanics of materials. Detailed worked examples and exercises are provided throughout the book to illustrate the application of the theory and the methodologies. Complete solutions to the end-of-chapter problems are presented in a *solutions manual*, which is available to instructors.

MAIN FEATURES OF THE BOOK

This book is an outgrowth of my experience in teaching an undergraduate course in mechanics of materials for large classes of students in mechanical, civil, manufacturing, materials and mineral engineering, and engineering physics and in teaching other courses in statics, dynamics, modeling, vibration, instrumentation, testing, and design. The book is targeted mainly toward engineering students, but the practical considerations, design issues, engineering techniques, and the simplified and snapshot-style presentation of more advanced theories and concepts renders the book useful for engineers and technicians, project managers, and other practicing professionals as well. This practical orientation is reinforced through my industrial experience at IBM Corporation, Westinghouse Electric Corporation, Bruel and Kjaer, and NASA's Lewis and Langley Research

Centers. To maintain clarity and focus and to maximize the usefulness of the book, the material in the book is presented in a manner that is convenient and useful to anyone with a basic background in engineering.

The book consists of 9 chapters and 4 appendices. The following are the main features of the book:

- The objectives and approaches are clearly stated and summarized at the beginning of each chapter.
- The material is presented in a progressive manner, first providing introductory material and then systematically leading to more advanced concepts and applications.
- Many worked examples are included throughout the book.
- Numerous problems and exercises are given at the end of each chapter.
- Worked examples and end-of-chapter problems are related to real-life situations and practical engineering applications.
- Analytical methods are presented using simple mathematics.
- Experimental techniques are presented in a concise manner.
- The main topics covered in each chapter are listed at the beginning of the chapter.
- Key issues presented in the book are summarized in a *summary sheet* at the end of each chapter for easy reference and recollection.
- Useful reference data and advanced material that cannot be conveniently integrated into the chapters are presented in a concise form as appendices at the end of the book. The inside covers contain a table of units and conversions, and a table of mechanical properties of common engineering materials, accessible for easy reference.
- The book is concise and avoids unnecessarily lengthy and uninteresting discussions, which makes it easier to comprehend.
- A *solutions manual* is available for the convenience of instructors.
- To access additional presentations and a variety of other resources (notes, course outline, exams, quizzes, homework problems, worked examples, etc.), go to http://www.crcpress.com/product/isbn/9781439877364, and follow the directions for accessing the author website.

Clarence W. de Silva

Acknowledgments

Many individuals have assisted in the preparation of this book, but it is not practical to acknowledge all such assistance here. First, I wish to recognize the contributions, both direct and direct, of my graduate students and research associates. Particular mention should be made of my lab manager and postdoctoral research associate, Roland H. Lang, and my research and teaching assistant, Edward Y. Wang. I am particularly grateful to Jonathan W. Plant, executive editor, CRC Press/Taylor & Francis Group, for his interest, enthusiasm, and strong support throughout the project. The smooth and timely production of this book is a tribute to many others at CRC Press and its affiliates as well, in particular senior project coordinator, Jessica Vakili; project editor, Richard Tressider; and project manager, Deepa Kalaichelvan of SPi Global. I have benefited from the knowledge and experience of the following colleagues who also teach mechanics of materials: Prof. Jon Mikkelsen, Prof. Farrokh Sassani, and Prof. Mohamed Gadala.

I am also grateful for the constructive comments on the book provided by the following experts on the subject: Prof. Marcelo H. Ang, National University of Singapore; Prof. Ruxu Du, The Chinese University of Hong Kong; Prof. Devendra P. Garg, Duke University; Prof. Piaras Kelly, University of Auckland; Prof. Chris K. Mechefske, Queen's University; Prof. J. N. Reddy, Texas A&M University; and Prof. Ronald E. Smelser, University of North Carolina–Charlotte.

Finally, my wife and children deserve appreciation and an apology for the unintentional "neglect" that they may have experienced during the preparation of this book.

Author

Dr. Clarence W. de Silva, PE, fellow ASME, fellow IEEE, is a professor of mechanical engineering at the University of British Columbia, Vancouver, Canada, and occupies the Senior Canada Research Chair professorship in mechatronics and industrial automation. Prior to that, he was the NSERC-BC Packers Research Chair in industrial automation since 1988. He has served as a faculty member at the Carnegie Mellon University (1978–1987) and as a Fulbright Visiting Professor at the University of Cambridge (1987–1988).

Dr. de Silva received his PhDs from the Massachusetts Institute of Technology (1978) and the University of Cambridge, England (1998), and an honorary DEng from the University of Waterloo (2008). He has also held the Mobil Endowed Chair Professorship in the Department of Electrical and Computer Engineering at the National University of Singapore; honorary professorship of Xiamen University, China; and honorary chair professorship of National Taiwan University of Science and Technology.

Other fellowships: Fellow, Royal Society of Canada; fellow, Canadian Academy of Engineering; Lilly Fellow at Carnegie Mellon University; NASA-ASEE Fellow; Senior Fulbright Fellow at Cambridge University; fellow of the Advanced Systems Institute of BC; Killam Fellow; and Erskine Fellow at the University of Canterbury.

Awards: Paynter Outstanding Investigator Award and Takahashi Education Award, ASME Dynamic Systems & Control Division; Killam Research Prize; Outstanding Engineering Educator Award, IEEE, Canada; Lifetime Achievement Award, World Automation Congress; IEEE Third Millennium Medal; Meritorious Achievement Award, Association of Professional Engineers of BC; and Outstanding Contribution Award, IEEE Systems, Man, and Cybernetics Society. He also made 32 keynote addresses at international conferences.

Editorial duties: Served on 14 journals, including *IEEE Transactions on Control System Technology* and *Journal of Dynamic Systems, Measurement and Control, Transaction of the American Society of Mechanical Engineers*; editor in chief, *International Journal of Control and Intelligent Systems*; editor in chief, *International Journal of Knowledge-Based Intelligent Engineering Systems*; senior technical editor, *Measurements and Control*; and regional editor, North America, *Engineering Applications of Artificial Intelligence—IFAC International Journal*.

Publications: 20 technical books, 18 edited books, 44 book chapters, 220 journal articles, and 250 conference papers.

Recent books: *Mechatronics: A Foundation Course* (2010); *Modeling and Control of Engineering Systems* (Taylor & Francis/CRC Press, 2009); *Sensors and Actuators: Control System Instrumentation* (Taylor & Francis/CRC Press, 2007); *Vibration: Fundamentals and Practice*, 2nd edn. (Taylor & Francis/CRC Press, 2007); *Mechatronics: An Integrated Approach* (Taylor & Francis/CRC Press, 2005); and *Soft Computing and Intelligent Systems Design: Theory, Tools, and Applications* (Addison-Wesley, 2004).

1 Mechanics of Materials

CHAPTER OBJECTIVES

- Introduce and define the subject
- Provide application areas and representative applications
- Indicate the significance of the subject
- Outline the history of the subject and its evolution
- Give typical problem scenarios and main steps of solution

1.1 WHAT IS MECHANICS OF MATERIALS?

Mechanics of materials deals with the internal effects (primarily, stresses and strains) in a deformable solid body or structure due to external loads (forces, moments, and torques) acting on it. The subject is also known as "strength of materials" or "mechanics of solids" or "mechanics of deformable bodies." In particular, this subject primarily studies

1. Strength (determined by stress at failure)
2. Deformation (determined by strain)
3. Stiffness (ability to resist deformation; load needed to cause a specific deformation; determined by the stress–strain "constitutive" relationship)
4. Stability (ability to avoid rapidly growing deformations caused by an initial disturbance, e.g., buckling)

In particular, the associated methodologies can be used to design structures, machinery, and other mechanical objects for "good" performance, and also to predict failure and to determine the failure conditions (i.e., loading that will cause failure) and the nature of failure (i.e., location, direction, surface orientation, and the corresponding stress and strain values) of these objects.

Consider an aircraft (Figure 1.1a). External loads applied on the aircraft body result in internal loads throughout the body. Imagine, for instance, the resulting loads in flight, at the wing joint (Figure 1.1b). Proper structural design of an aircraft can help mitigate structural damage and disaster. As an example that highlights this need, Flight 243 of Aloha Airlines Boeing 737 with 95 passengers and crew on board experienced structural damage and component loss in mid-air, on April 28, 1988. This resulted in a fatality (one crew member was sucked out through the created opening). However, the damaged aircraft (Figure 1.2) was able to land at Maui airport, Hawaii, without further loss of human life (there were eight serious injuries).

1.2 SUBJECT DEFINITION

External loads (forces, moments, and torques) applied on a body generate reaction loads at support locations (bearings, basements, mounts, anchors, suspension points, etc.) and internal loads throughout the body. Under static conditions, these loads satisfy "equations of equilibrium," which are typically studied in the subject of statics. Stress represents the "intensity" of a local force. Stresses in the body are caused by the internal loading and are a determining factor of the "strength" of the object. Strain represents the deformation per unit geometric element at a location. Hence, strains caused by loading are directly related to the "deflection" or "deformation" or

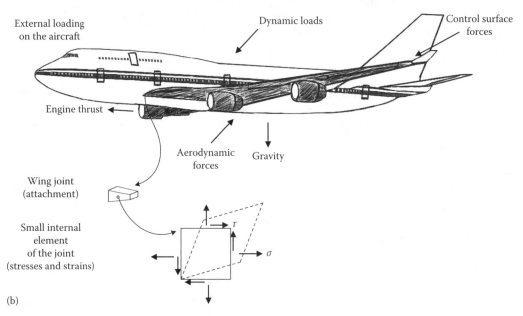

FIGURE 1.1 (a) An aircraft. (b) Loading on an aircraft.

"compatibility" or "integrity" of the object. The stress–strain relations (or "constitutive relations") determine the "stiffness" of an object and are governed by the physics of the object. Typically, however, stress–strain relations are determined experimentally, because analytical procedures for determining them using material physics can be quite complex if not impossible. As indicated in Figure 1.3, all these issues are studied in mechanics of materials. In addition to strength, deformation, and stiffness, the subject of mechanics of materials also concerns "stability," which studies the possibility of deformations that can grow suddenly and continuously without limit (in theory) when excited by a small disturbance.

Note: Deformations and strains due to thermal effects (heat transfer and temperature changes) are also considered in this subject.

Mechanics of Materials

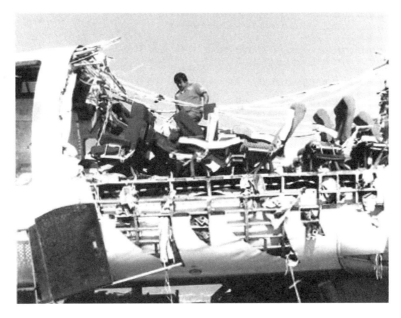

FIGURE 1.2 Mid-air structural damage of an aircraft (1 fatality out of 95 people on board).

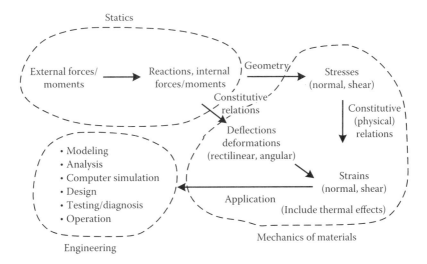

FIGURE 1.3 Subject definition.

1.3 APPLICATION OF THE SUBJECT

As indicated in Figure 1.3, the subject of mechanics of materials provides theory, formulas, and other information that are directly applicable in the modeling, analysis, simulation, design, operation, testing, and diagnosis of engineering systems (e.g., automobiles, airplanes, robots, machine tools, engines, bridges, elevated guideways, and buildings). The involved procedures are summarized in the following.

Modeling: Determine "equations" governing the stress–strain (or, load–deflection) behavior of an object/structure.

Analysis: Determine stresses and strains (based on internal loads and deformations) due to external loading. Numerical analysis is commonly done using the finite element method (FEM). In FEM, the analyzed structure is approximated by an interconnected set of small elements whose characteristics and parameter values are known. The numerical analysis is then carried out using the element equations and the interconnection equations, under specified external loading.

Simulation: Program a model of the system (possibly using both analytical and experimental equations and parameter values). Run the program under specified loading conditions. Determine stresses and strains (internal loads, deformations). The FEM is commonly used in computer simulations.

Design: Select materials, dimensions, and physical structures of a device to meet a set of "performance specifications" (related to strength, deformation, size, shape, cost, stability, safety, etc.). An "objective function" (or performance index) may be optimized with respect to the parameters that may be varied in the design process (this is known as optimal design), subject to a set of "constraints" or limiting values. A physical "prototype" of the system may be built based on the design, for subsequent testing and evaluation.

Testing: Apply a specified regime of loading (single or repetitive) to the physical system or prototype and measure the resulting deformations, or determine the loading that causes failure.

The importance of the subject is justified in many ways, for example

- Material optimization, energy efficiency, and size/compactness (lightweight) of modern designs of machinery and structures call for thin members, high flexibility, and complex geometry. Large deformations can lead to engineering problems (e.g., poor vehicle ride quality over guideways, bridges; undesirable contact between components causing wear, noise, sparks, hazard).
- Increased power levels and longer and various operating conditions of modern machinery will give rise to larger loading than usual. This will call for structural components with higher strengths.
- More stringent regulatory requirements on safety, architecture, and esthetics will require complex and more rigorous analysis, more precise design, and thorough testing and evaluation.

1.4 APPLICABLE ENGINEERING FIELDS

Mechanics of materials is useful in a variety of engineering areas including mechanical, civil, mining, materials, electrical, aerospace, and biomechanical engineering; for example:

Aeronautical and aerospace engineering: Design and development of aircraft and spacecraft.

Civil engineering: Design and evaluation of bridges, buildings, and roadways.

Electrical and computer engineering: Structural design of electronic and computer hardware (e.g., design of such components as hard-disk drives, switches, relays, and circuit-breakers); "product qualification" through analysis and testing for specialized applications (e.g., qualification of critical electrical/electronic components such as actuators and sensors for use in nuclear power plants for operation under seismic disturbances).

Manufacturing engineering: Machine component failure, tool wear, and breakage will lead to reduced productivity and product quality, and increased costs of operation and maintenance. Proper design of machine tools and components is essential in mitigating these problems.

Mechanical engineering: Design and testing of machinery, engines, vehicles, aircraft, robots, ships, etc.

Mining and mineral engineering: Design, development, and testing of mining machinery that operates under severe, risky, and hazardous conditions (e.g., human recue systems).

Some application examples are shown in Figure 1.4.

1.5 USEFUL TERMS

Force: A rectilinear load; has a magnitude and a direction (i.e., it is a vector); units: newton (N), 1 kN = 1000 N

Normal force: Force normal (perpendicular) to a considered area; tends to push/pull the body

Shear force: Force along the plane of a considered area; causes a shearing (or sliding deformation along the plane)

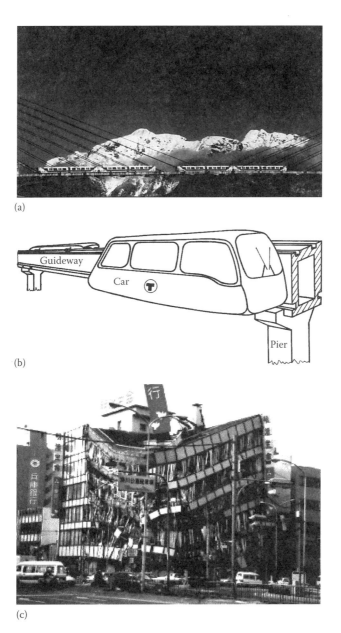

FIGURE 1.4 (a) High-speed ground transit (The Sky Train, Vancouver, Canada—A modern automated transit system). (b) Vehicle/guideway design, material optimization, cost, etc. (Torsional guideway transit system—TGT). (c) Seismic design for structural integrity, safety, etc. (Earthquake in Kobe, Japan, Magnitude 7.2, on January 17, 1995—Collapse of a bank building).

(*continued*)

Torque: A rotational (angular) load (i.e., torsional moment); tends to "twist" the object to which it is applied; has a magnitude and a direction (i.e., a vector); units: newton-meter (N·m)

Bending moment: A bending (i.e., flexural) load; tends to "bend" the object to which it is applied; has a magnitude and a direction (i.e., a vector); units: newton-meter (N·m)

Stress: Intensity of a force; force per unit area; not a vector (because the same force at a point will cause different stresses on areas of different orientation at that point)—it is a second-order tensor; units: N/m^2 (=pascal or Pa), 1 N/mm^2 = 1 MPa; normal stress is caused by a normal force component on the considered area, shear stress is caused by a shear force component on the considered sliding (shearing) area

(d)

(e)

(f)

FIGURE 1.4 (continued) (d) Building structure design (design of members, joints, configuration, etc., for structural integrity, safety, etc.)—the wooden roof structure of the Forest Sciences Center, University of British Columbia. (Pictures by C.W. de Silva.) (e) A window cleaning carriage boom. (Photo courtesy of C.W. de Silva.) (f) Cable stayed bridge in Incheon, South Korea. (Photo courtesy of C.W. de Silva.)

Mechanics of Materials 7

FIGURE 1.4 (continued)   (g) Adjustable tensioning rods and key joints of a structure at the airport, Vancouver, Canada. (Photo courtesy of C.W. de Silva.) (h) Failure of an exercise spring. (Photo courtesy of C.W. de Silva.) (Fatigue failure. Most likely due to shear in torsion, as the spring winds and unwinds, it experiences some twisting in addition to tension.) (i) Joints and connectors of machinery. (Courtesy of Siemens Gas Turbine SGT5-8000H.)

Strain: Intensity of deflection; deflection per unit length along the direction of the force (normal strain) or angle of shearing (sliding) deformation in radians (shear strain); dimensionless

Free-body diagram: "Virtually" separate the segment of interest from the rest of the object and mark the loads at the interface of separation

Homogeneous: Properties are uniform (do not change from point to point) in the body

Isotropic: Properties are nondirectional (do not vary with the considered direction at a given point)

1.6 HISTORY OF MECHANICS OF MATERIALS

Many scientists and engineers contributed to the development of the subject area. Some of the particularly noteworthy contributors with their main contributions to the subject are listed as follows.

Archimedes (287–212 BC): Statics, equilibrium of a lever
da Vinci (1452–1519): Concept of moments

Galileo (1564–1642): Effects of loads on beams and rods, virtual displacement
Newton (1642–1727): Foundation of mechanics
Bernoulli (1667–1748): Virtual displacement/work, beam bending
Hooke (1635–1703): Hooke's law relating stress and strain, Hooke's joint
Euler (1707–1793): Moment of inertia, beam bending, instability, column buckling, rigid body dynamics
d'Alembert (1717–1783): Inertia force (converts a dynamics problem into an equivalent statics problem)
Lagrange (1736–1813): Mechanics, energy methods
Coulomb (1736–1806): Friction (static and dynamic)
Laplace (1749–1827): Mechanics, etc.
Poisson (1781–1840): Lateral strain due to a normal load, Poisson's ratio
Saint-Venant (1797–1886): Strain distribution at abrupt changes in cross section, strain tensor, torsion
Castigliano (1847–1884): Analysis of structural loads and deflections by the energy method
Galerkin (1871–1945): Analysis of elastic plates, stresses in dams and retaining walls
Timoshenko (1878–1972): Theory of thick beams in bending

1.7 BASIC PROBLEM SCENARIOS

Consider an unstrained member schematically shown in Figure 1.5a. The grid lines schematically represent the shape of discrete segments of the member.

There are four basic scenarios of loading and deformation that are studied in this book.

1. *Axial loading*: Loads are forces (tensile or compressive) which are applied along the main axis of a member. The resulting deformations are primarily extensions or compressions, which occur along the loading axis even though deformations (strains) can occur perpendicular to this axis as well (Poisson effect). An example is given in Figure 1.5b.
2. *Shear loading*: Two equal and opposite parallel forces exist on two equal parallel areas of a member. The resulting deformation involves sliding (shearing) of one area with respect to the other along the direction of loading. An example is given in Figure 1.5c.
3. *Torsional loading*: The external loads are "torques" which tend to twist the member. An example is given in Figure 1.5d.
4. *Bending loading*: The external loads (forces and moments) result in bending deformations (i.e., flexure) of the member. An example is given in Figure 1.5e.

In practical problems, two or more of these basic scenarios may exist in combination.

1.8 PROBLEM SOLUTION

Solution of problems in mechanics of materials involves three basic considerations in general.

1. Statics (equations of equilibrium). This determines reactions at supports and internal loads, which determine stresses
2. Nature of deformation (nature of strains)
3. Stress–strain relations (i.e., constitutive relations or physical relations). These are needed to determine the strains (deformations) once the stresses are known (from the knowledge of internal loads)

The subject of statics uses equations of equilibrium, and does not consider deformations (or strains). It essentially treats the system as one or more rigid bodies. The underlying equations will determine

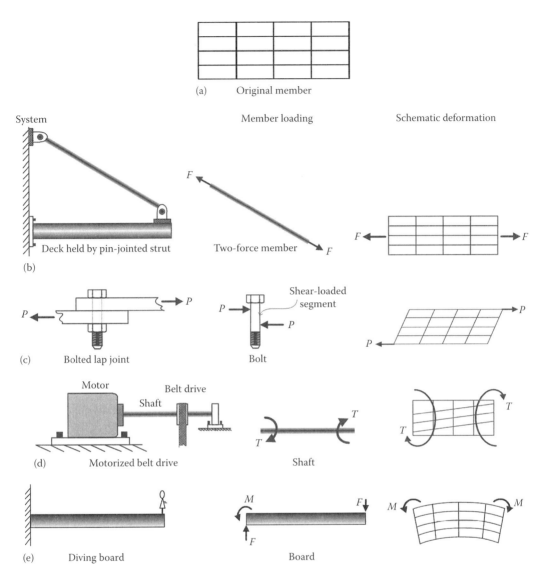

FIGURE 1.5 Representation of four problem scenarios: (a) original unstrained member, (b) axial loading, (c) shear loading, (d) torsional loading, and (e) bending loading.

the reactions at support locations. Furthermore, these considerations determine the internal loads (internal forces, internal torques, internal bending moments), which are necessary to determine the stresses at various locations and on various planes of the body. The consideration of deformations is not necessary to determine these unknowns (support reactions and internal loads) if the problem is *statically determinate*.

The nature (geometry) of deformation is an important consideration in the subject of mechanics of materials. Hence, in solving a problem, it is important to study how various locations in a body deform, what types of strains are generated, and what assumptions are needed to determine these deformations and strains.

To determine deformations (and strains) in a body, analytically or computationally, the governing stress–strain relationships must be known. They are called *constitutive relations*, and they depend on the physics of the problem and the material properties. Even though analytical representations of them are found using detailed theory (physics) and for simpler situations (e.g., under linear behavior), more commonly, the stress–strain relations are determined experimentally.

1.8.1 Problem Solution Steps

The use of a systematic approach is crucial for effective problem solving (this statement is not limited to mechanics of materials). The following main steps are suggested in solving problems in mechanics of materials:

1. Understand the problem (what has to be determined in the problem; what information is given; what are the assumptions and constraints; etc.)
2. Plan the solution (based on the understanding of the problem—Step 1 and what are the available approaches to solve the problem; decide which approach is the most appropriate one, unless the required approach is hinted in the problem)
3. Carry out the solution
4. Check the solution (e.g., for compatibility of units and dimensions, proper sign of the results, reasonability of the magnitudes, and against results from another approach)

1.9 ORGANIZATION OF THE BOOK

The topics covered in the book are organized as schematically represented in Figure 1.6.

The book consists of nine chapters and four appendices, which are intended for presenting the fundamentals of the subject of mechanics of materials, and their use in the solution of engineering problems. Illustrative examples and problems are presented throughout the book. A summary of the concepts covered in each chapter is given at the end of the chapter, along with useful definitions and formulas.

Chapter 1 introduces the subject of mechanics of materials. It provides application areas and representative applications, significance of the subject, history of the subject and its evolution, and typical problem scenarios and main steps of their solution.

Chapter 2 reviews the subject of statics, which is an important prerequisite for mechanics of materials. It outlines how to sketch and use proper *free-body diagrams* and write *equations of equilibrium*, which are necessary first steps in the solution of problems in mechanics of materials. It introduces various types of *supports* used in engineering structures. It describes the determination of *external reactions* (forces and moments) and *internal loads* (forces and moments), which are necessary to determine stresses and strains. It presents the analysis of structures (trusses/frames) consisting of *two-force members*. It shows the solution of problems containing *distributed forces*. At the end, it discusses the solution of problems that statics alone cannot solve. These are known as *statically indeterminate* systems, and are common in engineering applications.

Chapter 3 studies the subject of stress, which represents the intensity of an internal force.

It is emphasized that stress is not a vector but rather a second-order tensor (it has a magnitude and a direction, but also depends on the orientation of the section on which it acts). Two types of stress, normal stress and shear stress, are discussed. Normal stress in an axially loaded member (rod, bar, shaft, cable, etc.) is analyzed. Use of an axial force diagram in determining the stress distribution in an axially loaded member is illustrated. Analysis and design of joints (bolted, riveted, glued, welded, keyed, etc.) is shown. Shear stress and normal stress in bolts and the analysis of bolted joints in single shear and double shear are illustrated. The analysis of shear stress in a key, which connects a shaft to a disk, is given. The complementarity property of shear stress at a local element in an object—on a minute 2-D rectangular element, shear stresses are equal on all four sides; and equal and opposite on two opposite parallel sides—is presented. Stress transformation (i.e., determination of the stresses on a local section that is inclined to the section on which stresses are known) at a point is analyzed.

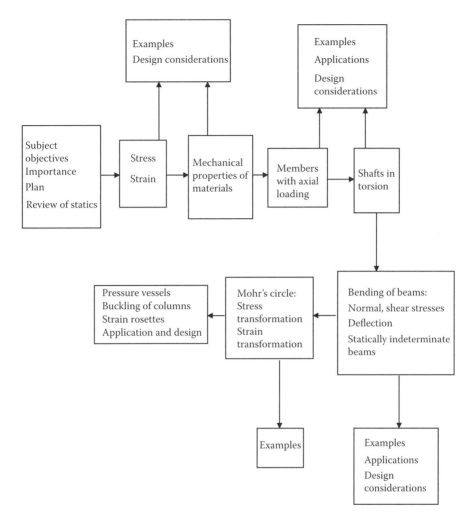

FIGURE 1.6 Organization of the book.

Chapter 4 studies the subject of strain, which represents the intensity of a local deformation, specifically, deformation of a reference element (of unit length or 90° corner). Two types of strain, normal strain and shear strain, are studied. Units of strain and the applicable sign convention are presented. The concepts of local strain and average strain are given. Strain caused by a temperature change (thermal strain) is studied. Measurement of strain using strain gauges is introduced.

Chapter 5 addresses the subject of constitutive relation, which represents the stress–strain behavior of a material. Tensile testing is introduced, which experimentally determines the stress–strain behavior (stress–strain curve) of a material. Hooke's law, which gives the linear stress–strain relationship, is presented. Associated material parameters, particularly, Young's modulus E, shear modulus G, and Poisson's ratio ν, are indicated. Various types of material behavior, linear, nonlinear, elastic, plastic, work hardening, creep, fatigue, and fracture, are discussed. Important material properties, homogeneity, isotropy, elasticity, rigidity, strength, elongation, resilience, toughness, and hardness, are outlined. Material types, particularly ductile, brittle, soft, hard, strong, and tough, are reviewed. Concepts of strain energy, and the associated modulus of resilience, and modulus of toughness are summarized.

Chapter 6 provides a detailed study of members under axial loading. The use of Saint-Venant's principle and equivalent loading to simplify their analysis is presented. The approach of determination of axial force diagram, which is an important step in solving this class of problems, is presented. The analysis of axially loaded members having (a) uniform cross section, (b) continuously varying nonuniform section, and (c) multiple segments of uniform cross section is studied. Application of the principle of superposition (valid when the force–deflection relation is linear) is given. The analysis of statically indeterminate structures (where statics—equilibrium equations—alone are not adequate to solve the problem) is studied. Thermal deformation due to temperature change and the associated strains and stresses are discussed. The incorporation of stress concentration due to geometric nonsmoothness and discontinuities, in solving problems of mechanics of materials, is reviewed.

Chapter 7 studies the analysis of pure torsion in uniform shafts of circular cross section. The underlying assumptions and their rationale are discussed. Sign convention used in the analysis is presented. Analysis of the linear elastic case is emphasized. Sketching the internal torque distribution, which is an important step in problem solution, is illustrated. Determination of the angle of twist is studied. Definition and use of the polar moment of area are illustrated. The analysis of statically indeterminate torsion members is presented. Torsion of members with noncircular cross section is introduced.

Chapter 8 explores the subject of beams in bending. The importance of the subject is outlined. The applicable sign convention is given. The use of shear diagram and moment diagram in solving this class of problems is shown. The graphical method of obtaining shear and moment diagrams is presented. The flexure formula for bending stress is derived and its use is illustrated. The analysis of composite beams by using the method of equivalent sections is presented. The concepts of transverse shear and shear flow at joints are presented. Determination of the deflection in beams is studied. The analysis of statically indeterminate beams is outlined.

Chapter 9 studies topics of stress transformation and strain transformation. The problem of plane stress (2-D problem) is emphasized. Stress transformation concerns finding stresses on different planes at the same location. The applicable formulas are derived and their application is illustrated. The use of Mohr's circle for plane stress in solving the associated problems is presented. The procedures in the determination of principal stresses, maximum in-plane shear stress, and absolute maximum shear stress are presented and illustrated. The analysis of thin-walled pressure vessels, particularly, cylindrical vessels and spherical vessels, is studied. The problem of plane strain and the associated methods of strain transformation—finding strains in different directions and corners at the same location—are studied. The applicable formulas are derived and their use is illustrated. The use of Mohr's circle for strain in solving problems of plane strain is shown. The methods of determination of principal strains, maximum in-plane shear strain, and absolute maximum shear strain are presented and illustrated. Theory of strain measurement using strain gauges is revisited. The use of strain-gauge rosettes in the measurement of multiple components of local strain is presented. Common theories of failure for ductile materials and brittle materials are presented.

Appendix A summarizes useful geometric properties of several planar shapes. Appendix B provides formulas for flexural deflection and slope of thin beams under different end conditions and loading. Appendix C outlines the analysis of buckling of columns, which is an important problem of structural instability and failure. Appendix D presents several advanced topics and theory of mechanics of materials (theory of elasticity, in particular).

PROBLEMS

1.1 What does the subject of mechanics of materials concern? What is the importance of this subject in engineering?

1.2 Several scientists and engineers who have made important contributions to the subject of mechanics of materials are listed in the following. Select a few of them (say, five) and for each of them

Mechanics of Materials

1. Write a short biographical sketch (one paragraph, not more then 10 lines)
2. Indicate the important contributions he made to the subject of mechanics of materials (one paragraph, not more than 7 lines)

You may obtain the necessary information through online search or literature review in library. *Note*: You should not simply copy what I have given in the chapter.

Archimedes (287–212 BC)
da Vinci (1452–1519)
Galileo (1564–1642)
Newton (1642–1727)
Bernoulli (1667–1748)
Hooke (1635–1703)
Euler (1707–1793)
d'Alembert (1717–1783)
Lagrange (1736–1813)
Coulomb (1736–1806)
Laplace (1749–1827)
Poisson (1781–1840)
Saint-Venant (1797–1886)
Castigliano (1847–1884)
Galerkin (1871–1945)
Timoshenko (1878–1972)

1.3 Structural wire mesh (e.g., for stucco reinforcement) is manufactured by joining steel wire using resistance welding. Weld strength is the primary consideration in the manufacturing process. An experimental setup for resistance welding of wire is shown in Figure P1.3. Two wires are placed on the bottom (stationary) electrode, one on top of the other at 90°. The top electrode is moved downward using a pneumatic cylinder to press the wires together, and the welding pressure is measured using the load cell. Welding controller is triggered to perform welding, by sending a large current through, and the root-mean-square (rms) welding current is recorded. Finally, the welded specimen is tested on a tensile testing machine, and the tensile force at break (of either the welded joint or the wire) is recorded.

Typically, it is required that the weld strength be slightly greater than the wire strength. What is the reason for this?

1.4 List several engineering applications where the subject of mechanics of materials is utilized. Select one of these applications and expand on how mechanics of materials may be used there.

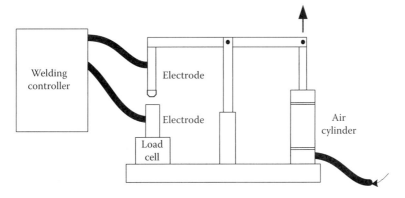

FIGURE P1.3 An experimental setup for resistance welding of cross-wire.

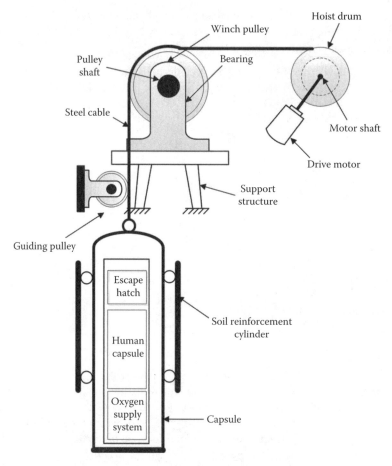

FIGURE P1.7 A rescue system for trapped miners.

1.5 In a typical application of mechanics of materials, it is required to maximize the material strength. What are some of the constraints associated with this optimization problem?

1.6 Various components are joined together when constructing such engineering systems as automobiles, buildings, construction machinery, and machine tools. List several types of joints that are commonly used in such fabrications.

1.7 Consider an emergency situation of collapsed mine where some miners are trapped in. Outline several applications of mechanics of materials in performing a rescue operation in this situation. A possible rescue mechanism is sketched in Figure P1.7.

2 Statics: A Review

CHAPTER OBJECTIVES

- Relevance of statics as a prerequisite for mechanics of materials
- A review of "statics"
- How to write equations of equilibrium
- Importance of recognizing the types of *supports* in problem solving
- Determination of *external reactions* (forces and moments)
- Determination of *internal loads* (forces and moments)
- Properties of *two-force members* and analysis of trusses/frames
- How to sketch and use proper *free-body diagrams* in effective problem solution
- Solution of problems with *distributed forces*
- Illustration that statics alone cannot solve *statically indeterminate* systems (another importance of mechanics of materials)

2.1 STATICS

With regard to statics, the following facts can be stated:

1. The subject of statics concerns bodies/structures in *equilibrium* (i.e., at rest or in uniform motion).
2. It does not take deformations of the components into account (i.e., it assumes that the components are *rigid*).
3. The basic equations in statics are the *equilibrium conditions*: force balance and moment (or torque) balance at any location and in any direction.
4. Solution of equations of equilibrium can determine
   a. Reaction loads (forces or moments) at support locations under known external loads (forces or moments)
   b. Internal loads (forces/moments) at any location and section of the body

Importance of statics in mechanics of materials: Internal loads determine the stresses at a given location and section in the body in the required directions. (*Note*: Stresses determine strains, both of which are important in mechanics of materials). Knowledge of the external loads and support reactions is needed to determine the internal loads. Hence, statics is an important prerequisite for Mechanics of Materials.

2.1.1 Equations of Equilibrium

In a problem of statics, a body is assumed to be in equilibrium. The basic equations in statics are then the equations of equilibrium for the body (or any segment of the body).

Note: The same equations of equilibrium hold even if the body is in motion, at constant velocity.

General 3-D case: In the three-dimensional (3-D) analysis of a body, the following equations of equilibrium hold

1. At any point, the forces should balance in any direction. Hence, forces should balance along three orthogonal (i.e., perpendicular) directions (x, y, and z directions in a Cartesian coordinate system):

$$\sum F_x = 0; \quad \sum F_y = 0; \quad \sum F_z = 0 \tag{2.1}$$

2. At any point, the moments should balance about any axis. Hence, moments should balance about three orthogonal axes (x, y, and z) in a Cartesian coordinate system:

$$\sum M_x = 0; \quad \sum M_y = 0; \quad \sum M_z = 0 \tag{2.2}$$

In this manner we get six equations for any segment of a body that is analyzed. Only six unknown external forces/moments (for the analyzed body segment) can be determined from them.

Rationale: If there is a nonzero resultant force in any direction, the body must move (accelerate) in that direction. If there is a nonzero moment about any axis, the body must rotate (accelerate) about that axis. These violate the condition of equilibrium.

Why use three orthogonal directions? Any direction (vector) in 3-D can be represented by (i.e., resolved into) three orthogonal vectors. Hence by using three orthogonal directions, we cover any direction in 3-D.

Note: If the unknown loads cannot be determined from the equilibrium equations alone, it is a "statically indeterminate system/problem."

2-D case (i.e., coplanar forces): In a two-dimensional (2-D) problem of statics, the forces lie on a single plane and the moments exist only about an axis perpendicular (normal) to this plane. Hence this is the case of co-planar forces. The applicable equations of equilibrium are the force balance along two orthogonal axes (x and y) on the plane and the moment balance about an axis normal to the plane (z axis):

$$\sum F_x = 0; \quad \sum F_y = 0; \quad \sum M_z = 0 \tag{2.3}$$

In this manner we get three equations (for the body or the segment of the body that is analyzed). Only three unknown external loads (forces/moments) can be determined from them, for the analyzed body segment.

Example 2.1

A cable of length $L = 5$ m is tied between points A and B, which are $c = 4$ m apart horizontally and $h = 0.6$ m apart vertically, as shown in Figure 2.1a. A mass of weight $Mg = 1.0$ kN is hung from a smooth and light pulley C, which can slide (without friction) on the cable. In the shown configuration, the pulley is stationary and the system is in equilibrium.

Determine
   a. The angles α and β of inclination of the two segments AC and CB, respectively, of the cable as measured from the vertical
   b. Tension T in the cable
   c. Length of the segment AC of the cable

Solution

   a. The equilibrium condition at C of the system is shown in Figure 2.1b.
   Equations of equilibrium at C in the horizontal and the vertical directions are (see Equation 2.3)

$$\rightarrow \sum F_x = 0 \rightarrow T\sin\alpha - T\sin\beta = 0 \rightarrow \alpha = \beta \tag{i}$$

Statics: A Review

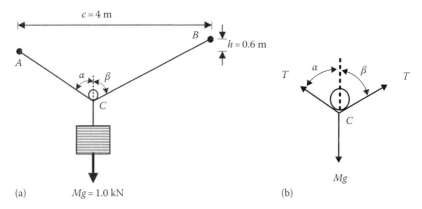

FIGURE 2.1 (a) An inextensible cable with a sliding pulley that carries a weight; (b) the state of equilibrium at the pulley.

$$\uparrow F_y = 0 \rightarrow T\cos\alpha + T\cos\beta - Mg = 0 \rightarrow T = \frac{Mg}{2\cos\alpha} \quad \text{(ii)}$$

Note: We have substituted (i) in getting the final result (ii).

From geometry

$$c = AC\sin\alpha + BC\sin\beta = (AC + BC)\sin\alpha \ \text{(From (i))}$$

$$\rightarrow c = L\sin\alpha \ (\textit{Note: Cable length} = L = AC + BC)$$

$$\rightarrow \sin\alpha = \frac{c}{L} = \frac{4}{5} \rightarrow \cos\alpha = \frac{3}{5}$$

$$\rightarrow \alpha = \beta = \sin^{-1}\frac{4}{5} = 53.1°$$

b. Substitute the angle in (ii)

$$T = \frac{1.0}{2 \times 3/5} \ \text{kN} = \frac{5}{6} \ \text{kN} = 0.833 \ \text{kN}$$

c. $BC\cos\beta - AC\cos\alpha = h \rightarrow BC - AC = \frac{h}{\cos\alpha}$ (*Note:* $\alpha = \beta$)

Substitute $BC + AC = L$ for BC

$$AC = \frac{1}{2}\left(L - \frac{h}{\cos\alpha}\right) = \frac{1}{2}\left(5 - \frac{0.6}{3/5}\right) = 2.0 \ \text{m}$$

Main Learning Objectives
1. Application of equilibrium equations to solve problems of statics
2. Proper use of geometry
3. The ideal case of "smooth" pulley (cable tension does not change around it)

■ **End of Solution**

2.2 SUPPORT REACTIONS

The reaction forces and reaction moments at the supports can be determined using the equations of equilibrium. The knowledge of the support loading is important in the complete solution of a problem in statics and hence in mechanics of materials. Note that

- Both magnitude and direction of the reactions are needed
- The type of support may determine the nature and direction of the support reaction

Some useful examples of supports and associated loading are shown in Figure 2.2.

Types of external forces: Surface forces (caused by contact with another body, e.g., cable pull; hand push)

Body forces (caused by another body without direct contact, e.g., weight; magnetic forces).

2.2.1 Free-Body Diagrams

Free a body of interest by cutting (virtual) at appropriate sections, removing (virtual) joints or supports, etc. On a sketch of this "free body," mark the loads (forces and moments) at these cut/released locations (in addition to other external loads). The result is a free-body diagram. It is useful in determining unknown forces/moments (e.g., external reactions; internal forces/moments).

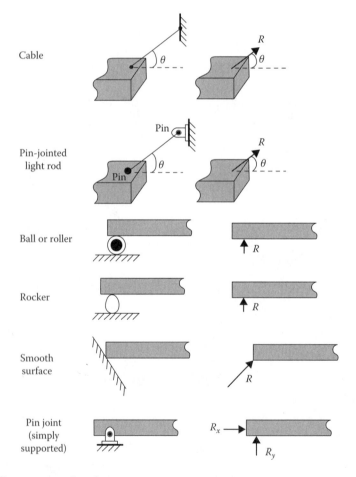

FIGURE 2.2 Common examples of supports.

Method

- Based on what unknowns are to be determined, isolate (free) the part (body) of interest from the overall system (using a "virtual" cut).
- Mark the forces/moments at the interface formed by the cut (at the correct location and in the correct direction—vectors).
- Mark all other external forces and moments present in the isolated body.
- Write equations of equilibrium (for forces and moments) and solve for unknowns.

Example 2.2

A structure consisting of two identical light beams AB and AC of length $L = 2$ m is pin-jointed at A, fixed at B, and supported on smooth rollers at C (Figure 2.3a). Beam AB is vertical and beam AC is horizontal. A cable is attached to the midpoint of AC at 45°, passed over a smooth pulley and carries a load of mass 20 kg, as shown.

Determine the external reaction loads at the supports B and C and the internal reactions at joint A of the two beam segments.

Note 1: Joint A does not have an external support.

Note 2: Neglect the beam thickness compared to the beam length.

Solution

The free-body diagrams of the two beam segments are shown in Figure 2.3b.

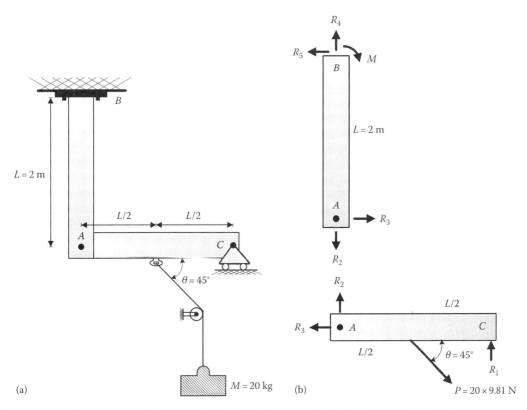

FIGURE 2.3 (a) A light structure carrying a heavy load; (b) free-body diagrams of the two beam segments.

Observations

1. Support B has a bending moment (M) as well as a shear force (R_5) in view of the fixed (i.e., clamped) support there.
2. Joint A is "smooth" and does not have a bending moment (because it is a pin joint). The reaction forces there are equal and opposite for the two beams (because there is no external load there).
3. There is only a vertical reaction force at C (because of smooth rollers → no horizontal force, no bending moment).

Equilibrium of AC (see Equation 2.3):

$$\rightarrow \sum F_x = 0 \rightarrow -R_3 + P\cos\theta = 0 \rightarrow R_3 = P\cos\theta$$

$$R_3 = 20 \times 9.81 \times \frac{1}{\sqrt{2}} \text{ N} = 138.7 \text{ N}$$

$$\curvearrowleft \sum M_A = 0 \rightarrow R_1 \times L - P\sin\theta \times \frac{L}{2} = 0 \rightarrow R_1 = \frac{P}{2}\sin\theta = \frac{1}{2} \times 20 \times 9.81 \times \frac{1}{\sqrt{2}} \text{N} = 69.35 \text{ N}$$

Note: We have neglected the beam thickness compared to the beam length. Otherwise, there would be an additional moment component in the ccw direction, about A (i.e., $P\cos\theta \times (h/2)$, where, h is the thickness of beam AC)

$$\uparrow \sum F_y = 0 \rightarrow R_2 - P\sin\theta + R_1 = 0 \rightarrow R_2 = P\sin\theta - R_1$$

$$\rightarrow R_2 = P\sin\theta - \frac{1}{2}P\sin\theta = \frac{1}{2}P\sin\theta = 69.35 \text{ N}$$

Equilibrium of AB:

$$\rightarrow \sum F_x = 0 \rightarrow R_3 - R_5 = 0 \rightarrow R_5 = R_3 = 138.7 \text{ N}$$

$$\uparrow \sum F_y = 0 \rightarrow -R_2 + R_4 = 0 \rightarrow R_4 = R_2 = 69.35 \text{ N}$$

$$\curvearrowleft \sum M_A = 0 \rightarrow -M + R_3 \times L = 0 \rightarrow M = R_3 \times L \rightarrow M = 138.7 \times 20 \text{ N·m} = 277.4 \text{ N·m}$$

Main Learning Objectives

1. Proper use of free-body diagrams
2. Nature of forces and moments at joints and support locations
3. Concepts of shear force and bending moment
4. Application of equilibrium equations to solve problems of statics

■ **End of Solution**

Example 2.3

A light uniform beam AE of length 4 m is clamped at end A and maintained horizontally, as shown in Figure 2.4a. A vertically downward load $P = 4$ kN is applied at end E. Also, the beam is supported by an inclined cable connected to the beam midspan at angle $\theta = 45°$ with tension $T = 5$ kN.

Statics: A Review

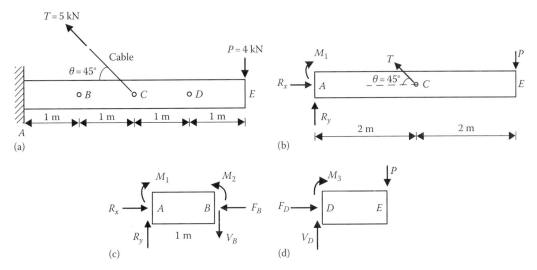

FIGURE 2.4 (a) A light uniform clamped beam carrying an end load and supported by a cable. (b) Free-body diagram of the beam. (c) Free-body diagram of beam segment AB. (d) Free-body diagram of beam segment DE.

a. What are the reaction loads at the clamped support A?
b. Determine the internal loading (forces and moments) present on the beam cross sections at B and D (at the ¼th and the ¾th, respectively, of the beam length from A)

Solution

a. Free-body diagram of the beam is shown in Figure 2.4b.

   Equilibrium equations:

$$\rightarrow \sum F_x = 0 : R_x - T\cos\theta = 0 \Rightarrow R_x = T\cos\theta = 5 \times \frac{1}{\sqrt{2}} = 3.54 \text{ kN}$$

$$\uparrow \sum F_y = 0 : R_y + T\sin\theta - P = 0 \Rightarrow R_y = P - T\sin\theta = 4 - 5 \times \frac{1}{\sqrt{2}} = 0.46 \text{ kN}$$

$$\circlearrowleft \sum M_A = 0: -M_1 + T\sin\theta \times 2 - P \times 4 = 0$$

$$\Rightarrow M_1 = 2T\sin\theta - 4P = 2 \times 5 \times \frac{1}{\sqrt{2}} - 4 \times 4 \text{ kN} \cdot \text{m} = -8.93 \text{ kN} \cdot \text{m}$$

   Note: R_x is the axial reaction (compressive)
   R_y is the shear force (+ve upward)
   M_1 is the bending moment (+ve when bending inward to form a "basin")

b. To determine the internal loading at X-section B, make a virtual cut across B and consider the segment AB. The resulting free-body diagram is shown in Figure 2.4c.

   Equilibrium equations:

$$\rightarrow \sum F_x = 0 : R_x - F_B = 0 \Rightarrow F_B = R_x = 3.54 \text{ kN}$$

$$\uparrow \sum F_y = 0: R_y - V_B = 0 \Rightarrow V_B = R_y = 0.46 \text{ kN}$$

$$\curvearrowleft \sum M_B = 0: M_2 - M_1 - R_y \times 1 = 0 \Rightarrow M_2 = R_y \times 1 + M_1$$

$$M_2 = 0.46 \times 1 - 8.93 \text{ kN m} = -8.47 \text{ kN m}$$

(*Note*: The negative sign of the result indicates that M_2 acts in the opposite direction to what is marked.)
R_x is the axial force (compressive) in the X-section at B
R_y is the shear force (downward) in the X-section at B
M_2 is the bending moment (+ve as shown in figure) in the X-section at B

To determine the internal loading at X-section D, make a virtual cut across D and consider the segment DE. The resulting free-body diagram is shown in Figure 2.4d.

Equilibrium equations:

$$\rightarrow \sum F_x = 0: F_D = 0$$

$$\uparrow \sum F_y = 0: V_D - P = 0 \Rightarrow V_D = P = 4.0 \text{ kN}$$

$$\curvearrowleft \sum M_D = 0: -M_3 - P \times 1 = 0 \Rightarrow M_3 = -P \times 1 = -4.0 \text{ kN} \cdot \text{m}$$

F_D is the axial force in the X-section at D
V_D is the shear force (upward) in the X-section at D
M_3 is the bending moment (+ve as shown) in the X-section at D

Main Learning Objectives

1. Proper use of free-body diagrams
2. Nature of forces and moments at support locations
3. Concepts of shear force and bending moment
4. Determination of internal loading
5. Sign convention for shear force and bending moment
6. Application of equilibrium equations to solve problems of statics

■ **End of Solution**

2.2.2 Principle of Transmissibility

The external reactions (at the supports) caused by a load are the same regardless of the point of application along its line of action. This is the principle of transmissibility and is illustrated in Figure 2.5.

Proof

The contribution of the force (*P*) to the equations of equilibrium (force balance and moment balance) will be the same in the two cases (i.e., same force and same moment of *P*), because the line of action is the same. So, in order to satisfy the conditions of equilibrium, the support reactions must be the same in the two cases.

Note 1: Body may deform differently (if not rigid), but the principle still holds for a problem of statics.

Note 2: This principle is not generally true for problems of dynamics.

Statics: A Review

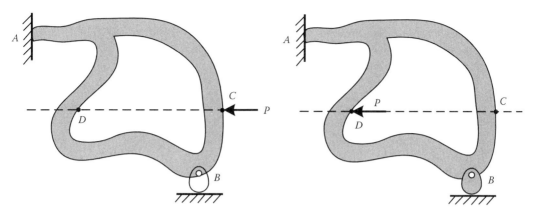

FIGURE 2.5 Illustration of the principle of transmissibility.

Example 2.4

A light L-shaped structural member is made by welding two straight segments AB (of length 0.4 m) and BC (of length 0.3 m) at right angles, as shown in Figure 2.6a. It is hung from a smooth pivot at end A, and a weight of mass $m = 50$ kg is hung from corner B. The end C is pulled by a straight cable in the direction AC, with a force $F = 200$ N.

a. Show that, under conditions of static equilibrium, the segment AB will remain vertical.
b. Determine the external reaction at pivot A. (*Note:* This reaction is exerted by the pivot on the structural member, in maintaining the equilibrium.)

Solution

a. From the principle of transmissibility (or common sense), the cable tension F may be considered to be applied at A of the L-shaped structure, in the direction AC.

If AB is not vertical, B will have a horizontal offset from A. Then, due to the weight mg of the mass hung at B, there will be a moment ($= mg \times$ horizontal offset) at A. This moment cannot be sustained because the support at A is a pin joint (which is smooth and cannot exert a moment). Hence, AB must be vertical (*Note:* This is a proof by contradiction).

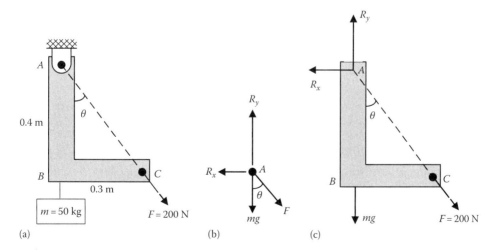

FIGURE 2.6 (a) An L-shaped structure carrying a weight and pulled by a cable; (b) loads at pivot A; (c) free-body diagram of the L-shaped component.

b. Isolate the pivot at A and consider the loads acting on it (which remains in equilibrium), as shown in Figure 2.6b. This is the "method of joints" as we will see later in this chapter.

Note: There is no moment (torque) there because of the smooth pin.
R_x is the horizontal component of the reaction (taken as +ve when acting to the left)
R_y is the vertical component of the reaction (taken as +ve when acting upward)
Equilibrium conditions (see Equation 2.3):

$$\rightarrow \sum F_x = 0 \rightarrow -R_x + F\sin\theta = 0 \rightarrow R_x = F\sin\theta = 200 \times \frac{0.3}{0.5} = 120\text{ N}$$

Note: From geometry

$$\sin\theta = \frac{0.3}{0.5};\ \cos\theta = \frac{0.4}{0.5}$$

$$\uparrow \sum F_y = 0 \rightarrow R_y - F\cos\theta - mg = 0 \rightarrow R_y = F\cos\theta + mg = 200 \times \frac{0.4}{0.5} + 50 \times 9.81 = 650.5\text{ N}$$

Alternative Approach

Draw the free-body diagram of the L-shaped structural element, as shown in Figure 2.6c. Write equations of equilibrium for the free body. We will get exactly the same equations as from the method of joints.

Main Learning Objectives

1. Ideal case of a "smooth" pivot
2. Nature of forces and moments at joints
3. Application of the principle of transmissibility
4. Application of equilibrium equations to solve statics problems
5. Introduction to the "method of joints"
6. Equivalence of the method of joints to the method of free-body diagrams

■ **End of Solution**

2.3 ANALYSIS OF TRUSSES

A truss is a structure that is made by joining together several straight members whose length is much larger than the cross-sectional dimensions. An ideal truss has light members (i.e., weight is negligible) that are rigid and joined using pins (i.e., frictionless joints, which cannot sustain moments or torques).

Learning Objective

To determine the loads in the members of a truss (e.g., bridge, building) and the external reactions.

Sign Convention: Tension in a member is represented by a +ve value → Compression is indicated by a −ve value.

In solving a truss problem, equilibrium conditions are written for carefully selected parts of the truss. Free-body diagrams are useful here. The method of joints and/or method of sections may be applied in solving a truss problem.

2.3.1 Method of Joints

- Isolate the joint of interest (using a virtual cut around the joint)—i.e., draw the free-body diagram of the joint
- Mark the forces acting at the joint by the members (rods)

Statics: A Review

- Write the equilibrium equations for the joint
- Solve for the unknowns

2.3.2 METHOD OF SECTIONS

- Separate (virtually) the truss into two segments by a sectioning line (i.e., by using a virtual cut across the truss).
- For the segment of interest isolated in this manner, mark (a) forces in the cut members; (b) external loads (known); and (c) reactions at the supports (unknown).
- Write the equilibrium equations for the isolated truss segment.
- Solve for the unknowns.

Consider the truss shown in Figure 2.7a. The application of the method of joints to joint A is illustrated in Figure 2.7b. The application of the method of sections for the left-hand side (LHS) segment that is isolated through the indicated virtual cut is shown in Figure 2.7c.

2.3.3 TWO-FORCE MEMBERS

These are light, rigid rods with pin-jointed ends. They are approximations of practical members. The pin joints are assumed frictionless. Hence they can support end forces only. End moments (bending moments, torques) cannot exist.

They are called two-force members in view of an important property, which can be established in view of the indicated assumptions.

Note:

- For equilibrium of the member: the two end forces must be equal and opposite; and they must be collinear with (i.e., they must fall along) the line joining the end joints (see Figure 2.8a).
- The shape of the member has no effect on the definition of a two-force member (see Figure 2.8b).
- Weight of the member or an external load applied on the member will violate the two-force condition (see Figure 2.8c). In particular, a heavy rod is not a two-force member.

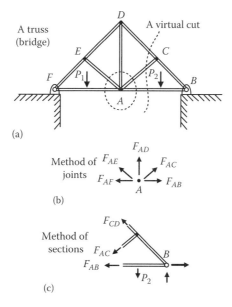

FIGURE 2.7 (a) A truss; (b) isolation of a joint; (c) isolation of a segment.

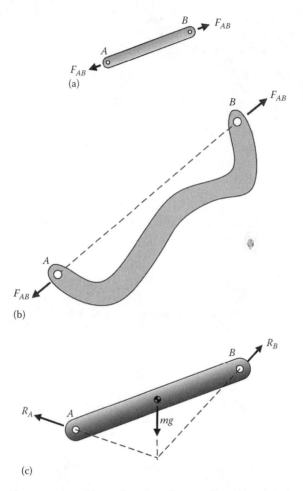

FIGURE 2.8 (a) A two-force member; (b) member shape has no effect; (c) weight (or external load) violates the requirement for a two-force member.

Example 2.5

The landing gear (see Figure 2.9a) is the structural component of an aircraft. A 2-D model of the landing gear is shown in Figure 2.9b. The links are assumed two-force members (light, pin-jointed). Also assume that the aircraft body applies two symmetric forces P on the landing gear, as shown. Determine the loads in the three structural links of the landing gear, and the ground reaction on the wheel.

Note: Assume a symmetrical structure.

Solution

Use the method of joints.
Equilibrium at joint A (Figure 2.9c):

$$\rightarrow \sum F_x = 0 \Rightarrow F_1 - P\cos\alpha + F_2\cos\theta = 0 \qquad \text{(i)}$$

$$\uparrow \sum F_y = 0 \Rightarrow F_2\sin\theta - P\sin\alpha = 0 \qquad \text{(ii)}$$

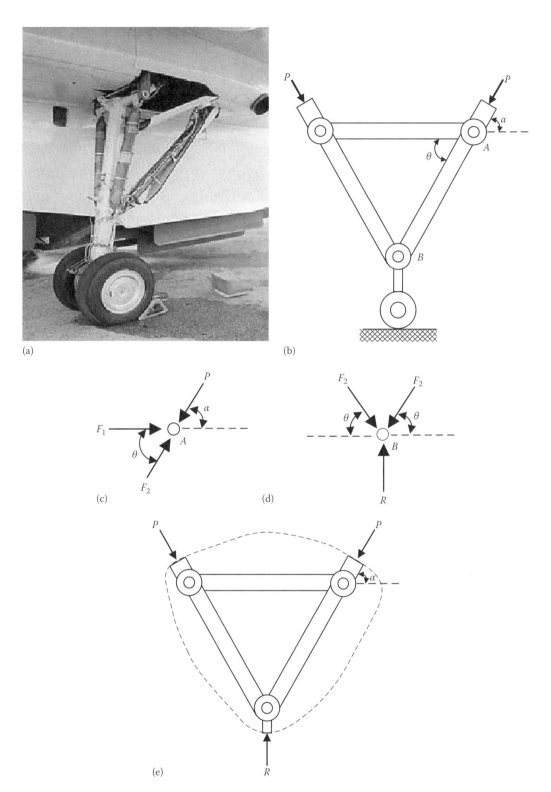

FIGURE 2.9 (a) An aircraft landing gear. (b) A 2-D model of the landing gear structure. (c) Forces at joint A. (d) Forces at joint B. (e) Application of the method of sections.

We get from (ii) $F_2 = P(\sin\alpha/\sin\theta)$
Substitute in (i)

$$F_1 = P\cos\alpha - F_2\cos\theta = P\cos\alpha - P\frac{\sin\alpha\cos\theta}{\sin\theta}$$

or

$$F_1 = P\left(\cos\alpha - \frac{\sin\alpha}{\tan\theta}\right)$$

Note: As marked, F_1 is the "compressive" force of the horizontal rod, and F_2 is the "compressive" force of the inclined rod (*AB*). That is, if their numerical values are +ve, they are compressive forces; if their numerical values are −ve, they are tensile forces. This is the opposite of the normal sign convention, but is used in this problem because we know that, in practice, *AB* must be in compression (i.e., F_2 must be +ve). However, F_1 can be +ve or −ve depending on the values of the angles α and θ.

Now assume a roller contact with the ground.
Equilibrium at joint *B* (Figure 2.9d):

$$\rightarrow \sum F_x = 0 \Rightarrow F_2\cos\theta - F_2\cos\theta = 0 \Rightarrow \text{Satisfied}$$

$$\uparrow \sum F_y = 0 \Rightarrow R - F_2\sin\theta - F_2\sin\theta = 0$$

$$\Rightarrow R = 2F_2\sin\theta \qquad (iii)$$

Substitute the previous result for F_2

$$R = 2P\frac{\sin\alpha}{\sin\theta}\sin\theta \rightarrow R = 2P\sin\alpha$$

Note: The result for the ground reaction *R* may be obtained using the method of sections as well, as shown now. Consider the isolated segment through the virtual cut shown in Figure 2.9e. Equilibrium of the cut segment:

$$\uparrow \sum F_y = 0 \Rightarrow -P\sin\alpha - P\sin\alpha + R = 0$$

Hence $R = 2P\sin\alpha$
This is the same result that we obtained from the method of joints.

Primary Learning Objectives
1. Determination of reactions and internal loads in a statics problem
2. Application of the methods of joints
3. Use of symmetry (of structure and loading)
4. Application of the method of sections
5. Simplifying the solution by selecting the proper free body for analysis

■ **End of Solution**

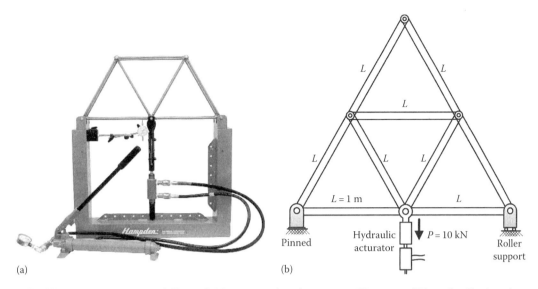

FIGURE 2.10 (a) A commercially available structural testing system. (Courtesy of Hampden Engineering, East Longmeadow, MA.) (b) An example of a structural test rig.

Example 2.6

Structural testing systems are commercially available and are widely used in testing for structural integrity under different loading and supporting conditions. A commercial system is shown in Figure 2.10a. Consider the truss shown in Figure 2.10b. Suppose that the test rig applies a vertical load $P = 10$ kN at the midspan of the truss using a hydraulic actuator. The members of the truss are light and identical, each of length $L = 1$ m. The joints of the rods are pinned (frictionless). One support of the structure is pinned while the other support has smooth rollers.

a. Determine the support reactions of the truss, and the loads in the members (take tensile loads as +ve).
b. If the vertical load P is moved from the midspan to the apex, determine the support reactions of the truss, and the loads in the members (rods).
c. Compare the results in the two cases.

Solution

a. The free-body diagram of the entire truss is shown in Figure 2.11a.

Note: No horizontal reaction at support D (because of the smooth rollers).

Equilibrium equations:

$$\rightarrow \sum F_x = 0 \rightarrow -R_x + 0 = 0 \rightarrow R_x = 0$$

$$\uparrow \sum F_y = 0 \rightarrow R_1 - P + R_2 = 0$$

$$\curvearrowleft \sum M_A = 0 \rightarrow -P \times L + R_2 \times 2L = 0 \rightarrow R_2 = \frac{P}{2}$$

Substitute

$$R_1 = \frac{P}{2}$$

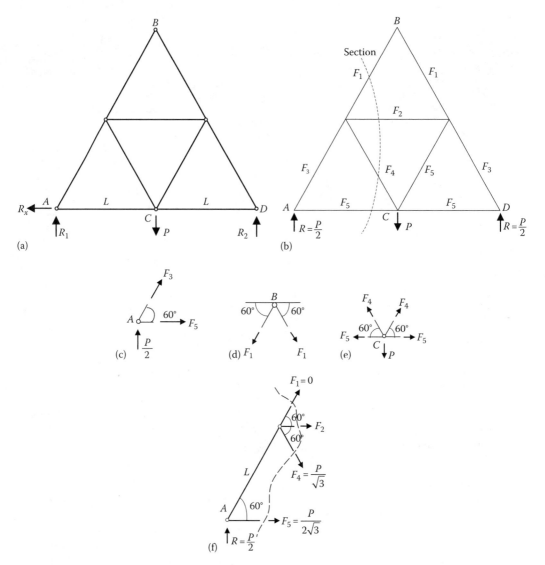

FIGURE 2.11 (a) Free-body diagram of the truss. (b) Symmetric loadings in the truss members. (c) Loading at joint A. (d) Loading at joint B. (e) Loading at joint C. (f) Loading on the sectioned segment.

Note: The truss and the external loading (including support reactions) are symmetric. Hence, the loads in the members are also symmetric.

The tensions in the members are denoted as in Figure 2.11b.

Apply the method of joints to the joints A, B, and C.

Joint A (Figure 2.11c):

$$\rightarrow \sum F_x = 0 \rightarrow F_3 \cos 60° + F_5 = 0 \rightarrow F_5 = -\frac{F_3}{2}$$

$$\uparrow \sum F_y = 0 \rightarrow F_3 \sin 60° + \frac{P}{2} = 0 \rightarrow F_3 \times \frac{\sqrt{3}}{2} + \frac{P}{2} = 0 \rightarrow F_3 = -\frac{P}{\sqrt{3}} \rightarrow F_5 = \frac{P}{2\sqrt{3}}$$

Joint B (Figure 2.11d):

$$\rightarrow \sum F_x = 0 \rightarrow -F_1 \cos 60° + F_1 \cos 60° = 0 \leftarrow \text{Satisfied}$$

$$\uparrow \sum F_y = 0 \rightarrow -F_1 \sin 60° - F_1 \sin 60° = 0 \rightarrow F_1 = 0$$

Joint C (Figure 2.11e):

$$\rightarrow \sum F_x = 0 \rightarrow -F_5 - F_4 \cos 60° + F_5 + F_4 \cos 60° = 0 \leftarrow \text{Satisfied}$$

$$\uparrow \sum F_y = 0 \rightarrow F_4 \sin 60° + F_4 \sin 60° - P = 0 \rightarrow 2F_4 \times \frac{\sqrt{3}}{2} - P = 0 \rightarrow F_4 = \frac{P}{\sqrt{3}}$$

Apply the method of sections to the LHS segment that results from the virtual section shown in Figure 2.11b (see Figure 2.11f).

$$\rightarrow \sum F_x = 0 \rightarrow F_5 + F_4 \cos 60° + F_2 + F_1 \cos 60° = 0 \rightarrow \frac{P}{2\sqrt{3}} + \frac{1}{2} \times \frac{P}{\sqrt{3}} + F_2 + 0 = 0 \rightarrow F_2 = -\frac{P}{\sqrt{3}}$$

$$\uparrow \sum F_y = 0 \rightarrow \frac{P}{2} - F_4 \sin 60° + F_1 \sin 60° = 0 \rightarrow \frac{P}{2} - \frac{P}{\sqrt{3}} \times \frac{\sqrt{3}}{2} = 0 \leftarrow \text{Satisfied}$$

$$\curvearrowleft \sum M_A = 0 \rightarrow -F_2 \times L \sin 60° - F_4 L \sin 60° = 0 \rightarrow F_2 + F_4 = 0$$

$$\rightarrow -\frac{P}{\sqrt{3}} + \frac{P}{\sqrt{3}} = 0 \leftarrow \text{Satisfied}$$

b. With the test load applied at B instead of C, denote the resulting loads using primes. The reactions are obtained using the free-body diagram of Figure 2.12a.

$$\rightarrow \sum F_x = 0 \rightarrow -R'_x + 0 = 0 \rightarrow R'_x = 0$$

$$\uparrow \sum F_y = 0 \rightarrow R'_1 - P + R'_2 = 0$$

$$\curvearrowleft \sum M_A = 0 \rightarrow -P \times L + R'_2 \times 2L = 0 \rightarrow R'_2 = \frac{P}{2} \rightarrow R'_1 = \frac{P}{2}$$

Joint A (Figure 2.12b):
As mentioned earlier (since the condition is identical)

$$R'_3 = -\frac{P}{\sqrt{3}}$$

32   Mechanics of Materials

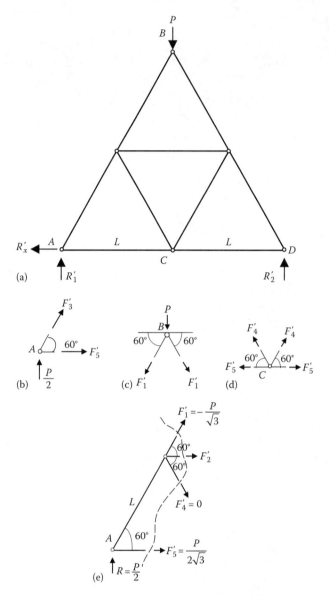

FIGURE 2.12 (a) Free-body diagram of the truss with modified external loading. (b) Loading at joint A. (c) Loading at joint B. (d) Loading at joint C. (e) Loading on the sectioned segment.

$$F_5' = \frac{P}{2\sqrt{3}}$$

Joint B (Figure 2.12c):

$$\rightarrow \sum F_x = 0 \rightarrow F_1'\cos 60° - F_1'\cos 60° = 0 \leftarrow \text{Satisfied}$$

$$\uparrow \sum F_y = 0 \rightarrow -P - F_1'\sin 60° - F_1'\sin 60° = 0 \rightarrow F_1' = -\frac{P}{\sqrt{3}}$$

Joint C (Figure 2.12d):

$$\rightarrow \sum F_x = 0 \rightarrow -F_5' - F_4' \cos 60° + F_5' + F_4' \cos 60° = 0 \leftarrow \text{Satisfied}$$

$$\uparrow \sum F_y = 0 \rightarrow F_4' \sin 60° + F_4' \sin 60° = 0 \rightarrow F_4' = 0$$

Apply the method of sections to the LHS segment that results from the virtual section shown in Figure 2.11b (see Figure 2.12e).

$$\rightarrow \sum F_x = 0 \rightarrow F_5' + F_4' \cos 60° + F_2' + F_1' \cos 60° = 0 \rightarrow \frac{P}{2\sqrt{3}} + 0 + F_2' - \frac{P}{\sqrt{3}} \times \frac{1}{2} = 0 \rightarrow F_2' = 0$$

$$\uparrow \sum F_y = 0 \rightarrow \frac{P}{2} - F_4' \sin 60° + F_1' \sin 60° = 0 \rightarrow \frac{P}{2} - 0 - \frac{P}{\sqrt{3}} \times \frac{\sqrt{3}}{2} = 0 \leftarrow \text{Satisfied}$$

$$\curvearrowleft \sum M_A = 0 \rightarrow -F_2' \times L \sin 60° - F_4' L \sin 60° = 0 \rightarrow -0 = 0 \leftarrow \text{Satisfied}$$

c.

Load	Case A	Case B
R	$\dfrac{P}{2}$	$\dfrac{P}{2}$
F_1	0	$-\dfrac{P}{\sqrt{3}}$
F_2	$-\dfrac{P}{\sqrt{3}}$	0
F_3	$-\dfrac{P}{\sqrt{3}}$	$-\dfrac{P}{\sqrt{3}}$
F_4	$\dfrac{P}{\sqrt{3}}$	0
F_5	$\dfrac{P}{2\sqrt{3}}$	$\dfrac{P}{2\sqrt{3}}$

Observation 1:

The reaction loads are identical in the two cases. This should be the case because the line of action of the applied external load is the same in the two cases (see the principle of transmissibility).

Observation 2:

The internal loading in the members is not identical in the two cases. In particular, the values of F_1, F_2, and F_4 have changed. *Note:* Those loads that are directly determined by the support reactions do not change.

Primary Learning Objectives

1. Method of joints
2. Method of sections
3. The use of symmetry (in the structure and the loading)
4. Directions of the reactions depend on the support characteristics (pinned, roller)
5. If the values and the action lines of the applied forces are the same, the support reactions will be the same (principle of transmissibility)

■ **End of Solution**

2.4 DISTRIBUTED FORCES

A distributed load may be replaced by an equivalent point load having the same total magnitude and acting at the centroid of the profile of the distributed load.
Then

- The resulting external reactions at supports will be the same.
- The effects at a sufficiently farther location from the distributed load will be (approximately) the same.

Proof (see Figure 2.13)

Note: By definition, the equivalent force must have the same contribution to the equations of equilibrium as that of the original distributed force (i.e., same resultant force in a specific direction; same resultant moment about a specific axis).

Denote the 1-D force profile (force per unit length along the x direction) by $f(x)$.

Note: In Figure 2.13a what is plotted is not the force but the "rate of change" of force with respect to the location (x). The force itself is very small when $f(x)$ is finite, and is given by $f(x) \cdot \delta x$ at location x, where δx is very small and approaches zero. Finite $f(x)$ is the typical case of distributed loading. For a point force, the force per unit length is infinite at the force location, and zero elsewhere (analytically, $f(x)$ of a point force is represented by a Dirac delta function). It follows that $f(x)$ is not the force itself but a representation of its "intensity." Another form of intensity of a force is "stress," which is force per unit area. This is discussed in Chapter 3.

By definition, the magnitude of the equivalent force must be equal to the resultant magnitude of the distributed force. Hence

$$F = \int f(x)dx = \text{Area of the force profile} \tag{2.4}$$

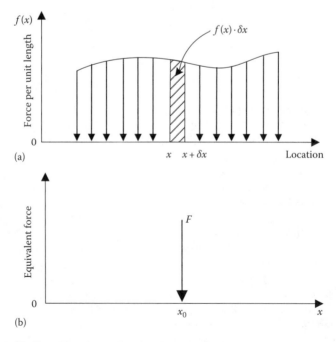

FIGURE 2.13 (a) A distributed load (1-D); (b) equivalent point load (acts at the centroid of the distributed load).

Statics: A Review

Let location of the equivalent point force be x_0 along the x axis (Figure 2.13b).

The moment of the forces about a normal axis must be the same for the two cases. (*Note*: This is true by definition of the equivalent force.)

We have

$$Fx_0 = \int x \cdot f(x) dx$$

Hence,

$$x_0 = \frac{\int x \cdot f(x) dx}{\int f(x) dx} \tag{2.5}$$

Equation 2.5 is the definition of the centroid of the force profile. Hence, the proof is complete.

Note: The support reactions must be the same for the two cases because their contribution to the equations of equilibrium must be the same in the two cases (because, by definition, contributions of the distributed force and the equivalent force to the equations of equilibrium are the same).

Example 2.7

A uniform beam *AB* of length $L = 2.5$ m and mass $m = 100$ kg is hung from point *O* using two pieces of rope attached to its two ends, as shown in Figure 2.14a. The lengths of the rope segments are $AO = 2.0$ m and $BO = 1.5$ m. When the system is in static equilibrium, determine the

a. Angle of inclination α of the beam to the horizontal, and the tensions T_A and T_B in the rope segments
b. Internal loading on the beam *X*-section at its centroid (midpoint) *G*

Solution

a. Since the beam is uniform, its weight *mg* is uniformly distributed along the length. Hence, in determining the end reactions (i.e., the tensions in the rope segments), the beam weight can be represented by a single point force *mg* acting downward through *G*, as shown in Figure 2.14b.

For equilibrium of the beam, the line of action of the weight *mg* should pass through *O*, where the tensions T_A and T_B meet. Otherwise, the three force vectors will not add up to zero (there will be a net moment), violating the equilibrium condition. This is further confirmed by the condition of force balance, as given by the triangle of force vectors drawn in Figure 2.14c.

Since $AO = 2.0$, $BO = 1.5$, and $AB = 2.5$, from Pythagoras' theorem, angle $AOB = 90°$. Then from geometry, a circle with diameter *AB* should pass through *O*. Hence, $OG = AG = GB = 2.5/2$ m = radius of the circle.

Then, $\angle GOB = \angle GBO = \theta$.

It is clear from Figure 2.14b that $\angle AOG + \angle A = \pi/2$

$$\rightarrow \left(\frac{\pi}{2} - \theta\right) + \left(\frac{\pi}{2} - \theta + \alpha\right) = \frac{\pi}{2}$$

$$\text{Hence } \alpha = 2\theta - \frac{\pi}{2} \tag{i}$$

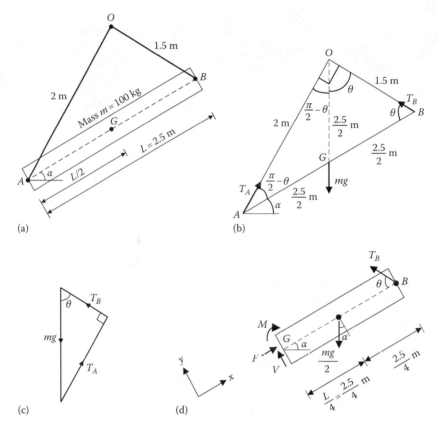

FIGURE 2.14 (a) A heavy uniform beam hung by two segments of rope. (b) Equilibrium configuration of the beam. (c) Triangle of forces (the condition of force balance). (d) Free-body diagram of the beam segment GB.

From trigonometry, for triangle AOB,

$$\sin\theta = \frac{2}{2.5} = 0.8 \Rightarrow \theta = 53.1°; \quad \cos\theta = \frac{1.5}{2.5} = 0.6$$

Hence, from (i)

$$\alpha = 2 \times 53.1° - 90° = 16.2°$$

Also, $mg = 100 \times 9.81$ N $= 981$ N
From Figure 2.14c

$$T_A = mg \sin\theta = 981 \times 0.8 = 785 \text{ N}$$

$$T_B = mg \cos\theta = 981 \times 0.6 = 589 \text{ N}$$

b. Make a virtual X-section at G and consider the beam segment GB. Its free-body diagram is shown in Figure 2.14d.

Note: Since the beam segment GB is uniform, its weight ($mg/2$) is uniformly distributed along the segment. Hence, in determining the end loads of the segment, the weight can be represented by a point force of magnitude $mg/2$ acting vertically downward through the midpoint of the segment (Figure 2.14d).

Statics: A Review

The internal loads on the X-section at G are
- F is the axial force (+ve when compressive, as marked)
- V is the shear force (+ve in the y direction, as marked)
- M is the bending moment (+ve when bending upward, as marked)

Equilibrium equations (note the x, y coordinates in Figure 2.14d):

$$\sum F_x = 0: F - \frac{1}{2}mg\sin\alpha - T_B\cos\theta = 0$$

$$\Rightarrow F = \frac{1}{2}mg\sin\alpha + T_B\cos\theta = \frac{1}{2}\times 981\times\sin 16.2° + 589\times 0.6\,\text{N} = 490\,\text{N}$$

$$\sum F_y = 0: V - \frac{1}{2}mg\cos\alpha + T_B\sin\theta = 0$$

$$\Rightarrow V = \frac{1}{2}mg\cos\alpha - T_B\sin\theta = \frac{1}{2}\times 981\times\cos 16.2° - 589\times 0.8\,\text{N} \simeq 0$$

$$\sum M_G = 0: -M - \frac{1}{2}mg\cos\alpha\times\frac{L}{4} + T_B\sin\theta\times\frac{L}{2} = 0$$

$$\Rightarrow M = T_B\frac{L}{2}\sin\theta - mg\frac{L}{8}\cos\alpha = 589\times\frac{2.5}{2}\times 0.8 - 981\times\frac{2.5}{8}\cos 16.2° = 295\,\text{N}\cdot\text{m}$$

Main Learning Objectives

1. Handling of distributed loading (distributed weight in particular)
2. Equivalent point load of a distributed load (representing weight in particular)
3. Concepts of shear force and bending moment
4. Equilibrium conditions for three forces (intersection at a common point; triangle of forces)

■ **End of Solution**

Example 2.8

Dams of water reservoirs have to be designed to resist hydrostatic pressure loading (see Figure 2.15a). Consider a 1 m horizontal segment of a uniform and vertical dam wall in a water reservoir, as shown in Figure 2.15b. Due to the water pressure, the distribution of the hydrostatic force (i.e., force per unit length or the line load) in the vertical direction of this wall segment varies from 0 at the top level of water according to the relation $w = 60\,z$ kN/m, where z is measured vertically down from the water surface. The depth of water is $L = 10$ m.

Determine
  a. The magnitude of the resultant hydrostatic force on the dam segment
  b. The location of the center of pressure on the dam segment. (*Note*: Center of pressure is the location where the resultant hydrostatic force acts on the considered surface.)

Note: Atmospheric pressure p_0 at the water surface has to be taken into account as well when determining the total force on the water side of the dam surface (i.e., p_0 has to be added to w given in the example). However, this is cancelled out by the atmospheric pressure on the other (dry) side of the dam. Hence, in this example, only the hydrostatic force (due to w) needs to be considered.

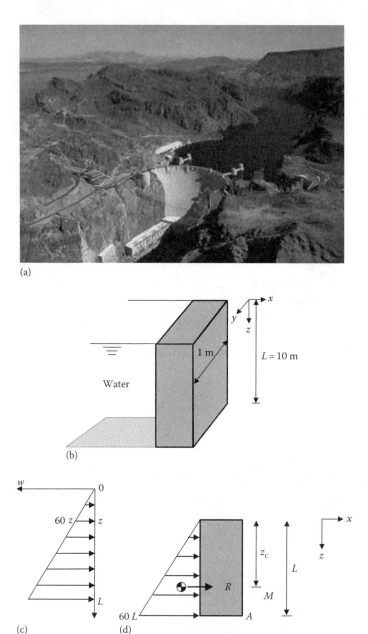

FIGURE 2.15 (a) Hoover Dam and reservoir of the Colorado River; (b) an idealized dam of a water reservoir; (c) line load on the dam wall due to hydrostatic pressure; (d) location of the center of pressure.

Solution

Loading on the 1 m wall segment (free-body diagram) from the water surface is shown in Figure 2.15c. Also shown are the depth z_c of the center of pressure and the resultant hydrostatic force R acting there.

a. Hydrostatic force acting on a strip of height δz of the dam at depth z from the water surface is $60z\delta z$.

Total resultant hydrostatic force $R = \int_0^L 60z\,dz = 60 \times (L^2/2) = 30L^2$

Statics: A Review

b. Moment of the hydrostatic force acting on the strip of height δz at a depth z, taken about the water surface, is $z \times 60z\delta z = 60z^2\delta z$.

Total moment

$$R \times z_c = \int_0^L 60z^2 dz = 60 \times \frac{L^3}{3} = 20L^3$$

Substitute the previous result for R

$$z_c = \frac{20L^3}{30L^2} = \frac{2}{3}L$$

Note: This result is confirmed by the fact that the centroid of a triangular loading profile is at $(2/3)L$ from its apex.

Main Learning Objectives

1. Handling of distributed loading (hydrostatic load in particular)
2. Magnitude and location of the equivalent point load of a distributed load (specifically a line load)
3. Center of pressure (location where the resultant hydrostatic force acts)

■ **End of Solution**

Example 2.9

A baggage trolley is made of rigid and light members, as shown in Figure 2.16a. The joints at B, E, and F use smooth pins. The base of the trolley is supported on rollers at its two ends A and C. The loading on the trolley is assumed to be uniformly distributed from D to E, at the rate w_0 per unit length, and linearly decreasing to zero from E to F, as shown.

Determine

a. The ground reactions on the rollers at A and C
b. The force in the two-force member BF and the internal loading at joint E
c. The internal loading at an X-section near (either side of) joint F
d. The internal loading at an X-section near (either side of) joint B

The relevant parameter values are $L = 1.0$ m, $w_0 = 500$ N/m.

Solution

For the purpose of determining the external reactions, the uniform load from D to E may be represented by a vertical point force $w_0 L$ acting at the midpoint of DE; and the linear load from E to F by a vertical point force $w_0 L/2$ acting at a distance $L/3$ to the right of E (i.e., $2L/3$ to the left of F). The corresponding free-body diagram of the trolley is shown in Figure 2.16b.

a. *Note*: There are no horizontal external forces. Hence, there are no horizontal reactions on the rollers at A and C (even though we have not neglected friction at these support points).

Equilibrium equations:

$$\uparrow \sum F_y = 0 : R_C + R_A - w_0 L - \frac{w_0 L}{2} = 0 \Rightarrow R_A + R_C = \frac{3w_0 L}{2} \quad \text{(i)}$$

$$\circlearrowleft \sum M_C = 0 : R_A \times 2L - w_0 L \times \frac{L}{2} - \frac{w_0 L}{2} \times \left(L + \frac{L}{3}\right) = 0$$

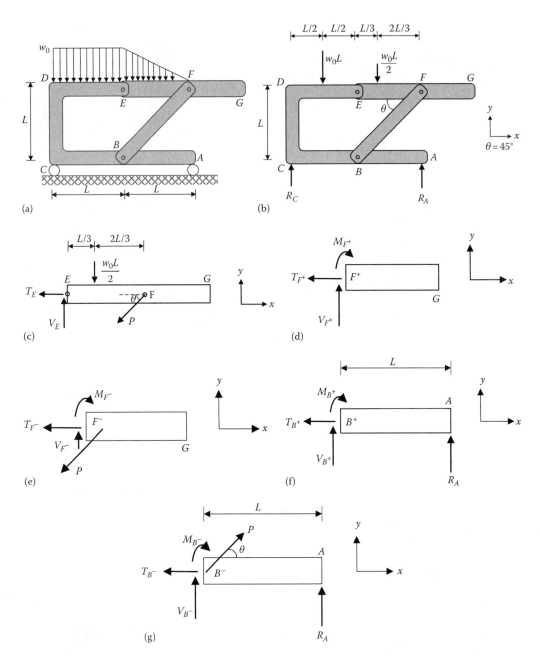

FIGURE 2.16 (a) A baggage trolley with a distributed vertical load. (b) Free-body diagram of the trolley. (c) Free-body diagram of EG. (d) Free-body diagram of F^+G. (e) Free-body diagram of F^-G. (f) Free-body diagram of B^+A. (g) Free-body diagram of B^-A.

$$\Rightarrow R_A = \frac{7}{12} w_0 L \qquad \text{(ii)}$$

Substitute (ii) in (i)

$$R_C = \frac{3}{2} w_0 L - \frac{7}{12} w_0 L = \frac{11}{12} w_0 L \qquad \text{(iii)}$$

Substitute numerical values.

ii.
$$R_A = \frac{7}{2} \times 500 \times 1 \, \text{N} = 1750 \, \text{N}$$

iii.
$$R_C = \frac{11}{12} \times 500 \times 1 \, \text{N} = 458 \, \text{N}$$

b. Virtually separate the joint *E* and joint *F* and consider the freed member *EG*. The corresponding free-body diagram is shown in Figure 2.16c.

   Note 1: Since *E* has a pin joint, there is no bending moment there.
   V_E is the shear force; T_E is the tension.

   Note 2: Load on the two-force member *BF* is *P* (along the direction of *BF*; taken +ve when in tension).

   Note 3: From geometry of the problem, $\theta = 45°$.

   Note 4: Since the entire length *EF* is retained in the free body, its distributed load is represented by the equivalent point force $w_0 L/2$ at the centroid of the distributed load, as for the previous free body.

   Equilibrium equations:

$$\rightarrow \sum F_x = 0 : -T_E - P\cos\theta = 0 \Rightarrow T_E = -P\cos\theta \tag{iv}$$

$$\uparrow \sum F_y = 0 : V_E - \frac{w_0 L}{2} - P\sin\theta = 0 \Rightarrow V_E = \frac{w_0 L}{2} + P\sin\theta \tag{v}$$

$$\Rightarrow P = -\frac{w_0 L}{6\sin\theta} \tag{vi}$$

$\Rightarrow$ Member *BF* is in compression.
Substitute (vi) in (iv)

$$T_E = \frac{w_0 L}{6\tan\theta} \tag{vii}$$

Substitute (vi) in (v)

$$V_E = \frac{w_0 L}{2} - \frac{w_0 L}{6} = \frac{w_0 L}{3} \tag{viii}$$

Substitute numerical values.

(vi) Load in the two-force member $P = -\dfrac{500 \times 1}{6 \times 1/\sqrt{2}} \, \text{N} = -118 \, \text{N}$

$\rightarrow$ *P* is a compressive force.

(vii) Joint tension $T_E = \dfrac{500 \times 1}{6 \times 1} = 83 \, \text{N}$

(viii) Joint shear force $V_E = \dfrac{500 \times 1}{3} = 167\,\text{N}$

c. Make a virtual X-section just right of F and consider the resulting free body up to G. The corresponding free-body diagram is shown in Figure 2.16d.

Equilibrium equations:

$$\rightarrow \sum F_x = 0 : -T_{F^+} = 0$$

$$\uparrow \sum F_y = 0 : V_{F^+} = 0$$

$$\curvearrowleft \sum M_{F^+} = 0 : -M_{F^+} = 0$$

Make a virtual X-section just to the left of F (also remove the two-force member at F) and consider the resulting free body up to G. The corresponding free-body diagram is shown in Figure 2.16e.

Equilibrium equations:

$$\rightarrow \sum F_x = 0 : -T_{F^-} - P\cos\theta = 0 \Rightarrow T_{F^-} = -P\cos\theta$$

Substitute (vi)

$$T_{F^-} = \dfrac{w_0 L}{6\tan\theta} \qquad (ix)$$

$$\uparrow \sum F_y = 0 : V_{F^-} - P\sin\theta = 0 \Rightarrow V_{F^-} = P\sin\theta$$

Substitute (vi)

$$V_{F^-} = -\dfrac{w_0 L}{6} \qquad (x)$$

$$\curvearrowleft \sum M_{F^-} = 0 : -M_{F^-} = 0$$

Substitute numerical values into (ix) and (x)

$$T_{F^-} = \dfrac{500 \times 1}{6 \times 1} = 83\,\text{N}$$

$$V_{F^-} = -\dfrac{500 \times 1}{6} = -83\,\text{N}$$

d. Make a virtual X-section just right of B and consider the resulting free body up to A. The corresponding free-body diagram is shown in Figure 2.16f.

Equilibrium equations:

$$\rightarrow \sum F_x = 0 : -T_{B^+} = 0$$

Statics: A Review

$$\uparrow \sum F_y = 0 : V_{B^+} + R_A = 0 \Rightarrow V_{B^+} = -R_A$$

Substitute (ii)

$$V_{B^+} = -\frac{7}{12} w_0 L \qquad (xi)$$

$$\curvearrowleft \sum M_{B^+} = 0 : -M_{B^+} + R_A \times L = 0 \Rightarrow M_{B^+} = R_A L$$

Substitute (ii)

$$M_{B^+} = \frac{7}{12} w_0 L^2 \qquad (xii)$$

Substitute numerical values.
xi.

$$V_{B^+} = -\frac{7}{12} \times 500 \times 1 \mathrm{N} = -292 \mathrm{N}$$

xii.

$$M_{B^+} = \frac{7}{12} \times 500 \times 1^2 \mathrm{N} \cdot \mathrm{m} = 292\ \mathrm{N} \cdot \mathrm{m}$$

Make a virtual X-section just to the left of B (also remove the two-force member at B) and consider the resulting free body up to A. The corresponding free-body diagram is shown in Figure 2.16g.

Equilibrium equations:

$$\rightarrow \sum F_x = 0 : -T_{B^-} + P\cos\theta = 0 \Rightarrow T_{B^-} = P\cos\theta$$

Substitute (vi)

$$T_{B^-} = -\frac{w_0 L}{6 \tan\theta} \qquad (xiii)$$

Note: Member is in compression.

$$\uparrow \sum F_y = 0 : V_{B^-} + P\sin\theta + R_A = 0 \Rightarrow V_{B^-} = -P\sin\theta - R_A$$

Substitute (vi) and (ii)

$$V_{B^-} = \frac{w_0 L}{6} - \frac{7}{12} w_0 L = -\frac{5}{12} w_0 L \qquad (xiv)$$

$$\curvearrowleft \sum M_{B^-} = 0 : -M_{B^-} + R_A \times L = 0 \Rightarrow M_{B^-} = R_A L$$

Substitute (ii)

$$M_{B^-} = \frac{7}{12} w_0 L^2 \qquad \text{(xv)}$$

Substitute numerical values.

xiii.

$$T_{B^-} = -\frac{500 \times 1}{6 \times 1} \text{N} = -83 \text{N (compression)}$$

xiv.

$$V_{B^-} = -\frac{5}{12} \times 500 \times 1 \text{N} = -208 \text{N}$$

xv.

$$M_{B^-} = \frac{7}{12} \times 500 \times 1^2 \text{Nm} = 292 \text{Nm}$$

Main Learning Objectives
1. Handling of distributed loading
2. Equivalent point load of a distributed load
3. Concepts of shear force and bending moment
4. Determining joint loads and support reactions
5. Determining internal loading
6. Use of proper free-body diagrams
7. Handling of frames and trusses having a mixture of two-force members and general members

■ **End of Solution**

2.5 STATICALLY INDETERMINATE STRUCTURES

Thus far we solved problems of statics (i.e., determined external reactions, internal loads, etc.) by using the "equations of equilibrium" alone. These are "statically determinate" problems/systems. We state the following regarding this topic:

- For some systems, equilibrium equations alone are not adequate to solve the loading problems (i.e., determination of support reactions, internal loads).
- These are "statically indeterminate" problems/systems.
- To solve a statically indeterminate problem, additional equations on how the system deforms will be necessary.
- These additional relations are typically "compatibility conditions," which may be expressed as: Deformation at a given point is the same regardless of which segment of the structure is analyzed to determine it.

In future chapters, we will study how to solve statically indeterminate problems. Now we will simply present an example of a statically indeterminate problem.

Example 2.10

A circular barrel of mass M with smooth surface is being pulled up a set of steps using a rope and a pulley (having smooth surface and frictionless axle). Some flexible and smooth material (modeled as a spring) is placed against the riser of the upper step in order to cushion the barrel against possible damage (see Figure 2.17a).

a. Write equations of equilibrium for the barrel.
b. Discuss solution for the unknown reactions on the barrel, assuming that M and the rope tension P are known.

Solution

a. Free-body diagram of the barrel is shown in Figure 2.17b.

Note: In view of the smooth surface of the barrel, reaction R_1 is vertical and reaction R_2 is radial, both passing through the center of the barrel. Reaction R_3 of the cushioning material is assumed to be horizontal (smooth contact) and passing through the center of the barrel.

Equilibrium equations:

$$\rightarrow \sum F_x = 0 \rightarrow P\cos\theta - R_3 - R_2 \cos\alpha = 0 \rightarrow R_3 + R_2 \cos\alpha = P\cos\theta \quad \text{(i)}$$

$$\uparrow F_y = 0 \rightarrow -Mg + R_1 + R_2 \sin\alpha + P\sin\theta = 0 \rightarrow R_1 + R_2 \sin\alpha = Mg - P\sin\theta \quad \text{(ii)}$$

Note: The moment equation is already satisfied since all the external forces pass through a common point (center of the barrel).

b. M, P, θ, and α are known.
There are three unknowns (R_1, R_2, R_3) and two Equations (i) and (ii). Hence, this is a statically indeterminate system.

The third equation, which is needed to solve for the unknowns, will come from the force–deflection relation of the cushioning material. This is a geometric compatibility condition.

Main Learning Objectives

1. Establishment of the correct directions of the support reactions
2. Introduction to statically indeterminate systems
3. The need for geometric compatibility relations (involving displacements and deformations) in solving statically indeterminate problems

■ **End of Solution**

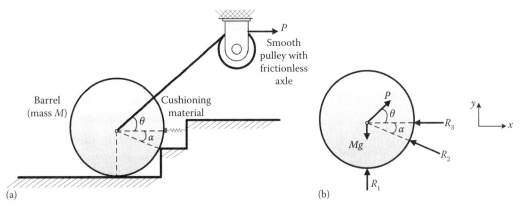

FIGURE 2.17 (a) A barrel being pulled up a set of stairs; (b) free-body diagram of the barrel.

SUMMARY SHEET

Statics: Addresses the determination of support reactions (forces and moments) and internal loads (forces and moments) for a given set of applied loads (forces and moments) on a mechanical system (body, structure, etc.). Equilibrium equations are the basis of statics.

Relevance to mechanics of materials: Internal load distribution determines the stress distribution, deformation, and the strain distribution, which are the issues addressed in mechanics of materials.

Equilibrium equations:
Sum of forces $(F_i) = 0$ in any direction
Sum of moments $(M_i) = 0$ about any point (or any axis)
Two-dimensional (2-D) case (with Cartesian axes x, y, z):

$$\sum F_x = 0; \quad \sum F_y = 0; \quad \sum M_z = 0$$

Three-dimensional (3-D) case:

$$\sum F_x = 0; \quad \sum F_y = 0; \quad \sum F_z = 0; \quad \sum M_x = 0; \quad \sum M_y = 0; \quad \sum M_z = 0$$

Support reactions: Depend on the nature of the support.

Common types of supports and their reactions:
Cable: Tensile force along the cable length
Pin-jointed Light Rod: Tensile or compressive force along the rod
Ball or Roller: Radial force through contact point
Smooth Surface: Force normal to the surface
Pin-jointed Support (or, Simple Support): Only reaction forces; no moments
Clamped End: Provides both forces and moments
Ball Bearing: No loading in the axial direction; no moments
Journal Bearing: No loading in the axial direction; can provide both forces and moments in the other directions.

Free-body diagram: Isolate a desired part of the system by means of a virtual cut. Mark the externally applied loads (forces and moments) and the loads at the cut locations.

Principle of transmissibility: Support reactions are the same if an externally applied load is moved to any point along its line of action. (*Note*: The internal loads will change in general.)

Two-force member: A pin-jointed light member with no external loading along it. Its two end forces will be collinear and equal and opposite.

Method of joints: Isolate a joint (using a virtual cut around it); mark the forces acting on the point (joint); and write equations of force balance.

Method of sections: Separate a segment of interest of a mechanical system (structure) by means of a virtual cut; mark the loads at the cut locations; and write equilibrium equations (for both forces and moments).

Equilibrium conditions for three nonparallel forces: Must pass through a common point; must be representable (in magnitude and direction) by the three sides of a triangle, in the same direction (cw or ccw).

Distributed load: Represented by the load density (e.g., force per unit length), unlike a point (concentrated) load.

For 1-D force density profile $f(x)$, along the x-axis,

Magnitude of the equivalent force: $F = \int f(x)dx$ = area of the force density profile
Location of the equivalent force = centroid of the distributed force profile:

$$x_0 = \frac{\int x \cdot f(x)dx}{\int f(x)dx}$$

Statically indeterminate systems: Systems for which equilibrium equations alone are not adequate to solve the loading problems (i.e., determination of support reactions, internal loads, etc.). Additional equations on how the system deforms (i.e., compatibility conditions) will be necessary.

PROBLEMS

2.1 A uniform disk of radius $r = 300$ mm and mass $m = 20$ kg is hung from point A of a vertical wall using a massless cable of length $L = 500$ mm, which is attached to center O of the disk (Figure P2.1). The disk rests against the wall at point B. Under equilibrium, determine the tension T of the cable and the reaction R from the wall on the disk at the point of contact B. *Note*: No need to assume smooth disk or smooth wall.

2.2 A traffic light of mass $m = 50$ kg is hung at point C from two overhead points A and B, which are 5 m apart at the same horizontal level, using two light cables of length $CA = 3$ m and $CB = 4$ m, respectively (Figure P2.2). Under static conditions, determine the tensions T_A and T_B in the two cables.

2.3 A rigid yet light handle stick of length $L = 1$ m is firmly attached to a disk of radius $r = 0.5$ m and mass $m = 50$ kg from its center O. The disk rests on a rough floor at point A and against a smooth wall at point B with the handle making angle $\theta = 30°$ above the horizontal (Figure P2.3). A force $P = 200$ N is applied at the end of the handle and normal to it. Determine the reaction forces at A and B on the disk.

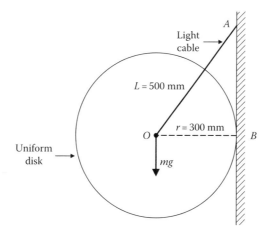

FIGURE P2.1 A disk hung by a light cable from a wall.

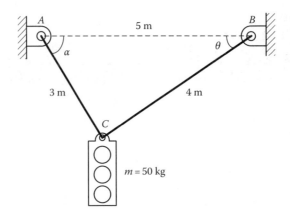

FIGURE P2.2 An overhung traffic light.

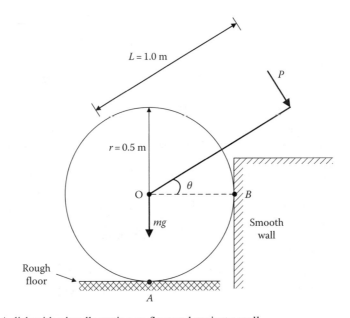

FIGURE P2.3 A disk with a handle resting on floor and against a wall.

2.4 A rigid and light shaft is supported horizontally by two ball bearings at points A and C with $AC = L = 1.0$ m. A light pulley of radius $r = 0.2$ m and carrying a load of mass $m = 20$ kg is firmly mounted at B along the shaft where $AB = L_1 = 0.6$ m. This load is supported by vertical force P applied to the end D of a handle of length $d = 0.5$ m, which is firmly attached to the end of the shaft beyond C (Figure P2.4). The axial location of the handle is given by $L_2 = 1.2$ m. Determine P and the reactions at the two bearings.

2.5 A rigid and light frame structure is made of a horizontal segment of length $AB = L = 1.0$ m, a vertical segment of length $h = 0.8$ m, and another horizontal segment of length $L/2$, as shown in Figure P2.5. End A of the structure is supported on a smooth pin while the corner B is supported on a smooth roller. A force $P = 200$ N is applied at end C, outward, making an angle $\theta = 60°$ below the horizontal there.
Determine
  a. The support reactions at A and B
  b. The internal loading on the X-section at D and E, where $AD = L/2$ and $BE = h/2$

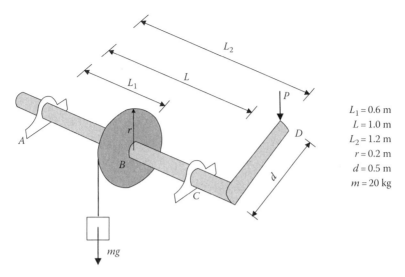

FIGURE P2.4 A pulley carrying a load and supported on a shaft with handle.

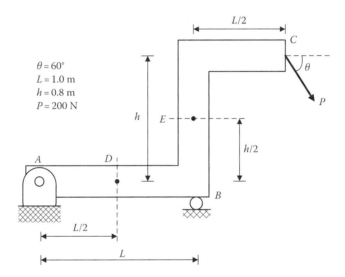

FIGURE P2.5 A rigid frame structure supported on a pin joint and a smooth roller.

2.6 Consider the two loading situations of a truss, as shown in Figure P2.6. Members $BC = 1$ m and $AC = 2$ m are light and pin-jointed at A, B, and C.
Case (a): A vertically downward force $P = 20$ kN is applied at joint C.
Case (b): A horizontal force $P = 20$ kN acting to the right is applied at joint C.
Determine the forces in BC and AC (magnitudes and directions) and the reactions at A and B (magnitudes and directions) in each of the two cases.

2.7 Determine the reactions at A and B and the loads in the members of the boom (truss) of the crane shown in Figure P2.7. All joints are frictionless (pin joints). The pulley surface is frictionless (hence cable tension does not change around it).

2.8 A pipe gripper is sketched in Figure P2.8. It has a symmetric configuration. The gripping points of the pipe are A and A'. The force P applied to the handle is taken as concentrated at G and G', and normal to the handle. The joints B, C, C', D, D', and E are assumed to contain smooth pins. The links BC, BC', and BE may be taken as two-force members.

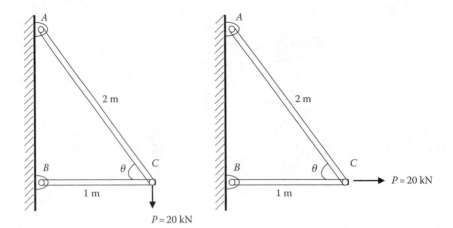

FIGURE P2.6 Two cases of loading of a truss.

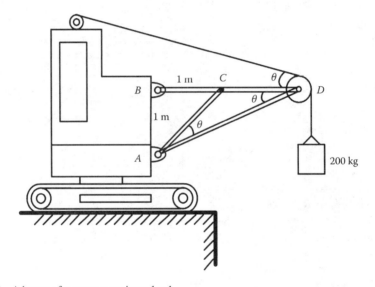

FIGURE P2.7 A boom of a crane carrying a load.

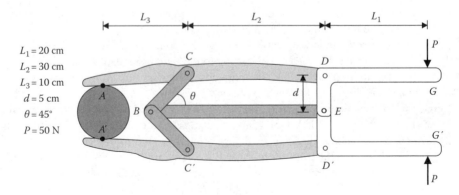

$L_1 = 20$ cm
$L_2 = 30$ cm
$L_3 = 10$ cm
$d = 5$ cm
$\theta = 45°$
$P = 50$ N

FIGURE P2.8 A pipe gripper (symmetric).

Statics: A Review

For a handle force of $P = 50$ N, determine the gripping force at A (and A'), when $\theta = 45°$. The values of the member dimensions indicated in the figure are

$$DG = L_1 = 20 \text{ cm}, \quad DC = L_2 = 30 \text{ cm}, \quad CA = L_3 = 10 \text{ cm}, \quad ED = d = 5 \text{ cm}.$$

Note 1: Assume that ACD and $A'C'D'$ are parallel straight lines.
Note 2: Do not neglect possible friction at A and A'.

2.9 Consider an inclined wall of a water tank as shown in Figure P2.9. The depth of water in the tank is $h = 2.0$ m. The point where the water surface touches the wall is at a horizontal distance $d = 1.5$ m from the base of the wall. Determine the magnitude, direction, and location (center of pressure) of the resultant hydrostatic force acting on a 1 m horizontal segment (in the z-direction) of the wall.
Note: Since atmospheric pressure acts on all sides of the wall and balances out, it does not have to be considered in this problem.

2.10 A uniform rectangular steel plate of length $L = 1.0$ m and height $h = 0.8$ m has a circular area of radius r removed centrally with its center at a distance $0.6 L$ from one edge. The plate is supported using a frictionless hinge at corner A, and its longitudinal edge is maintained horizontally by means of a vertical cable at corner B (see Figure P2.10). Area density of the

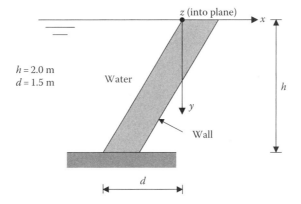

FIGURE P2.9 Water loading on an inclined wall.

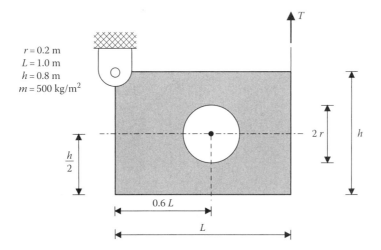

FIGURE P2.10 A uniform steel plate suspended from a hinge and a cable.

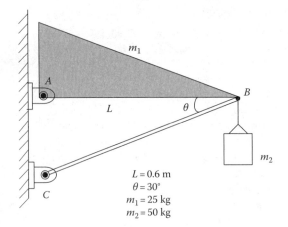

FIGURE P2.11 Nonuniform arm carrying a load.

plate is $m = 500$ kg/m². Determine the reaction at hinge A and the tension T in the cable under conditions of static equilibrium.

2.11 Consider the nonuniform arm with load, as shown in Figure P2.11. All joints (A, B, C) are pin joints (no friction; no moments). Determine the reactions at the joints and the force in the rod BC.

Note: BC is a two-force member. Hence, its end forces will be along BC.

2.12 An overhung platform is modeled as a light beam with distributed loading. It is supported by three light truss members (see Figure P2.12).

Determine the reaction loads; forces in the truss members; and the shear force, normal force, and bending moment at the beam sections through F and G.

2.13 A uniform vertical column of rectangular cross section and height $L = 10$ m is fixed at the base (see Figure P2.13a). A distributed force is applied on one side of the column, which varies from 0 at the top of the column according to $w = 60\,z$ kN/m, where z is measured vertically downward from (see Figure P2.13b). Determine

a. The vertical distribution of the shear force in the column X-section
b. The vertical distribution of the bending moment in the column X-section
c. The magnitude and the location of the maximum shear force and of the maximum bending moment in the column X-section

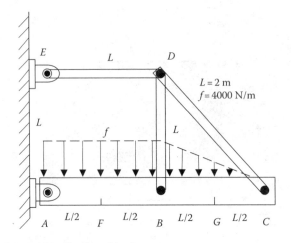

FIGURE P2.12 A platform with a distributed load.

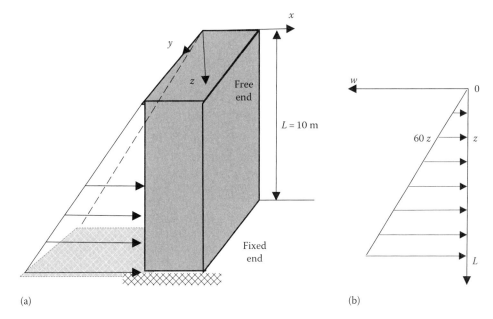

FIGURE P2.13 (a) Rectangular column with a distributed load; (b) distributed load on the column.

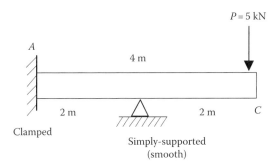

FIGURE P2.14 A cantilever beam with a peg support and an end load.

2.14 Cantilever beam of length 4 m is clamped at end A, simply supported (using a smooth peg) at midpoint B, and loaded with a vertically downward force $P = 5$ kN at free end C (Figure P2.14).
  a. Draw a free-body diagram for the beam and mark all the external loads (including the reactions at A and B).
  b. Write the possible independent equations for determining the unknowns in part (a).
  c. Is this a statically determinate beam? Why?

2.15 An example of a torque arm is shown in Figure P2.15. A vertical force $P = 100$ N is applied at the handle, which is maintained horizontally. Determine the loads at the "torquing point" C, ball bearing B, and the fixed (clamped) joint A.

Further Review Problems

2.16 The truss in Figure P2.16 is made of three two-force members. It is attached to a wall using a smooth hinge at A and a smooth hinge on smooth rollers at C. A vertically downward load P is applied at B.

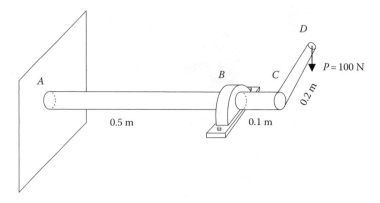

FIGURE P2.15 A torquing arm.

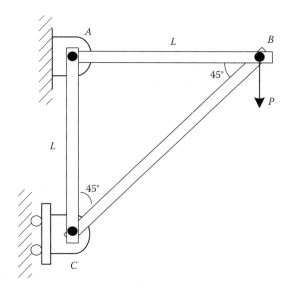

FIGURE P2.16 A simple truss carrying a load.

  a. Draw a free-body diagram of the truss and determine the support reactions at A and C.
  b. Determine the internal tensions in the two-force members.

2.17 Figure P2.17 shows a truss that is made of five two-force members. It is supported on a rigid wall using a smooth hinge at A and a hinge on rollers at D. All the joints use smooth pins. The length $AD = L = 2$ m. The angles of the members are as shown in the figure.

A vertical downward load $P = 10$ kN is applied at B. Determine the internal forces (tensions) in the five members using
  a. The method of joints.
  b. The method of sections.
  c. If the vertical downward load P is applied at C instead of B, determine the support reactions and the member tensions then.

2.18 A horizontal beam AC is hinged (smooth) at end A and simply supported (on a smooth peg) at its midspan B. A uniformly distributed downward load of value w_0 per unit length is applied from B to C. Length of the beam $= 2L$ (see Figure P2.18).

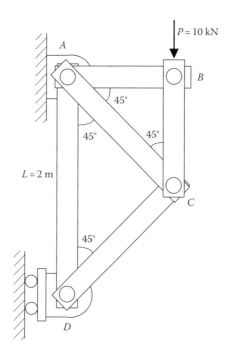

FIGURE P2.17 A truss carrying a vertical load.

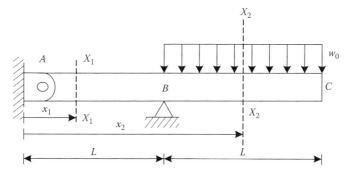

FIGURE P2.18 A beam with a distributed load.

   a. Determine the reactions at A and B.
   b. Determine the shear force and bending moment on a cross section X_1 and on a cross section X_2 (Figure P2.18).
   c. Sketch the shear force diagram and the bending moment diagram of the beam.
2.19 Beam AB of length $L = 2$ m is clamped (cantilevered) at end A and maintained horizontally using a cable of tension $T_C = 5$ kN at the other end. The cable is attached to hook at C, which is at a height $h = 0.1$ m from the beam axis, and inclined at $\theta = 45°$, as shown in Figure P2.19.
   a. Determine the loading at support location A of the beam.
   b. On a general location cross section X–X of the beam, located at a distance x from end A, determine the shear force V and the bending moment M. Using these results, sketch the shear force diagram and the bending moment diagram of the beam.
   Note: Neglect the weight of the beam.

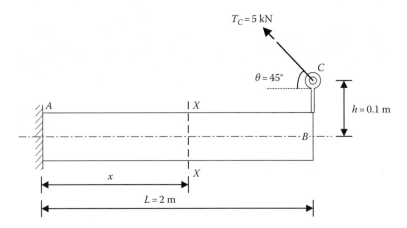

FIGURE P2.19 An end cable attached to a light cantilever.

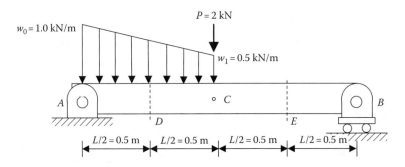

FIGURE P2.20 A beam with a point load and a distributed load.

2.20 A beam AB of length $2L = 2.0$ m is supported horizontally using a smooth hinge at end A and smooth hinge on smooth rollers at end B, as shown in Figure P2.20. A point force $P = 2$ kN acts vertically downward at the midspan C of the beam. Also, a trapezoidal distributed load acts vertically downward, which varies from $w_0 = 1.0$ kN/m at A to $w_1 = 0.5$ kN/m at C, as shown. Determine
  a. The reactions of the supports at A and B
  b. The shear force V and bending moment M on cross sections at D and E, where D is the midpoint of AC and E is the midpoint of CB

Note: Neglect the weight of the beam.

3 Stress

CHAPTER OBJECTIVES

- Understanding the meaning of stress (intensity of an internal force)
- Interpretation of stress as a second-order tensor, not a vector (it has a magnitude and a direction, but also depends on the orientation of the plane on which it acts)
- Two types of stress: normal stress and shear stress
- Normal stress in an axially loaded member (rod, bar, shaft, cable, etc.)
- Use of axial force diagram in determining the stress distribution in an axially loaded member
- Analysis and design of joints (bolted, riveted, glued, welded, keyed, etc.)
- Shear stress and normal stress in bolts
- Bolted joints with single shear and double shear
- Shear stress in a key, which connects a shaft to a disk
- Complementarity property of shear stress at a local element in an object (on a minute 2-D rectangular element, shear stresses are equal on all four sides, and equal and opposite on two opposite parallel sides)
- Stress transformation (determination of the stresses on a plane that is inclined to the plane on which stresses are known) at a point

3.1 INTRODUCTION

Mechanics of materials concerns the internal effects in a deformable solid body or a structure due to external loads (forces, moments, and torques) acting on it. Particularly useful internal effects are stresses and strains at various critical locations in the body. Stress is the "intensity" of a local force that acts on an area (a plane or a section) at a location. Strain is the normalized deformation (change in length or change in angle of a corner in a unit geometric element) at a location. Stress is a crucial consideration when designing a mechanical system such as a structure or machinery, because permanent deformation of a component occurs when its stress exceeds some limiting value (yield stress) and the failure (fracture) occurs when its stress exceeds another limiting value (ultimate stress).

In this chapter, we will study stresses. There are two types of stresses, normal stress and shear stress, and we will study both. The scenarios of axial loading and shear loading are specifically addressed in this chapter. Normal stress and shear stress both can occur under axial loading, depending on the orientation of the considered plane (section) in the member. Strains due to axial loading and shear loading, and stresses and strains due to torsional loading and bending loading will be treated in other chapters. In those chapters, we will observe that shear stresses are of primary importance in a member in torsion (e.g., in a shaft of a motor; pump; or a gear) while both normal (bending) stresses and shear stresses are of significance in a member subjected to bending (e.g., in a bridge; deck; or tall building).

Consider again the familiar example of an aircraft in flight (Figure 3.1). Under various loading conditions, we are interested in determining the types, locations, orientations, and values of the stresses in the aircraft, for instance, the stresses at the wing joints.

The stresses that an object can withstand will determine the strength of the object. In particular, *ultimate strength* (or ultimate stress) is the stress at which the material fails (fractures or ruptures); *yield strength* (or yield stress) is the stress at which the material undergoes permanent deformation (from which it cannot fully recover); and *allowable stress* is the stress level for which an object or

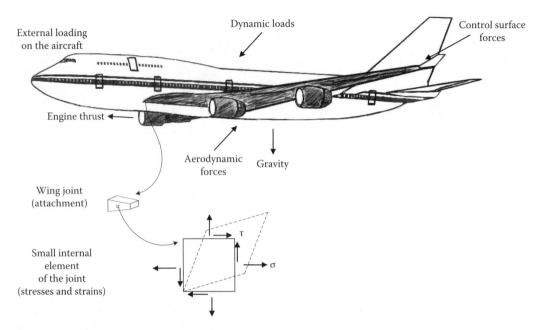

FIGURE 3.1 Stresses in an aircraft.

a component is designed so that under conditions of normal (good) operation, this level of stress will not be exceeded. (*Note*: A *factor of safety* is applied either to the ultimate strength or the yield strength in order to arrive at the allowable stress.)

3.2 DEFINITION OF STRESS

1. Stress is a force per unit area at a specific location and on a specific plane (section) at that location in a body.
2. Hence, stress is the local intensity of the force at that section. The following should be clear:
   a. For a given force, the intensity (stress) increases when the area of application decreases.
   b. For a given area of application, the intensity increases with the force.

Consider a general body under a set of external loads (see Figure 3.2a). Make a section (virtual or imaginary cut) across it (see Figure 3.2b). There will be a specific distribution of forces acting on this section, which will maintain the sectioned body (free body) in equilibrium.

Consider a small area element ΔA_z on the cut section (whose normal is defined to be along the z direction). Forces on this element may be combined into a single elemental (i.e., very small) resultant force ΔF (a vector with a magnitude and a direction). Then it may be resolved into three orthogonal elemental forces ΔF_x, ΔF_y, ΔF_z along the orthogonal directions x, y, and z in a three-dimensional (3-D) Cartesian coordinate system at the element.

Note: ΔF_z acts normal to the element plane ΔA_z, along the z direction (i.e., it is a *normal force*); ΔF_x and ΔF_y act along the element plane along the x and y directions, respectively (i.e., they are *shear forces*).

The normal force on the elemental area generates a local normal stress there, as given by

$$\sigma_z = \lim_{\Delta A_z \to 0} \frac{\Delta F_z}{\Delta A_z} \tag{3.1}$$

Stress

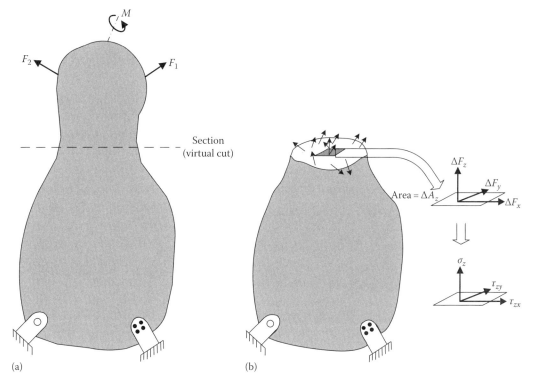

FIGURE 3.2 (a) A general body under general loading; (b) a virtually sectioned segment of the body (a free body).

Similarly, the shear forces ΔF_x and ΔF_y on the elemental area generate the local shear stresses

$$\tau_{zx} = \lim_{\Delta A_z \to 0} \frac{\Delta F_x}{\Delta A_z} \qquad (3.2)$$

$$\tau_{zy} = \lim_{\Delta A_z \to 0} \frac{\Delta F_y}{\Delta A_z} \qquad (3.3)$$

Note: The first subscript denotes the plane on which stress acts, and the second subscript denotes the direction of the stress.

Note: Even though a stress has a "magnitude and a direction," it is not a vector (unlike a force or a moment). The reason is that stress also depends on the "orientation" of the section or plane on which it acts. Specifically, one coordinate axis (first subscript of stress) should define the direction of the plane on which the stress acts (by giving the normal—perpendicular—direction to the plane), and the other coordinate axis (second subscript of the stress) should define the direction of the stress on the plane. Hence, stress is a "second-order tensor" similar to a "matrix."

- *Normal stress*: related to pulling (tensile), pushing (compressive), or bending (flexural) actions in objects
- *Shear stress*: related to "shearing," "sliding," or "twisting" actions in objects

3.3 NORMAL STRESS UNDER AXIAL LOADING

Consider a prismatic bar (i.e., a bar whose area of cross section is uniform—identical—along its length). See Figure 3.3 (shows a square X-section).

$$\text{Area of } X\text{-section} = A$$

The bar is in equilibrium under equal and opposite axial forces P at the two ends (i.e., the end forces act along the main axis of the bar).

Note: For equilibrium, the two end forces must be equal and must fall on the same straight line (i.e., *collinear*).

Make a section (a virtual cut) at C, normal to the axis of the bar.

By equilibrium, the resultant force at the X-section must be equal to P and also must be collinear with the axial end forces.

$$\text{Average normal stress at the } X\text{-section } \sigma = \frac{P}{A} \tag{3.4}$$

If the stress distribution is uniform (i.e., σ is constant) across the section, the resultant force P must act at the centroid of the section.

Proof

Take moments about the x and y axes through the point of action of the resultant force. These moments must be zero because the resultant force passes through these axes. Hence,

$$M_x = \int y\sigma \, dA = \sigma \int y \, dA = 0$$

$$M_y = \int x\sigma \, dA = \sigma \int x \, dA = 0$$

Note: σ can be taken out of the integral since it is a constant.

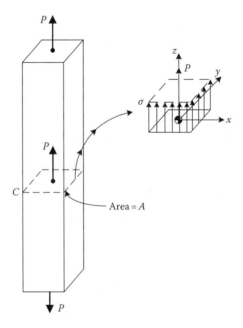

FIGURE 3.3 A prismatic rod under axial loading.

Stress

We get

$$\int y\,dA = 0; \quad \int x\,dA = 0$$

These are the required (i.e., necessary) and sufficient conditions for the centroid of a cross section.

Note: Even when the stress is not uniform, we may assume it to be uniform on sections located sufficiently far from the two ends of the rod where the loads are concentrated (see Saint-Venant's principle in Chapter 6).

Observation

As pointed out earlier, the stress at a point depends not only on the location of the point but also on the plane at the point on which the stress acts (i.e., stress is a 2-D tensor, not a vector). To emphasize, consider Figure 3.4a, which revisits the tensile loading scenario of Figure 3.3. In Figure 3.4b, we have made a virtual section on the rod perpendicular to the loading axis (such a "perpendicular" section is called a cross section or *X*-section) at point *X*. As observed earlier, in view of equilibrium

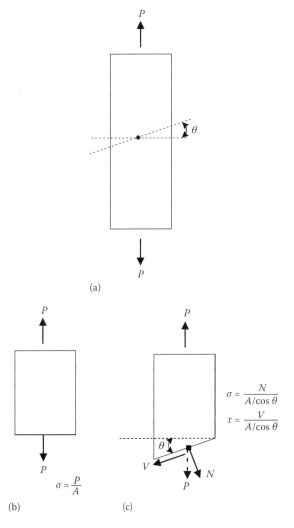

FIGURE 3.4 Section dependence of stress at a location: (a) a rod with tensile loading; (b) stress on a normal section (*X*-section) at *X*; (c) stresses on an inclined section at *X*.

of the cut portion of the rod, the only force at X is P, and it acts normal to the section. Hence, the normal stress on the sectional plane is $\sigma = P/A$, and there is no shear stress (since there is no shear force along the plane of the section, according to the equilibrium condition).

Alternatively, in Figure 3.4c, we have made a virtual section at the same point X, but this time at an angle θ to the original X-section. The area of the new section at X is $A/\cos\theta$. As mentioned earlier, for equilibrium of the isolated segment of the rod (free body), the resultant force there should be equal to P and it should fall along the same axis as the external force P that is applied at the two ends of the rod. The normal force component N and the shear force component V of this force P on this inclined section are given by $N = P \cos\theta$ and $V = P \sin\theta$. It is clear then (see Figure 3.4c) that the normal stress on the new section at X is $\sigma = N/(A \cos\theta)$, which is different from the normal stress on the previous section (X-section) at X. Furthermore, on the new section, there exists a shear stress as given by $\tau = V/(A \cos\theta)$.

3.3.1 Solution Steps for Stress Problems

1. Establish what stresses are to be determined (in what components of the body, at what locations, and on what planes—sections)
2. Solve the problem of "statics" (see Chapter 2) to determine the "internal" forces at those locations
3. Establish the areas on which these forces act
4. Force divided by the area gives the stress (at the appropriate location, on the appropriate plane, in the appropriate direction)

Note: In this manner, what we determine is the "average" stress, unless the considered area is infinitesimally small or the stress is uniform (i.e., constant) across the area.

Example 3.1

Beam Supported by Hanger Rods

Consider a uniform beam with vertical support hangers and carrying an end load (see Figure 3.5a). Assume pin joints (i.e., smooth) at A and B. The parameter values are indicated in the figure.

Determine the tensile stress in the hanger rod at A.

What should be the area of cross section of the hanger rod at B so that the magnitudes of the normal stresses in the two hanger rods would be equal?

Solution

Free-body diagram (FBD) of the beam is shown in Figure 3.5b.
Equations of equilibrium:

$$\uparrow \sum F_y = 0 \Rightarrow R_1 + R_2 - mg - P = 0 \tag{i}$$

$$\curvearrowleft \sum M_A = 0 \Rightarrow -mg \times 2L + R_2 \times 3L - P \times 4L \tag{ii}$$

$$\Rightarrow R_2 = \frac{2mg + 4P}{3}$$

Substitute (ii) into (i)

$$R_1 = mg + P - \frac{2mg + 4P}{3} = \frac{mg - P}{3} \tag{iii}$$

Stress

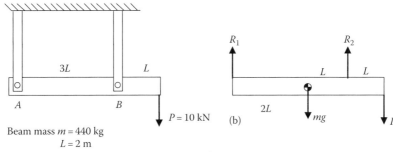

Beam mass $m = 440$ kg
$L = 2$ m
Area of X-section of hanger rod at $A = 500$ mm^2

(a)

FIGURE 3.5 (a) A horizontal beam supported by vertical hangers. (b) Free-body diagram of the beam.

Substitute numerical values

(iii) $$R_1 = \frac{400 \times 9.81 - 10{,}000}{3} \text{ N} = -2025 \text{ N}$$

(ii) $$R_2 = \frac{2 \times 400 \times 9.81 + 4 \times 10{,}000}{3} \text{ N} = 15{,}949 \text{ N}$$

Normal stress in rod at A: $\sigma_A = -\dfrac{2025}{500}$ N/mm^2 = -4.05 MPa (compressive)

Normal stress in rod at B: $\sigma_B = \dfrac{15{,}949}{a}$ N/mm^2 = $\dfrac{15{,}949}{a}$ MPa (tensile)

where a is the area of cross section of hanger rod at B.

We require: $\dfrac{15{,}949}{a} = 4.05$

$$\rightarrow a = \frac{15{,}949}{4.05} \text{ mm}^2 = 3938.0 \text{ mm}^2$$

Primary Learning Objectives

1. Solution of the problem of statics of a structure to determine the internal loading at appropriate locations
2. Determination of stresses on planes at the appropriate locations
3. Designing of structural members to maintain uniform stress

■ **End of Solution**

Example 3.2

Support Column of a Water Tank

An overhead water storage tank of mass $M = 100 \times 10^3$ kg is supported on a uniform steel column having a square box section of side a and constant thickness $t = 2$ cm, as shown in Figure 3.6a.
Height of the column $h = 5$ m
Ultimate strength of structural steel = 400 MPa

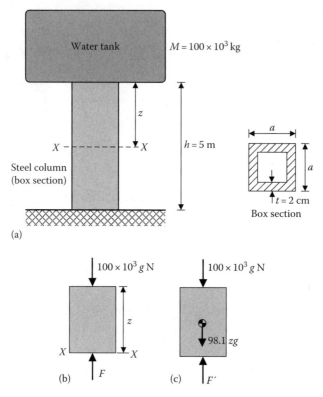

FIGURE 3.6 (a) An overhead tank and support column; (b) FBD of column segment (weight neglected); (c) FBD of column segment (weight included).

a. Neglecting the weight of the column, design a suitable section for it (i.e., select a suitable value for a).
b. If the density of structural steel is 10×10^3 kg/m³, what is the maximum average normal stress on the column X-section and what is its location?
How does this value compare with that in part (a) where the column was assumed to be light?

Note: Use a factor of safety = 4, which will take into account possible uncertainty of the true value of maximum stress (e.g., uncertainties in external loading; stresses due to bending moments and axial loads, which may result from other external loads such as seismic disturbances, wind gusts, rain, and snow; errors and uncertainties of the analysis).

Solution

a. Make a normal section (virtual) in the column at a depth z from the base of the tank, and consider the free-body diagram (FBD) of the resulting column segment above this section (Figure 3.6b):

$$R = \text{axial force at the X-section}$$

Note: There cannot be any other loads (e.g., shear forces or bending moment) at this X-section, in view of equilibrium of the considered column segment under the only other external force applied to it, which is the axial force due to the weight of the tank:

$$\text{Equilibrium equation}: \uparrow \sum F_z = 0 \Rightarrow$$

$$-Mg + F = 0 \Rightarrow F = Mg = 100 \times 10^3 \times 9.81 \, \text{N}$$

Area of X-section of the column

$$A = a^2 - (a-2t)^2 = 4at - 4t^2$$

Average stress at the X-section

$$\sigma_{AV} = \frac{F}{A} = \frac{Mg}{A} = \frac{Mg}{4at - 4t^2}$$

Allowable stress for the column design

$$\sigma_{ALW} = \frac{\sigma_{ULT}}{\text{Safety factor}} = \frac{400}{4} \text{ MPa} = 100 \text{ MPa}$$

We need $\sigma_{AV} = \sigma_{ALW}$ for safe performance.

$$\Rightarrow \frac{Mg}{A} = 100 \Rightarrow \frac{100 \times 10^3 \times 9.81 \, (N)}{A \, (mm^2)} = 100 \text{ MPa} \Rightarrow A = 9.81 \times 10^3 \text{ mm}^2$$

$$\Rightarrow 4at - 4t^2 = 9.81 \times 10^3 \Rightarrow a = \frac{9.81 \times 10^3}{4t} + t$$

$$\Rightarrow a = \frac{9.81 \times 10^3}{4 \times 20} + 20 \text{ mm} = 122.6 + 20 = 142.6 \text{ mm} \sim 5.6 \text{ in.}$$

b. *Note*: Weight per unit length × Length = Weight = Mass density × Volume × g

→ Weight per unit length of the column = Mass density × A × g

$$= 10 \times 10^3 \, (kg/m^3) \times g(m/s^2) \times 9.81 \times 10^3 \times 10^{-6} \, (m^2) = 98.1 \, gN/m$$

The new FBD of the column segment is shown in Figure 3.6c.
Equation of equilibrium: $\uparrow \sum F_z = 0 \Rightarrow F' - 100 \times 10^3 \times 9.81 - 98.1z \times 9.81 = 0$

$$\Rightarrow F' = 100 \times 10^3 \times 9.81 + 98.1z \times 9.81 \, N$$

It is clear that the internal axial force in the column X-section increases linearly from top to bottom.

$\Rightarrow$ Maximum internal load occurs at the base where $z = h = 5$ m

$$\Rightarrow F'_{max} = 100 \times 10^3 \times 9.81 + 98.1 \times 5 \times 9.81 \, N$$

Corresponding average normal stress

$$\sigma_{max} = \frac{F'_{max}}{A} = \frac{100 \times 10^3 \times 9.81 + 98.1 \times 5 \times 9.81 \, N}{9.81 \times 10^3} \text{ MPa}$$

$$= 100 + 0.49 \text{ MPa} = 100.49 \text{ MPa}$$

This corresponds to only a very slight increase (about 0.5%) over the value in part (a).

Primary Learning Objectives

1. Normal stress analysis of axially loaded members
2. Column design for normal stress
3. Accommodation of distributed axial loading, particularly due to weight, in stress analysis

■ **End of Solution**

3.3.2 Axial Force Diagram

In determining the normal stress (average) at a cross section along an axially loaded member, we first determine the "internal" axial load at that X-section. This is done by applying the axial force balance (equilibrium condition) of the free body created through virtual sectioning of the original member. (This procedure was followed in Example 3.2.) The internal axial force, as determined in this manner, may vary along the length of the member. The variation of the internal axial force may be shown by a sketch on a Cartesian coordinate frame, where the axial location is given by the horizontal axis and the internal axial force is given by the vertical axis. This is called the *axial force diagram*.

In part (c) of Example 3.2, for example, the axial force varied continuously and linearly along the member (column) due to its weight distribution. Axial force diagrams are particularly useful, however, in determining the variation of the normal stress along the axis when the external axial forces are point forces rather than distributed forces. Then the internal axial force undergoes "step changes" at the points of application of the external forces. Examples of situations that create such step changes in axial loading are shafts with collars, thrust bearings, helical gears, etc.; towers with multiple platforms along the height; turbine rotors with multiple blade stages along the rotor shaft; and cutting tools (boring, drilling, etc.) with multiple blades along the axis.

Example 3.3

Tower with Cable-Anchored Deck

A tower is made of three uniform cylindrical segments AB, BC, and CD of diameters d_1, d_2, and d_3, respectively in series (Figure 3.7a). It carries a load of mass $M_1 = 1000$ kg at the top (location A), a platform (deck) at B, which is anchored using cables with a net vertical upward force $P = 500g$ N, and another load of mass $M_2 = 2000$ kg at C.

   a. Determine the internal axial force distribution in the three segments of the tower and sketch the corresponding axial force diagram. What is the ground reaction (at D) on the tower?
   b. If $d_2 = 0.2$ m, determine d_1 and d_2 such that the average axial normal stress is uniform (i.e., constant) along the tower.

Note: Assume the tower segments and the platform to be relatively light.

Solution

   a. Cut (virtual) a segment of the tower by sectioning it (at X_1) between A and B, and draw the FBD of the segment above the cut (Figure 3.7b).

$$\text{Equilibrium}: -M_1 g + F_1 = 0$$

Force along AB: $F_1 = M_1 g = 1000g$ N (compressive)
   Cut a segment of the tower by sectioning it (at X_2) between B and C, and draw the FBD of the segment above the cut (Figure 3.7c).

$$\text{Equilibrium}: -M_1 g + P + F_2 = 0$$

Stress

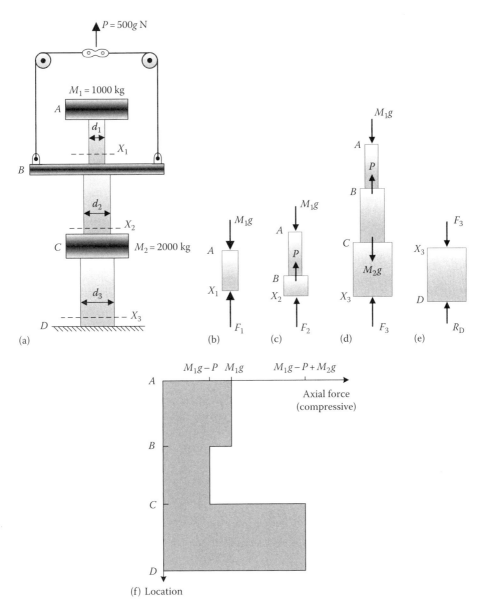

FIGURE 3.7 (a) Tower with an anchored deck; (b) FBD above section X_1; (c) FBD above section X_2; (d) FBD above section X_3; (e) FBD below section X_3; (f) axial force diagram.

→ Force along BC:

$$F_2 = M_1g - P = 1000g - 500g = 500g \text{ N (compressive)}$$

Cut a segment of the tower by sectioning it (at X_3) between C and D, and draw the FBD of the segment above the cut (Figure 3.7d).

$$\text{Equilibrium}: -M_1g + P - M_2g + F_3 = 0$$

→ Force along CD:

$$F_3 = M_1g - P + M_2g = 1000g - 500g + 2000g = 2500g \text{ N (compressive)}$$

Consider the segment below the cut at X_3 instead (Figure 3.7e).

$$\text{Equilibrium}: -F_3 + R_D = 0 \rightarrow R_D = F_3 = 2500g \text{ N(upward)}$$

The axial force diagram is shown in Figure 3.7f.

b. We require

$$\frac{F_3}{(\pi/4)d_3^2} = \frac{F_2}{(\pi/4)d_2^2} = \frac{F_1}{(\pi/4)d_1^2} \rightarrow \frac{2500g}{d_3^2} = \frac{500g}{d_2^2} = \frac{1000g}{d_1^2}$$

$$\rightarrow d_1 = \sqrt{2}d_2 = \sqrt{2} \times 0.2 \text{ m} = 0.283 \text{ m}$$

$$\rightarrow d_3 = \sqrt{5}d_2 = \sqrt{5} \times 0.2 \text{ m} = 0.450 \text{ m}$$

Primary Learning Objectives
1. Use of the axial force diagram to determine the axial stress distribution
2. Tower design to achieve uniform average normal stress under several external point loads along the length (and step changes in the internal axial force)

■ **End of Solution**

3.4 BEARING STRESS

When a member rests on another member or when two members are joined using such connecting devices as bolts, rivets, and pins, then forces are generated at the contact surface between the two components that are in contact. As a consequence, stresses are produced on the common contact surface called the *bearing surface*. These stresses can be quite complex and depend on many factors including the material, geometry of the contact surface, force at the contact surface, and so on. An approximate and rather simple representation of this contact stress is often made using an equivalent stress called the bearing stress.

$$\text{Bearing stress } \sigma_b = \frac{\text{Resultant force at the contact surface}}{\text{Projected area of the contact surface normal to the resultant force}} \quad (3.5)$$

Usually, this representation of bearing stress is adequate for purposes of component design.

Note: Bearing stress is an average compressive stress.

As an example, consider a bar connected to a gusset plate using a single bolt, as shown in Figure 3.8. Axial force applied to the bar = P.

From equilibrium, the resultant contact force from the bolt on the surface of the hole in the bar is also equal to P, as shown.

The projected area of the contact surface, normal to the direction of P, is $t \times d$ (see Figure 3.8), where

t is the thickness of the bar (depth of the bolt hole in the bar)
d is the diameter of the bolt (hole)

Hence,

$$\text{Bearing stress on the bar } \sigma_b = \frac{P}{td} \quad (3.6)$$

Stress

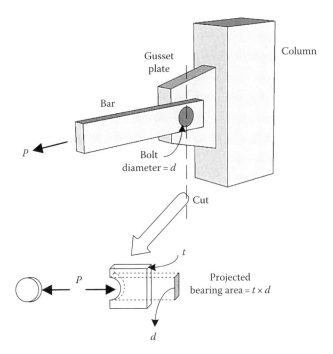

FIGURE 3.8 Bearing stress due to the contact force between a bolt and a bar.

Note 1: The bearing stress on the gusset place is obtained from the same formula (3.6), where *t* now is the thickness of the gusset plate. (Clearly, *P* and *d* are the same as before.)

Note 2: Bearing stress is an average "compressive" normal stress.

Note 3: Another situation of bearing stress is when a loaded (or heavy) member rests on a surface (e.g., horizontal plate, floor), as in the case of a drop-in anchor.

Example 3.4

Bearing Stress between a Post and the Supporting Floor

A cylindrical post of radius *R* carries a total downward load *P* and rests on a horizontal surface at its base (see Figure 3.9a).

  a. What is the bearing stress σ_b at the supporting surface?
  b. Now suppose that the local normal stress (compressive) on the bearing surface remains constant from radius $r=0$ to $r=R/2$ and then drops linearly to zero at $r=R$ (see Figure 3.9b). Determine the maximum normal local stress on the bearing surface (compressive).

Suggest a suitable factor of safety when using σ_b to represent the normal compressive stress on the bearing surface.

Solution

  a. Bearing area = $\pi R^2 \Rightarrow \sigma_b = P/\pi R^2$
  b. Consider an annular area of elemental width δr at radius *r* of the bearing surface (see Figure 3.9c).

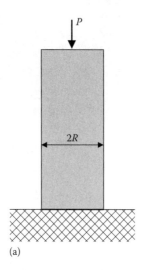

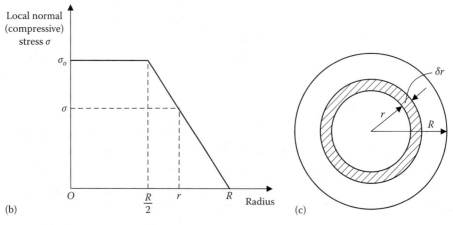

FIGURE 3.9 (a) A loaded post resting on a floor; (b) normal stress distribution on the bearing surface; (c) elemental annular area of the bearing surface.

Elemental area $= 2\pi r \cdot \delta r$
Bearing force on the elemental area $= \sigma \times 2\pi r \cdot \delta r$
where

$$\sigma = \sigma_o \quad \text{for } r = 0 \text{ to } R/2$$

$$= \left(\frac{R-r}{R/2}\right)\sigma_o \quad \text{for } r = \frac{R}{2} \text{ to } R$$

Total bearing force

$$P = \int_0^R \sigma \times 2\pi r \cdot dr = \int_0^{R/2} \sigma_o \times 2\pi r \cdot dr + \int_{R/2}^R \frac{R-r}{R/2}\sigma_o \times 2\pi r \cdot dr$$

$$= \sigma_o \pi r^2 \Big|_0^{R/2} + \frac{4\pi\sigma_o}{R}\left[\frac{Rr^2}{2} - \frac{r^3}{3}\right]_{R/2}^R$$

$$= \pi \frac{R^2 \sigma_o}{4} + \frac{4\pi\sigma_o}{R}\left[\frac{R}{2}\left(R^2 - \frac{R^2}{4}\right) - \frac{1}{3}\left(R^3 - \frac{R^3}{8}\right)\right]$$

$$= \frac{\pi R^2}{4}\sigma_o\left[1 + 16 \times \frac{1}{2} \times \frac{3}{4} - 16 \times \frac{1}{3} \times \frac{7}{8}\right] = \frac{7}{12}\pi R^2 \sigma_o$$

$$\Rightarrow \sigma_o = \frac{12}{7}\frac{P}{\pi R^2} = \frac{12}{7}\sigma_b$$

$\Rightarrow$ A suitable factor of safety would be $\dfrac{\sigma_o}{\sigma_b} = \dfrac{12}{7} \sim 2$

Primary Learning Objectives
1. Determination of bearing stress
2. Nonuniformity of normal stress distribution on a bearing surface
3. Its relevance in structural design

$$\text{An interpretation}: \quad \left[\text{Factor of safety}\right] = \frac{\left[\text{Maximum stress}\right]}{\left[\text{Average stress}\right]} \tag{3.7}$$

■ **End of Solution**

3.5 SHEAR STRESS

- Shear stress (or, shearing stress) acts "along" the plane of a section. (Contrast: Normal stress acts normal to the plane.)
- It causes a tearing, shearing, sliding, or twisting (torsion) effect on the material.
- Ductile material is weaker in shear than in tension. Hence, under some loading conditions (e.g., pure torsion), ductile material tends to fail when shear stress exceeds its ultimate value.
- Brittle material is weaker in tension. Hence, typically it tends to fail when normal stress exceeds its ultimate value.
- The primary loading on rivets, bolts, pins, shafts, etc., is shear.

3.5.1 AVERAGE SHEAR STRESS

Typically, the shear stress varies along the section on which it acts (e.g., in a laterally loaded beam, shear stress is zero at the outer edges of an X-section and maximum somewhere within the section—see Chapter 8; in a member under torsion, the shear stress is zero at corners and the center of an X-section, and is maximum somewhere on the outer edge, in between the corners—see Chapter 7). Average shear stress is the uniform shear stress that gives the same resultant shear force as that from the actual shear stress distribution.

As an illustration, consider a horizontal cantilever beam subjected to vertical force P at the free end (Figure 3.10). Take a cross section (virtual) at point X along the beam.

For equilibrium of the cut segment, there must be a resultant shear force P and a bending moment M at the X-section (also, see Chapter 2).

$$\text{Average shear stress } \tau = \frac{P}{A} \tag{3.8}$$

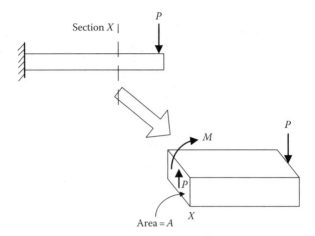

FIGURE 3.10 Shear stress in a cantilever X-section.

Example 3.5

Shear Stress in a Tennis Racquet

A tennis racquet is held firmly and horizontally. A tennis ball hits the racquet horizontally, as shown in Figure 3.11a, creating a horizontal force of 250 N. The point of impact of the ball from the gripped location of the racquet handle is given by the coordinates $a = 500$ mm and $b = 100$ mm, as shown.

a. Determine the internal loading at a vertical cross section through the handle slightly to the left of the gripped location.
b. If the area of cross section of the handle is 10 cm², what is the average shear stress there?

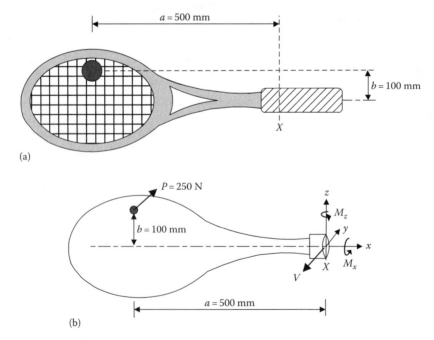

FIGURE 3.11 (a) Impact of a tennis ball on a racquet; (b) FBD showing the internal loading at a handle X-section.

Stress

Solution

a. Make a vertical cut (virtual) at X, which is slightly to the left of the gripped location (Figure 3.11b). The x-axis is along the central axis of the handle. The z-axis is in the vertical direction. Since the external loading P is in the y direction, as shown, the only nonzero loading components at the X-section are
 V = shear force in the negative y direction
 M_x = bending moment component about the x-axis
 M_z = bending moment component about the y-axis
Equations of equilibrium:
 Force balance in y direction: $P - V = 0$
 Moment balance about x-axis: $M_x - P \times b = 0$
 Moment balance about z-axis: $M_z - P \times a = 0$
We have

$$V = P = 250 \text{ N}$$

$$M_x = P \times b = 250 \times 100 \times 10^{-3} \text{ N·m} = 25.0 \text{ N·m}$$

$$M_z = P \times a = 250 \times 500 \times 10^{-3} \text{ N·m} = 125.0 \text{ N·m}$$

b. Average shear stress

$$\tau_{av} = \frac{V}{A} = \frac{250 \text{ (N)}}{10 \times 10^2 \text{ (mm}^2)} = 0.25 \text{ N/mm}^2 = 0.25 \text{ MPa}$$

Note: Also, there is a further shear stress distribution at the X-section due to the torque (M_x) and moment M_z there. But, their average value will be zero. Shear stress due to torque is studied in Chapter 7. Shear stress distribution of a beam cross section is studied in Chapter 8.

Primary Learning Objectives

1. Determination of shear force and bending moment (internal loading) at a cross section of a practical structure due to some external loading
2. Determination of average shear stress
3. Noncontribution of internal torques/moments to the average shear stress

■ **End of Solution**

Example 3.6

Glued Joint of a Tube

A plastic tube of external diameter $d = 5$ cm is glued inside another plastic tube of internal diameter slightly greater than 5 cm (see Figure 3.12a). The two glued tubes are expected to carry a maximum axial force of $P = 5$ kN.
 Determine a suitable glued length L for the two tubes, with a factor of safety of 4.
 Ultimate shear strength of glue (after setting—drying) $\tau_u = 1$ MPa.
Note: Assume that the failure is likely to occur in the set glue rather than in the tube material.

Solution

Consider the FBD of the segment of internal tube as shown in Figure 3.12b.
Equation of equilibrium: $V - P = 0 \rightarrow V = P$
where V is the shear force at the glued interface around the segment of axial length L

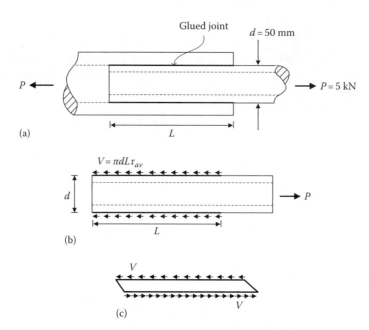

FIGURE 3.12 (a) A glued joint of two tubes; (b) FBD of internal tube segment; (c) loading on the glue layer.

Area of shear in the glued segment = $\pi d L$

Average shear stress in the glued joint $\tau_{av} = \dfrac{V}{\pi d L} = \dfrac{P}{\pi d L}$

We need $\tau_{av} = \dfrac{\tau_u}{4}$ for a factor of safety 4

Hence, $\dfrac{P}{\pi d L} = \dfrac{\tau_u}{4} \rightarrow \dfrac{5 \times 10^3 \,(\text{N})}{\pi \times 50 \times L\,(\text{mm}^2)} = \dfrac{1}{4}\,\text{MPa}$

$\rightarrow L = \dfrac{4 \times 5 \times 10^3}{\pi \times 50}\,\text{mm} = 127.3\,\text{mm} \sim 130\,\text{mm}$

Note 1: It should be clear from Figure 3.12c that the force on the glue layer is shear, not tension.

Note 2: In practice, joints are designed to be slightly stronger than the individual components that are joined. In this manner, we allow a slightly larger factor of safety for the joint than to the individual components. This takes into consideration that, typically, the joint strength has greater uncertainty than the strength of the joined components, because practices of manufacturing and quality control of components are more uniform than those of constructing the joints.

Primary Learning Objectives

1. Analysis of a glued joint
2. Shear force and its orientation in a glued joint
3. Average shear stress in a glued joint
4. Design considerations of a glued joint

■ **End of Solution**

Example 3.7

Vibration Mount

A vibration mount of an engine is sketched in Figure 3.13a. Each engine post is anchored to a support base using a bolt and a nut. There is packing material to dampen vibration at the point on which the post rests in the support base. Let

Engine load on the post $P = 10$ kN
Compressive force in the damping material $P_d = 2$ kN
Ultimate shear strength of the bolt $\tau_u = 200$ MPa

Determine a suitable diameter d for the bolt so that it is able to support the shear loading, at a factor of safety of 4.

Solution

FBD of the engine post is shown in Figure 3.13b.

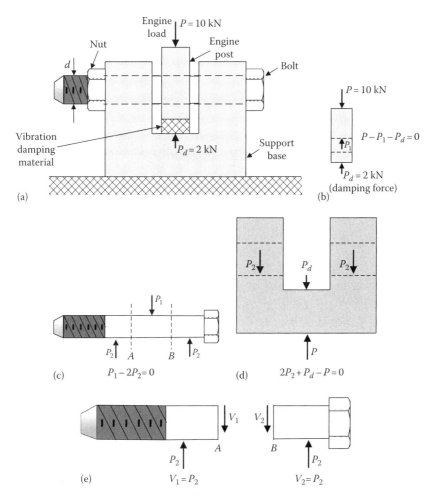

FIGURE 3.13 (a) A vibration mount; (b) FBD of engine post; (c) FBD of bolt; (d) FBD of mount base; (e) FBD of a left segment and a right segment of bolt.

Equilibrium condition:

$$P - P_1 - P_d = 0 \qquad \text{(i)}$$

$$\Rightarrow P_1 = P - P_d = 10 - 2 = 8 \text{ kN}$$

FBD of the bolt is shown in Figure 3.13c. It is assumed that loading in the support base is symmetric. Hence, there are two upward reactions P_2 from the two support points of the base, acting on the bolt.

Equilibrium condition:

$$P_1 - 2P_2 = 0 \qquad \text{(ii)}$$

$$\Rightarrow P_2 = \frac{1}{2}P_1 = \frac{1}{2} \times 8 \text{ kN} = 4 \text{ kN}$$

Note: By considering the FBD of the entire unit, the reaction on the support base from the ground may be determined to be equal to P. This is so because, all the forces that were considered before, other than P, are internal within this overall unit.

The FBD of the support base is shown in Figure 3.13d.

Equilibrium condition:

$$2P_2 + P_d - P = 0 \qquad \text{(iii)}$$

It is seen that (iii) = −(i) − (ii). Hence, (iii) is not a new result, but it further confirms the previous observation that the ground reaction on the support base is P.

To determine the maximum shear force in a cross section of the bolt, make a virtual section at A in Figure 3.13c and consider the left segment of the bolt (see Figure 3.13e).

Equilibrium condition: Shear force $V_1 = P_2$

Similarly, make a virtual section at B in Figure 3.13c and consider the right segment of the bolt (see Figure 3.13e).

Equilibrium condition : Shear force $V_2 = P_2$

It follows that the maximum shear force = P_2

Average shear stress in a bolt X-section, $\tau_{av} = \dfrac{P_2}{(\pi/4)d^2}$

We require $\tau_{av} = \dfrac{\tau_u}{4} \Rightarrow \dfrac{P_2}{(\pi/4)d^2} = \dfrac{200}{4} \Rightarrow \dfrac{4 \times 10^3}{(\pi/4)d^2} = \dfrac{200}{4}$

$$\Rightarrow d = \sqrt{\dfrac{320}{\pi}} \text{ m} = 10.1 \text{ mm}$$

Primary Learning Objectives
1. Choice of proper FBD in stress analysis
2. Shear loading and shear stress in a bolt
3. Design considerations of bolts
4. An introduction to engine (vibration) mounts

■ **End of Solution**

3.5.2 SINGLE SHEAR AND DOUBLE SHEAR CONNECTORS

- Connectors such as bolts, pins, and rivets may be loaded in single shear or double shear.
- In a bolt under single shear, the stem of the bolt has a single section by which the entire load is supported.
- In a bolt under double shear, the stem of the bolt has two sections by which the load is supported. Hence, the shear force (and hence the average shear stress) on the bolt is halved. In this manner, double shear provides a stronger connector.

Consider the two bolted joints shown in Figure 3.14. Axial load on the two connected members = P (equal and opposite, for equilibrium)

Diameter of the bolt = d

$$\text{Area of } X\text{-section of the bolt} = \frac{\pi}{4}d^2$$

In single shear (Figure 3.14a): There is a single shearing section that carries a shear force P.

$$\text{Average shear stress at bolt section (in single shear) } \tau = \frac{4P}{\pi d^2} \tag{3.9}$$

In double shear (Figure 3.14b): There are two shearing sections, each carrying a shear force $P/2$

$$\text{Average shear stress at a bolt section (in double shear) } \tau = \frac{2P}{\pi d^2} \tag{3.10}$$

Note: Example 3.7 gives a practical situation of "double shear" in a bolted connection.

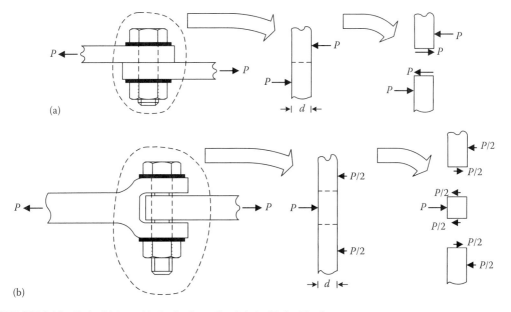

FIGURE 3.14 Bolted joints: (a) single shear (lap joint); (b) double shear.

3.5.3 SHEAR STRESS IN A KEY

- Keys are used to connect wheels, disks, pulleys, handles, etc., to shafts.
- The primary force on a key is a shear force, which is acting between the two connecting components.
- The average shear stress on a key is the main design consideration for the key.

As an example, consider a lever arm of length a connected to a shaft of radius r, using a key of thickness t and length L (see Figure 3.15a).

A force P is applied at the free end of the handle.

From the FBD of the handle (Figure 3.15b), the reaction force (downward) V from the key on the handle unit, at the shaft surface, is obtained using the equation of equilibrium:

$$\circlearrowleft \Sigma \text{ Moments about shaft center } O = 0 \Rightarrow V \times r - P \times a = 0 \Rightarrow V = \frac{aP}{r}$$

In return, an upward force of V acts on the key from the handle, at the contact surface. Hence, an equal and opposite force should act on the key from the shaft, to maintain the equilibrium of the key (see Figure 3.15c).

Note: There will be a bending moment as well on the key, to balance the "couple" created by the two equal and opposite forces V. However, since the distance between these two forces is small, the bending moment can be ignored. Even if this bending moment is included in the FBD shown in Figure 3.15c, it will not affect the shear force in the marked section (see shear line drawn as a broken line) of the key.

Shear force on the key = V
Shearing area of the key = $t \times L$
Hence, the average shear stress on the key is

$$\tau_{key} = \frac{V}{tL} = \frac{aP}{rtL} \tag{3.11}$$

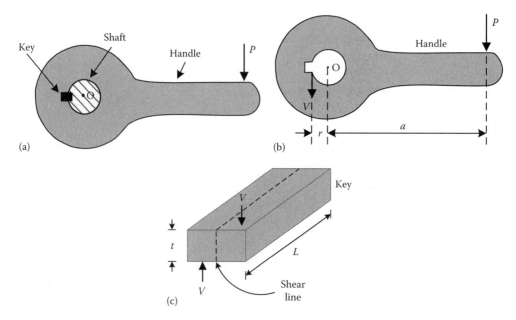

FIGURE 3.15 (a) A handle connected to a shaft using a key; (b) FBD of the handle; (c) FBD of the key.

Example 3.8

Motor Shaft and Pulley with Key

A payload of mass $M = 1000$ kg is hoisted at steady speed using a dc servo motor with a pulley and belt drive, as shown in Figure 3.16a. The pulley is attached to the motor shaft using a key of rectangular cross section. The following parameters are known.
 Radius of the pulley $R = 0.2$ m
 Radius of the motor shaft $r = 0.05$ m
 Thickness of the key $t = 0.01$ m
Determine a suitable length L for the key if the allowable shear stress for the key = 50 MPa.

Solution

The FBD of the pulley is shown in Figure 3.16b.
 P = force from the key on the pulley
 Since the hoisting conditions are steady, there is no inertia loading. Hence, the system can be treated as static. The moment balance of the pulley is (Figure 3.16b) $r \times P = R \times Mg$

$$\Rightarrow P = \frac{R}{r} \times Mg$$

Substitute numerical values: $P = \dfrac{0.2}{0.05} \times 1000 \times 9.81 \, \text{N} = 4 \times 9.81 \times 10^3 \, \text{N}$

Shear area of the key = $L \times t$

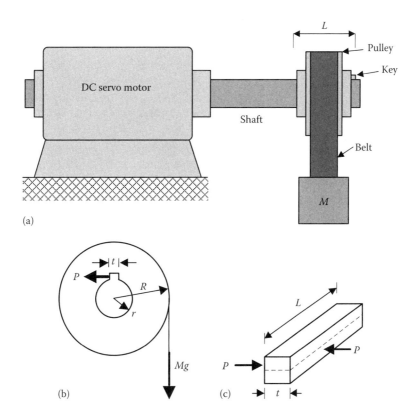

FIGURE 3.16 (a) Steady hoisting of a package by a motor-driven pulley; (b) FBD of pulley; (c) FBD of key.

$$\text{Average shear stress in the key} = \frac{P}{L \times t}$$

$$\text{We require } \frac{P}{L \times t} = 50 \text{ MPa}$$

$$\text{Substitute for } P \text{ and } t: \frac{4 \times 9.81 \times 10^3 \text{ (N)}}{L \times 0.01 \times 10^3 \text{ (mm}^2)} = 50 \text{ MPa}$$

$$\Rightarrow L = \frac{4 \times 9.81}{0.5} \text{ mm} = 78.5 \text{ mm} \approx 80 \text{ mm}$$

Primary Learning Objectives
1. Proper use of FBD in the analysis/design of keys
2. Shear stress in a key
3. Design of a key

■ **End of Solution**

3.5.4 Complementarity Property of Shear Stress

- The shear stresses on two perpendicular sides of a minute element of a body should be equal. They both should be either directed toward or directed away from the corner of intersection of the two sides.
- The shear stresses on two parallel sides of a minute element of a body should be equal and opposite.
- These two facts directly follow from equilibrium of the body.

Proof (2-D Case or Plane Stress Case)

Consider a point O of interest in the body and mark a Cartesian coordinate system there. Draw a minute rectangular solid element of dimensions $\Delta x \times \Delta y \times \Delta z$ at O. Mark the shear stress components on the orthogonal faces of this element, as shown in Figure 3.17a.

Note 1: What is shown is the case of *plane stress* because all the stresses are directed along a single plane (x–y plane; there are no stresses in the z direction).

Note 2: As usual, the first subscript of stress denotes the plane on which stress acts (i.e., the direction normal to the plane) and the second subscript denotes the direction of the stress.
Equilibrium of the small element:

$$\rightarrow \sum F_x = 0 \Rightarrow \tau_{yx} \times \Delta x \times \Delta z - \tau'_{yx} \times \Delta x \times \Delta z = 0 \Rightarrow \tau'_{yx} = \tau_{yx}$$

$$\uparrow \sum F_y = 0 \Rightarrow -\tau'_{xy} \times \Delta y \times \Delta z + \tau_{xy} \times \Delta y \times \Delta z = 0 \Rightarrow \tau'_{xy} = \tau_{xy}$$

$$\circlearrowleft \Sigma \text{ Moments about } z\text{-axis at } O = 0$$

$$\Rightarrow -\tau_{yx} \times (\Delta x \times \Delta z) \times \Delta y + \tau_{xy} \times (\Delta y \times \Delta z) \times \Delta x = 0 \Rightarrow \tau_{yx} = \tau_{xy}$$

Stress

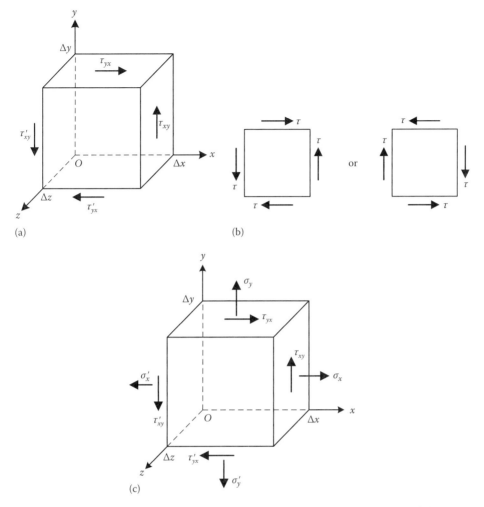

FIGURE 3.17 (a) State of plane stress under pure shear; (b) complementarity property of shear stress; (c) general plane stress.

Hence,

$$\tau_{yx} = \tau_{xy} = \tau'_{yx} = \tau'_{xy} = \tau \tag{3.12}$$

The only two possible configurations of shear stress under equilibrium are shown in Figure 3.17b. This is called the complementarity property of shear stresses.

Note: The second case is obtained by reversing all the stresses in the first case.

It is seen that, in addition to equality of the magnitudes of shear stress on perpendicular planes and parallel planes, the following rule applies:

> On adjacent planes (perpendicular), the shear stresses are either directed toward each other or away from each other.

It can be easily verified that the complementarity property of shear stress holds even under a general state of stress. In particular, consider the case of general plane stress shown in Figure 3.17c. Since the opposite horizontal planes of the element are so close (because $\Delta y \to 0$), the state of stress on these two planes must be equal. Hence, $\tau_{yx} = \tau'_{yx}$ and $\sigma_y = \sigma'_y$. Similarly, since the opposite vertical planes of the element are so close (because $\Delta x \to 0$), the state of stress on these two planes also must be equal. Hence, $\tau_{xy} = \tau'_{xy}$ and $\sigma_x = \sigma'_x$. These observations guarantee the balance of forces separately, in the two directions. Now write the moment balance about the axis through the center of the element and parallel to z (*Note*: There is no moment contribution from the normal stresses because the corresponding forces pass through the axis about which the moment is taken.). As obtained earlier, we get $\tau_{yx} = \tau_{xy}$.

Note: It can be shown that the previous results (complementarity property, in particular) hold true even in the 3-D situation (where stresses are present in the z direction as well).

3.6 STRESS TRANSFORMATION IN A BAR UNDER AXIAL LOADING

- The stresses at a location depend on the orientation of the local plane on which the stresses act.
- Given the stresses on one local plane, determination of the stresses on another plane (a plane inclined to the first plane) at the same location is called *stress transformation*.
- This subject will be studied in more detail in Chapter 9.
- The present section considers the special case of an axially loaded member.

Consider a bar in tension under an axial force P at its ends (Figure 3.18a).

Note: The X-section needs neither be rectangular nor uniform. However, unless the area of the X-section is infinitesimally small, what is considered is the "average stress" over the area.

Let us determine the stresses on a section (plane) inclined at angle θ to the normal section (section normal to the loading axis or X-section).

Note: Even though the inclined plane is shown to have rotated in the counter-clockwise (ccw) direction from the original section, a cw rotation will give the same final result, except the shear stress will be in the opposite direction.

FBD of the segment to the left when sectioned by the inclined plane is shown in Figure 3.18b. Equilibrium of this segment requires that the resultant force on the inclined section is also P and is collinear (uniaxial) with the end force. This force (shown by a dotted arrow) can be resolved into a component N normal to the inclined plane (normal force) and a component V along the inclined plane (shear force), as shown. We have

Normal force $N = P\cos\theta$
Shear force $V = P\sin\theta$
Let area of cross section $= A$
Then, area of the inclined section $= A/\cos\theta$
It follows that

$$\text{Normal stress on the inclined section } \sigma = \frac{P\cos\theta}{A/\cos\theta} = \frac{P}{A}\cos^2\theta = \frac{1}{2}\frac{P}{A}(1+\cos 2\theta) \quad (3.13)$$

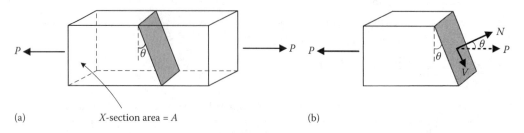

(a) X-section area = A   (b)

FIGURE 3.18 (a) An axially loaded bar with a general inclined section indicated; (b) FBD of the bar segment separated by the inclined section.

$$\text{Shear stress on the inclined section } \tau = \frac{P\sin\theta}{A/\cos\theta} = \frac{P}{A}\sin\theta\cos\theta = \frac{1}{2}\frac{P}{A}\sin 2\theta \qquad (3.14)$$

Note: We have used the familiar trigonometric identities in obtaining the final results of (3.13) and (3.14). It is seen from (3.13) that the maximum normal stress occurs when $\theta = 0°$ (i.e., at a normal section or cross section) and its value is

$$\sigma_{max} = \frac{P}{A} \qquad (3.15)$$

It is seen from (3.14) that the maximum shear stress occurs on a section that makes an angle of 45° with the normal (cross) section and its value is

$$\tau_{max} = \frac{1}{2}\frac{P}{A} = \frac{1}{2}\sigma_{max} \qquad (3.16)$$

Hence we may write

$$\sigma = \frac{1}{2}\sigma_{max}(1+\cos 2\theta) \qquad (3.17)$$

$$\tau = \frac{1}{2}\sigma_{max}\sin 2\theta \qquad (3.18)$$

The variation of the transformed stresses is sketched in Figure 3.19.

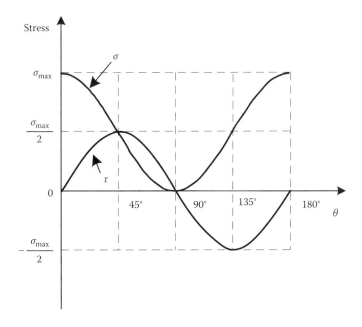

FIGURE 3.19 Variation of the transformed stresses.

Example 3.9

Failure of a Glued Joint

Two segments of a bar are glued together along a section at an angle 60° to the normal section (cross section), as shown in Figure 3.20a.

For the glued joint (for set—dried—glue)
Ultimate normal stress $\sigma_{ult} = 20$ MPa
Ultimate shear stress $\tau_{ult} = 12$ MPa
Area of normal section of the bar $A = 100$ mm²

Will the joint fail in shear or in normal detachment?
What is the maximum axial load which the glued joint is able to carry?

Note: Assume that the failure is likely to occur in the set glue rather than in the material of the bar.

Solution

Angle of inclination of the glued section (from the X-section) $\theta = 60°$ in cw direction (i.e., $-60°$ in ccw direction)

From the usual results on stress transformation (equations 3.17 and 3.18),

$$\text{Normal stress on the glued surface } \sigma = \frac{1}{2}\sigma_{max}(1 + \cos 2 \times (-60°)) = \frac{1}{4}\sigma_{max} \quad \text{(i)}$$

$$\text{Shear stress on the glued surface } \tau = \frac{1}{2}\sigma_{max} \sin 2 \times (-60°) = -\frac{\sqrt{3}}{4}\sigma_{max} \quad \text{(ii)}$$

This state of stress is shown in Figure 3.20b.

Note: To understand the reason for the minus signs in (ii), compare the +ve directions shown in Figure 3.18b with the directions shown in Figure 3.20b.

From (i) and (ii) we have, in magnitude,

$$\left|\frac{\sigma}{\tau}\right| = \frac{1}{\sqrt{3}} < \frac{\sigma_{ult}}{\tau_{ult}} = \frac{20}{12}$$

It follows that when the magnitude of τ reaches τ_{ult}, then the magnitude of σ will be less than σ_{ult}.
→ Failure of the joint will be due to *shear* along the glued surface.

$$\text{At failure, } \sigma_{max} = \frac{P_{max} \text{ (N)}}{100 \text{ (mm}^2)} \text{ MPa}$$

$$\text{Hence from (ii), at failure, } |\tau| = \frac{\sqrt{3}}{4}\frac{P_{max}}{100} = \tau_{ult} = 12$$

$$\rightarrow \text{Maximum axial load } P_{max} = \frac{400 \times 12}{\sqrt{3}} \text{ N} = 2770 \text{ N}$$

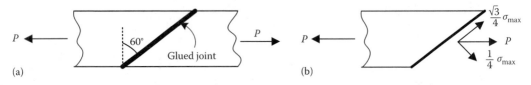

FIGURE 3.20 (a) A glued joint at an inclined section; (b) stresses on the glued section.

Stress

Primary Learning Objectives

1. Stress transformation
2. Recognition of correct directions of transformed stresses (sign convention)
3. Design of glued joints

■ **End of Solution**

Example 3.10

Stress Transformation in a Truss Member

A two-member frame structure with pin joints supports a vertical load $P = 40$ kN as shown in Figure 3.21a. The member AB is horizontal and the member CB makes an angle θ with the vertical. They are uniform two-force members having areas of X-section A_1 and A_2.

  a. For $\theta = 60°$, determine the forces F_1 and F_2 in the members AB and CB.
  b. For $\theta = 60°$, determine suitable A_1 and A_2 if the allowable normal stress for each of the two members is 100 MPa.
  c. For a general value of θ, obtain expressions for the normal stress and the shear stress on a horizontal section (X in Figure 3.21a) in the member CB. Compute these two stresses for angles $\theta = 30°$, 45°, 60°, for A_2 as determined in part (b).
  Note: In part (c), as θ changes the shape of the structure itself changes, not just the angle of the considered section.

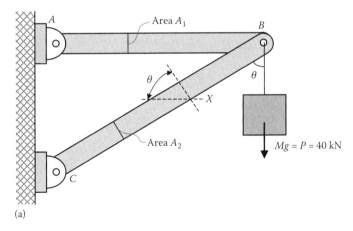

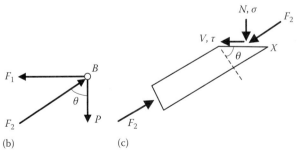

FIGURE 3.21 (a) A two-member frame structure; (b) forces at Joint B; (c) FBD of sectioned segment of member CB.

Solution

a. Consider the equilibrium of joint B (Method of joints—see Figure 3.21b).

We have $-F_1 + F_2 \sin\theta = 0$; $F_2 \cos\theta - P = 0$

Hence,

$$F_2 = \frac{P}{\cos\theta}, \quad F_1 = P\tan\theta$$

Substitute $P = 40$ kN and $\theta = 60°$

$$F_1 = 40\tan 60° \text{ kN} = 40\sqrt{3} \text{ kN}$$

$$F_2 = \frac{40}{\cos 60°} \text{ kN} = 80 \text{ kN}$$

b. Average normal stress in AB

$$\sigma_{1av} = \frac{F_1}{A_1} = \frac{P\tan\theta}{A_1}$$

Average normal stress in CB

$$\sigma_{2av} = \frac{F_2}{A_2} = \frac{P}{A_2 \cos\theta}$$

We require $\sigma_{1av} = \sigma_{2av} = 100$ MPa

$$\Rightarrow \frac{40\sqrt{3} \times 10^3 \text{ [N]}}{A_1 \text{ [mm}^2\text{]}} = 100 \quad \text{and} \quad \frac{80 \times 10^3 \text{ [N]}}{A_2 \text{ [mm}^2\text{]}} = 100$$

$$\Rightarrow A_1 = 400\sqrt{3} \text{ mm}^2 = 693 \text{ mm}^2$$

$$A_2 = \frac{80 \times 10^3}{100} \text{ mm}^2 = 800 \text{ mm}^2$$

c. From the standard results on the state of stress on an inclined section in an axially loaded member, we have
Normal stress on the horizontal section (Equation 3.13)

$$\sigma = \frac{1}{2}\frac{F_2}{A_2}(1+\cos 2\theta)$$

Note: Since F_2 is compressive, σ will be compressive. The marked directions of the force and stresses in Figure 3.21c take this into account.

Shear stress on the horizontal section (Equation 3.14)

$$\tau = \frac{1}{2}\frac{F_2}{A_2}\sin 2\theta$$

TABLE 3.1
Average Stresses on a Horizontal Section of CB

θ (°)	σ (MPa)	τ (MPa)
30	43.3	25.0
45	35.4	35.4
60	25.0	43.3

Substitute for F_2

$$\sigma = \frac{1}{2}\frac{P}{A_2 \cos\theta}(1+\cos 2\theta)$$

$$\tau = \frac{1}{2}\frac{P}{A_2 \cos\theta}\sin 2\theta$$

Compute the values with $P = 40$ kN and $A_2 = 800$ mm². The results are tabulated in Table 3.1.

Note: Since θ determines not only the orientation of the section but also the orientation of the member CB (see Figure 3.21a), it cannot be assigned arbitrarily. Specifically we must have $0 < \theta < 90°$.

Primary Learning Objectives

1. Stress transformation
2. Recognition of the correct directions of transformed stresses
3. Stresses and transformed stresses in members of a structure depend on the geometry of the structure
4. Design of truss members

■ **End of Solution**

SUMMARY SHEET

Stress:

- Intensity of an internal force at a specific location in the body and on a specific internal plane
- Force per unit area
- Units: 1 N/m² = 1 Pa; 1 lbf/in.² = 1 psi ≈ 6.95 kPa
- Defined by the magnitude and the direction of the force (a vector), and the size and orientation of the area (another vector). Hence, stress is a "second-order tensor," not a vector. Its definition needs two subscripts: one to define its direction and the other to define the plane on which the stress acts (or the unit vector normal to the plane)

Normal stress (σ): Acts normal to the plane (hence, needs only one subscript, which defines the normal to the plane; related to pulling (tensile), pushing (compressive), or bending (flexural) actions in objects; tensile is positive, compressive is negative.

Local normal stress: $\sigma_z = \lim_{\Delta A_z \to 0} (\Delta F_z / \Delta A_z)$ in the z direction.

Average normal stress: $\sigma_z = F_z / A_z$; area A_z is not infinitesimal.

Shear stress (τ): Acts along the plane (hence, needs two subscripts: one to define the orientation of the plane and the other to define the direction of the stress); related to "shearing," "sliding," or "twisting" actions in objects.

Local shear stress: $\tau_{zy} = \lim_{\Delta A_z \to 0} (\Delta F_y / \Delta A_z)$ in the y direction on a plane whose normal is in the z direction.

Average shear stress: $\tau_{zy} = F_y / A_z$; area A_z is not infinitesimal.

Bearing Stress:

$$\sigma_b = \frac{\text{Resultant force at the contact surface}}{\text{Projected area of the contact surface normal to the resultant force}}$$

Note: Bearing stress is an average stress.

Single shear (in a bolt): There is a single shearing section carrying the entire shear force P.

$$\text{Average shear stress on bolt section } \tau = \frac{4P}{\pi d^2}$$

Double shear (in a bolt): There are two shearing sections, each carrying half the force $P/2$.

$$\text{Average shear stress on bolt section } \tau = \frac{2P}{\pi d^2}$$

Shear stress in a key: $\tau_{key} = V/tL$
where
  V = shear force
  t = key thickness
  L = key length

Complementarity property of shear stress: At a given point in a body, the shear stresses along two perpendicular planes should be equal. → Shear stress is symmetric; in 3-D there are nine stress components, six of which are shear stresses. Three of these are equal to the remaining three: $\tau_{xy} = \tau_{yx}$; $\tau_{yz} = \tau_{zy}$; $\tau_{zx} = \tau_{xz}$.

Stress transformation: For a bar of X-section A and axial force P,
Normal stress on a plane inclined at θ with normal plane:

$$\sigma = \frac{1}{2}\frac{P}{A}(1 + \cos 2\theta) = \frac{1}{2}\sigma_{max}(1 + \cos 2\theta)$$

$$\text{Shear stress on the inclined plane } \tau = \frac{1}{2}\frac{P}{A}\sin 2\theta = \frac{1}{2}\sigma_{max}\sin 2\theta$$

PROBLEMS

3.1 Consider the truss shown in Figure P3.1. All truss members are light and all joints are frictionless (i.e., pin joints).
Given: Ultimate strength of the truss material = 400 MPa; factor of safety = 5
Determine a suitable area of cross section for the truss members (i.e., design the truss).

3.2 A uniform shaft AE with axial end loads is shown in Figure P3.2. Area of cross section of the shaft = 2500 mm².
Thrust bearings are located at B, C, and D, which exert axial loads as shown in the figure.
Sketch the axial force diagram for the shaft. Determine the maximum average normal stress and where it occurs along the shaft.

3.3 A square steel plate of thickness $t = 15$ m is firmly supported in horizontal orientation at its four edges (Figure P3.3a). A hole of diameter slightly larger than 10 mm is drilled at its center. A steel shaft of diameter 10 mm is supported vertically on the plate using an end cap of diameter 20 mm, which is machined at the top end of the shaft. It passes through the hole of the plate, as shown. The other end (bottom end) of the shaft carries a vertical load $P = 12$ kN.

a. Determine the average normal stress in the axial direction of the shaft in MPa.
b. Determine the bearing stress at the interface of the end cap and the plate.
c. What is the total shear force in the vertical direction in a cylindrical section of diameter d slightly greater than 20 mm in the plate around the end cap? What is the corresponding average shear stress in MPa (vertical) in the virtual section of the plate?
d. Repeat part (c) for a cylindrical section of diameter 40 mm in the plate around the end cap.
e. If the bearing stress in part (b) is not to exceed $125/\pi$ MPa, what should be the diameter of the end cap, for the same shaft as mentioned earlier?

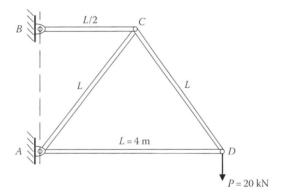

FIGURE P3.1 A planar truss carrying a load.

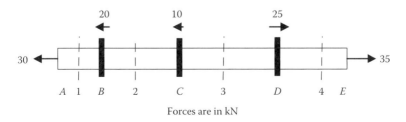

FIGURE P3.2 A uniform shaft with axial loading.

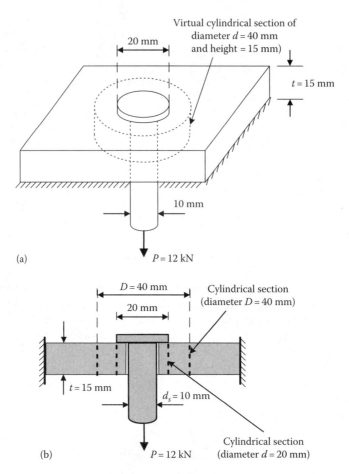

FIGURE P3.3 A horizontal plate carrying a vertical shaft with end load; (a) 3-D (perspective) view; (b) side sectional view.

A sectional view of the system, through the shaft, is shown in Figure P3.3b, which also indicates the cylindrical sections of parts (c) and (d).

Note: The answer may be expressed in terms of π.

3.4 A light rod with rectangular X-section ($a \times b$) is vertically supported at its head on a rigid and fixed platform. The square head has side dimensions $b \times h$, where b = width and h = thickness, so that it is flush with the stem of the rod on one side (Figure P3.4a). The stem of the rod passes through a hole in the platform without touching it, as shown in the front sectional view of Figure P3.4b. The rod supports a vertical load P as shown.

The yield strength of the rod in normal stress is $\sigma_Y = 400$ MPa.

The yield strength of the rod in shear stress is $\tau_Y = 160$ MPa.

Determine the maximum value of the dimension ratio a/h so that, when sufficiently loaded, the rod will fail in tension in its stem rather than in shear in its head.

Use the same factor of safety for the two types of failure.

Hint: First obtain expressions for normal stress (average) in the stem of the rod and the shear stress (average) on a vertical section of the head, in terms of P, a, b, and h. Equate the normal stress of the stem to the allowable normal stress (i.e., normal yield strength divided by the safety factor) and determine the corresponding P. For this value of P, make the shear stress in the head less than the allowable shear stress (i.e., shear yield strength divided by the safety factor).

Stress

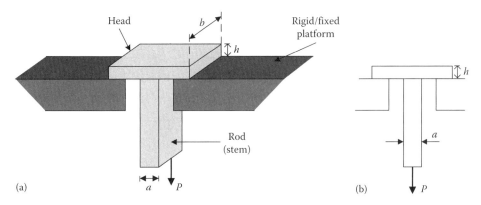

FIGURE P3.4 A rectangular rod supported on a platform and carrying an end load: (a) perspective view; (b) front view.

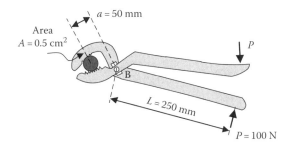

FIGURE P3.5 A pair of pipe pliers.

3.5 Forces $P = 100$ N are applied by hand to the handles of a pair of pipe pliers (Figure P3.5). A pipe is held by the jaws of the pliers through this applied force. The area of X-section of the jaw segment of the pliers, near the point of pipe contact, is $A = 0.5$ cm². Some relevant dimensions are shown in the figure. Determine
   a. The holding force R applied to the pipe
   b. The loading on a section of the jaw just to the right of its contact point with the pipe and parallel to the direction of the pipe reaction R, and the corresponding average shear stress
   c. The loading on a section of the jaw just to left of its contact point with the pipe and parallel to the direction of R

   Note: Assume a smooth joint (B) for the two plier segments.

3.6 A rotating shaft is connected to another rotating shaft using a U-jaw in the second shaft and a pin, as shown in Figure P3.6.
   Diameter of the first shaft $d_1 = 50$ mm
   Diameter of the pin $d_2 = 20$ mm
   Torque applied to the first shaft (which is transmitted to the second shaft) = 50 N·m
   Determine the average shear stress in a cross section of the pin between the first shaft (right shaft) and the jaw of the second shaft (left shaft).
   Note: Assume that the only load transmitted along the shaft is the torque.

3.7 Two identical strips are riveted for carrying a tensile load P, as shown in Figure P3.7.
   The dimensions of the strip cross section: width = 100 mm; thickness = 20 mm
   Diameter of a rivet = 15 mm
   Maximum allowable normal stress of the strip material = 150 MPa
   Maximum allowable shear stress of the rivet material = 120 MPa

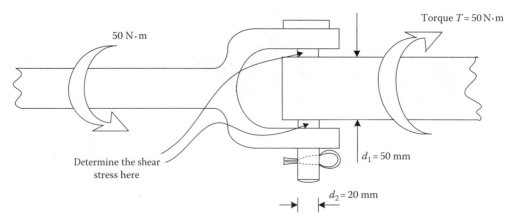

FIGURE P3.6 Two shafts joined by a jaw and a pin and transmitting a pure torque.

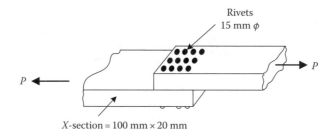

FIGURE P3.7 A riveted joint of two rectangular strips.

Determine
   a. The maximum load P that can be carried by the device
   b. The minimum number of rivets needed

3.8 Consider again the vibration mount sketched in Figure P3.8.

Engine load on the post $P = 10$ kN

Compressive force in the damping material $P_d = 2$ kN. (*Note*: This is determined by the deflection and the stiffness of the damping material. It cannot be determined by the equilibrium equations. This is a *statically indeterminate* problem.)

Diameter of the bolt $d = 10$ mm

Allowable bearing strength of the support base $\sigma_b = 60$ MPa

Select suitable dimensions L_1 and L_2 for the support base on the basis of its allowable bearing strength.

3.9 An anchoring cable is secured to a concrete base through a steel end plate and an identical pair of bolts and nuts (Figure P3.9). The cable tension P is adjusted by turning the nuts. There is a steel washer between each nut and the end plate. Suppose

   d_c = cable diameter = 25 mm
   d_b = bolt diameter = 20 mm
   d_w = washer diameter = 30 mm
   t = end plate thickness = 5 mm
   P = cable tension = 50 kN

Determine the following:

   σ_c = average tensile stress in the cable
   σ_{bolt} = average tensile stress in a bolt

Stress

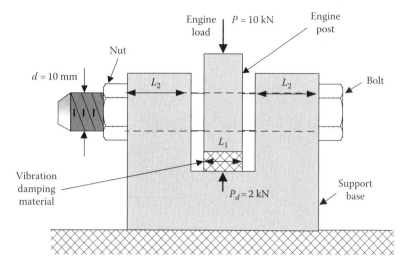

FIGURE P3.8 A vibration mount.

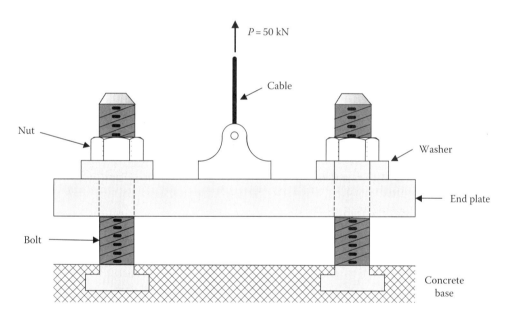

FIGURE P3.9 Bolt-secured end plate of an anchoring cable.

σ_b = bearing stress between the washer and the end plate
τ_p = average shear stress in a cylindrical section of the end plate just outside the washer
Note: Assume that the conditions in the two bolts are identical.

3.10 In Problem 3.9, for the given anchoring cable, determine the necessary bolt diameter so that the average normal stress in a bolt is equal to the average normal stress in the cable.

3.11 By definition, bearing stress is the average normal stress (compressive) on the bearing area. However, the actual normal stress on the bearing area is not uniform.
In Problem 3.9, assume that the normal stress σ in the bearing area under a washer on the end plate varies linearly with $\sigma = 0$ at diameter d_b and maximum $\sigma = \sigma_o$ at diameter d_w (see Figure P3.11).

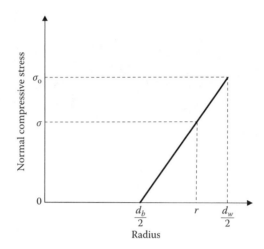

FIGURE P3.11 Normal stress distribution under a washer.

For a bearing load of P/2 as described earlier, determine σ_o. By comparing this result with the bearing stress σ_b obtained in Problem 3.9, suggest a suitable factor of safety when using the "average" bearing stress in designing this anchoring system.

3.12 A rod subjected to axial loading is made of a material with ultimate normal stress 100 MPa and ultimate shear stress 75 MPa.
  Will this member be more likely to fail due to normal stress or shear stress?
  Across what section of the rod is the failure likely to occur?
  If the cross-sectional area of the rod is 50 mm², estimate the axial load under which the failure occurs.

3.13 Two segments of a bar are glued together along a section at an angle 45° to the normal section (cross section), as shown in Figure P3.13.
  For the glued joint (for set glue), the following are given:
   Ultimate normal stress $\sigma_{ult} = 12$ MPa
   Ultimate shear stress $\tau_{ult} = 18$ MPa
   Area of cross section of the bar $A = 100$ mm²
  Will the joint fail by shearing along the glued surface or by detachment of the two segments of the bar normal to the glued surface?
  What is the maximum axial load which the glued joint is able to carry?
  Note: Assume that the material of the bar is stronger than the set glue.

3.14 Two identical strips of wood, each of thickness $t = 20$ mm, are joined together by bonding two identical flat pieces of wood symmetrically on the two sides of the joint, using a suitable adhesive (Figure P3.14). Design the length (2L) of the flat pieces, given that
   Ultimate shear strength of the set (dried) glue $\tau_{ug} = 2.0$ MPa
   Ultimate tensile strength of the wood strips $\sigma_{uw} = 14.0$ MPa

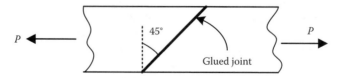

FIGURE P3.13 A glued joint at an inclined section.

Stress

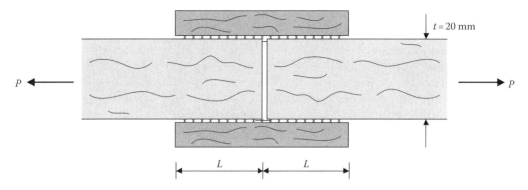

FIGURE P3.14 A bonded joint of two strips of wood.

Note: In this design, we should make the maximum shear force of the set glue (there are two sides) equal to the maximum tensile force in a wood strip (this is the ideal, "optimal" design). In practice, however, the former (i.e., glue strength) should be slightly higher than the latter (i.e., the wood strength), in view of the greater uncertainty and nonuniformity of the former (i.e., it is rather difficult to produce a glued joint that has uniform strength throughout it). Similar arguments apply to other types of joints as well (e.g., welded joints).

3.15 A uniform brass nail of diameter $d = 5$ mm and length L has been driven into a block of wood. Now, it is being pulled out using a claw hammer (Figure P3.15). Material properties are

Ultimate tensile stress of the nail $\sigma_u = 240.0$ MPa

Maximum frictional force per unit area at the interface of wood and nail $\tau_f = 4.0$ MPa

Determine the maximum possible length of the nail L_{max} so that it would not break before being pulled out.

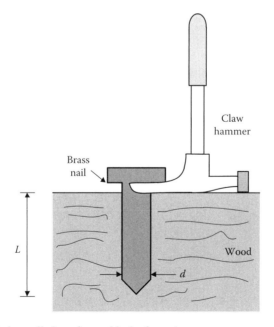

FIGURE P3.15 A nail being pulled out from a block of wood.

Further Review Problems

3.16 A uniform steel block of height $h=4$ m and uniform square cross section of side $a_1=0.5$ m is placed on a platform. A drop-in steel rod of uniform square cross section with side dimension a_2 is welded to the bottom of the steel block, through a central hole of the platform. The rod carries a mass m_1 of weight $m_1 g = 150$ kN. A hydraulic press applies a force P_0 downward, at the top of the steel block (see Figure P3.16).

Neglect the weight of the drop-in rod in comparison with other loads and weights of the system. Specific weight of steel, $\rho_s g = 80$ kN/m³. Yield stress of steel, $\sigma_Y = 240$ MPa.

a. Draw the axial force diagram of the system.
b. Determine the maximum P_0 that may be applied on the steel block such that the block will not yield, with a factor of safety 4.
c. Determine the minimum cross-sectional dimension a_2 of the drop-in rod such that when the maximum P_0 is applied, the rod will not yield with a factor of safety 4.
d. How would the result in part (c) change if the weight of the drop-in rod ($m_3 g = 10$ kN) is taken into account?

3.17 In Problem 3.16, suppose that the hole in the platform is square with side $a=0.1$ m.

a. Determine the bearing stress σ_b at the interface of the steel block and the platform.
b. If the allowable bearing stress is 75 MPa, what is the largest possible size of the square hole (a_{max}) in the platform?

Note: Include the weight of the drop-in rod. Also, P_{0max} applies on the steel block.

3.18 A pair of grip pliers in a specific configuration of gripping an object is shown in Figure P3.18. A holding force of $P = 100$ N is applied to each handle of the pliers. The gripping force F at the jaw makes an angle $\theta = 45°$ with P, as shown.

If the allowable shear stress of the pin at A is $\tau_{allow} = 20$ MPa, determine a suitable diameter for the pin.

3.19 A washer punching press punches out washers from a steel plate of thickness $t = 2$ mm. The internal radius and the external radius of the punching tool are $r_i = 1$ cm and $r_o = 2$ cm,

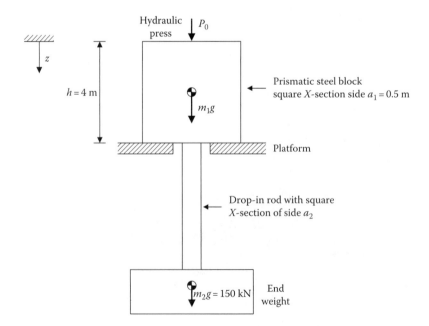

FIGURE P3.16 A steel block with a drop-in rod carrying a load and a top load.

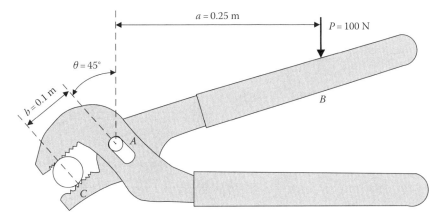

FIGURE P3.18 A pair of grip pliers.

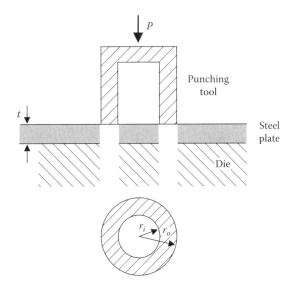

FIGURE P3.19 A punching press for steel washers.

respectively (Figure P3.19). If the ultimate shear stress of the steel plate is $\tau_{ult} = 275$ MPa, determine the force P that has to be generated by the punching press for producing the washers.

3.20 A sluice gate of a water reservoir is operated (moved up and down) by turning a horizontal two-bar handle. The central hub of the handle is attached to the vertical shaft of the gate using a single rectangular key of width $t = 10$ mm and length (along the hub) $L = 40$ mm. The location of the key-hole in the hub from shaft axis is $a = 20$ mm (this is slightly greater than the radius of the shaft, or the cylindrical hole of the hub), as shown in Figure P3.20. The equivalent hand force on each side of the handle is P, which acts at a distance $b = 0.5$ m from the shaft center. If the allowable shear stress for the key is $\tau_{allow} = 25$ MPa, determine the maximum value of P that may be applied on the handle.

3.21 Two pieces of wood are joined by cutting complementary wedges of angle $\theta = 60°$ on them, drilling a pass-through central hole perpendicular to the wedge face X–X, and tightening with

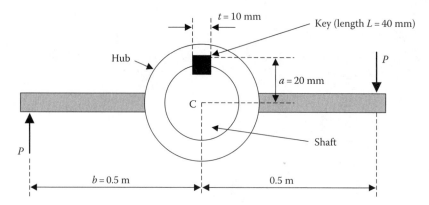

FIGURE P3.20 Handle of a sluice gate (top view).

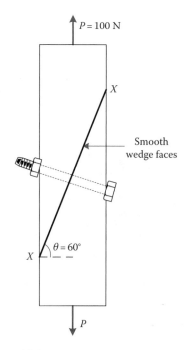

FIGURE P3.21 A single-bolted wood joint.

a bolt and nut (see Figure P3.21). Axial end forces $P = 100$ N are applied at the two ends of the member.

Determine the normal stress and the shear stress of the bolt on the cross section at the common wedge face.

Diameter of the bolt = 5 mm.

Note: Assume that the faces of the wedge are very smooth.

4 Strain

CHAPTER OBJECTIVES

- Understanding the meaning of strain: intensity of a local deformation; deformation of a reference element (of unit length or 90° corner)
- Two types of strain: normal strain and shear strain
- Units of strain and sign convention
- Local strain and average strain
- Thermal strain
- Measurement of strain using strain gauges

4.1 INTRODUCTION

Strain is caused by material "deformation" (not by a body "movement" without deformation). This is an important consideration in the design and usage of a structure or a machine. There are design requirements that limit the degree of deformation (e.g., lateral deformation of a tall building or a bridge can cause discomfort or fear in the users of these structures even when the structures themselves may not be damaged by the deflections; deformations in machine tools will affect the quality of the machined products and will also accelerate the degradation and failure of the machine tool; shaft deflections can damage the bearings and will also generate undesirable noise and vibration). Also there are deformations that are needed for proper operation of engineering devices (e.g., deflections are needed in the operation of electrical circuit breakers, relays, and switches; deflections are needed in vehicle suspension systems in order to assure proper ride quality; allowance for deflection is needed to release stresses such as thermal stresses in bridges and overhead vehicle guideways).

The solution of problems in mechanics of materials involves three basic considerations in general.

1. Statics (equations of equilibrium). This determines support reactions and internal loads. The latter determine stresses
2. Nature of deformation (nature of strains)
3. Stress–strain relations (i.e., constitutive relations or physical relations). These are needed to determine the strains (deformations) once the stresses are known (from the knowledge of internal loads), or to determine the stresses and loading corresponding to a specified limit on deflections or strains

The subject of statics is reviewed in Chapter 2. The subject of stress is treated in Chapter 3, which requires the knowledge of internal loads, as determined using equilibrium equations (statics). The exception is the case of statically indeterminate problems where all the associated loads cannot be determined using statics alone. Then, the deformations (compatibility conditions) have to be used. The nature (geometry) of deformation (hence strains) is the subject of the present chapter. Stress–strain relations are studied in Chapter 5. It is clear how strain is an important and integral part of Mechanics of Materials.

4.1.1 TYPES OF STRAIN

Some important characteristics of strain are listed as follows:

- Strain represents the *"intensity" of deformation* (of a straight-line segment or a right-angled corner) at a point in a body. (Compare: Stress represents the intensity of an internal force.)
- Strains are caused by deformations due to stresses or other factors such as changes in temperature.
- Strain deformations may be represented by change in length of a straight-line segment (i.e., *normal strain*) or change in the angle between two perpendicular line segments at a corner (i.e., *shear strain*).
- For expressing normal strains, the reference is a unit original length in a specified direction.
- For expressing shear strains, the reference is a corner of original angle 90°.
- The stress–strain relationship is a "physical" relationship governed by the material properties of the body. It is also called a "constitutive relation."

Note: A rigid body does not deform and hence does not exhibit strains (even though it may experience stresses). Deflections of a rigid body are "rigid body motions" and they are not deformations.

Normal strain: Deformation per unit length of a straight-line segment in a specified direction.

Note: Per unit length does not mean we have to consider a line segment of unity length. We just divide the extension by the considered length (ideally, the considered length is minute—infinitesimal).

Sign convention: Elongation is positive—it is called a *tensile strain*; contraction is negative—it is called a *compressive strain*.

Symbol: ε (Greek lowercase epsilon)

Unit: m/m or 1 unit of strain or 1 ε (*Note*: Strain is a dimensionless quantity.)

1 microstrain = 1 $\mu\varepsilon$ = 1 μm/m = 1 × 10^{-6} m/m = 1 × 10^{-6} ε

Note: To convert m/m into $\mu\varepsilon$, multiply by 10^6

1% strain = 1/100 strain = 0.01 m/m = 0.01 ε

Note: To convert m/m into % strain, multiply by 100

Shear strain: Change in angle of a specified corner of angle $\pi/2$

Sign convention: Angle reduction is positive; angle increase is negative

Symbol: γ (Greek lowercase gamma)

Unit: 1 rad (dimensionless)

$$1\,\mu\text{rad} = 1 \times 10^{-6}\text{ rad}$$

4.2 NORMAL STRAIN

4.2.1 LOCAL NORMAL STRAIN AND AVERAGE NORMAL STRAIN

Strains at a given location in a body are defined by considering an infinitesimal (extremely small) element at that location, as the size of the element approaches zero. These are called *local strains*.

Consider the elemental line segment OP of length δx along some direction (x-axis) in Figure 4.1a. When subjected to stresses, suppose that it deforms to $O'P'$. Let $\delta x'$ be the new length of the element.

Strain

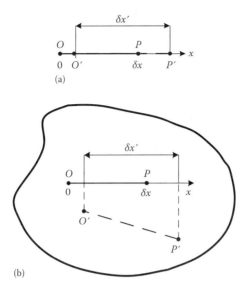

FIGURE 4.1 Definition of normal strain: (a) uniaxial case; (b) general case.

Note: This is the uniaxial (one-dimensional) case where both OP and $O'P'$ will fall along the same axis (x-axis in the figure).

The normal strain of the element OA is given by

$$\varepsilon_x = \frac{\text{Extension}}{\text{Original length}} = \frac{\delta x' - \delta x}{\delta x}$$

The local normal strain at O in the x direction is given by the limit

$$\varepsilon_x = \lim_{\delta x \to 0} \frac{\delta x' - \delta x}{\delta x} = \lim_{\delta x \to 0} \frac{\delta u}{\delta x} = \frac{du}{dx} \quad (4.1a)$$

Note: Here, δu is the extension of the element in the x direction.

Conversely, extension δu in the x direction of an element of original length δx is given by $\varepsilon_x \delta x$. The total extension u of the member in the x direction, from 0 up to x, is obtained by adding these elemental extensions from 0 to x. In the limit, this overall extension at x is expressed by the integral

$$u = \int_0^x \varepsilon_x \, dx \quad (4.2)$$

Note: The results (4.1a) and (4.2) allow for the general case where strain is not constant, but varies along the member.

In summary,

- Normal strain is the derivative of the deflection, with respect to the distance.
- Extension is given by the integral of the normal strain, with respect to the distance.

In the uniaxial case (1-D) case, as shown in Figure 4.1a, both OP and $O'P'$ will fall along the same axis. Now consider the general case as shown in Figure 4.1b, where the deformed element $O'P'$ does not fall along the same line as the original element *OP*. Then, the previous uniaxial results (4.1a) and (4.2) still hold, except that now $\delta x'$ is the new length of the element in the *x* direction (i.e., projection of $O'P'$ on the *x*-axis) as shown in Figure 4.1b.

4.2.1.1 Average Normal Strain

Typically, the strain varies from point to point in a body. Then, an average strain may be used to represent the state of strain of the body. For the case of normal strain, consider a 1-D member of length *L*. When subjected to an axial load, suppose that it stretches through δ. Then, the average normal strain in the member is

$$\varepsilon_{avg} = \frac{\delta}{L} \tag{4.1b}$$

Example 4.1

A wooden post of circular cross section has radius $r = 0.7$ m. A reinforcing steel wire is wrapped around it (Figure 4.2). Due to moisture, the radius of the post increases by $\Delta r = 0.005$ m. Determine the strain (normal) in the wire, assuming that initially it was not under strain.

Solution

Consider only one loop of the wire.
Length of the wire loop $L = 2\pi r$
Due to swelling of the post, the new length of the wire loop $= 2\pi(r + \Delta r)$
Change in wire length $\Delta L = 2\pi(r + \Delta r) - 2\pi r = 2\pi \Delta r$
Normal strain (average) in the wire $\varepsilon = \dfrac{\Delta L}{L} = \dfrac{2\pi \Delta r}{2\pi r} = \dfrac{\Delta r}{r}$

Note: We get the same answer even if we consider multiple loops of wire (check this).

Substitute the given data

$$\varepsilon = \frac{0.005}{0.7} = 0.007$$

$$\varepsilon = 0.7\%$$

Primary Learning Objectives

1. Solution of a problem of normal strain
2. The equivalence of considering the radius and the circumference of a circular object when analyzing normal strain
3. The convenience of using a single loop when analyzing the strain in multiple loops of wound wire

■ **End of Solution**

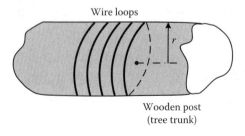

FIGURE 4.2 Wire-reinforced wooden post.

Strain

Example 4.2

A rigid beam hinged at end O is kept in the horizontal position using two identical hanger rods 1 and 2 attached at A and B on the beam using pin joints (see Figure 4.3a). The hanger rods are initially unstrained (neglect the weights of the members). The following dimensions are given:

Lengths of the hanger rods are the same at 3 m (i.e., $CA = DB = 3$ m)
Also, in the beam, $OA = AB = 2$ m

When a vertical load is applied at the free end of the beam, the beam location B moves down through a vertical distance $\delta_B = 3$ mm.

a. Determine the normal strains in the two hanger rods after the load is applied.
b. Suppose that there is a vertical clearance of 0.5 mm at joint A (see Figure 4.3b). Hence, the initial vertical movement of the beam through 0.5 mm at A will not strain the hanger at this joint. As joint B moves down through $\delta_B = 3$ mm when the load is applied, determine the resulting strains in the hanger rods.
c. Suppose that the vertical clearance is at joint B, not at joint A. Determine the strains in the hanger rods after the load is applied, if the beam movement at B is $\delta_B = 3$ mm as before.

Solution

a. Suppose that A moves to A' and B moves to B' when the load is applied (see Figure 4.3c).
Let θ = angle of rotation of the beam (very small).
Since the beam is rigid, $OA = OA'$ and $OB = OB'$.

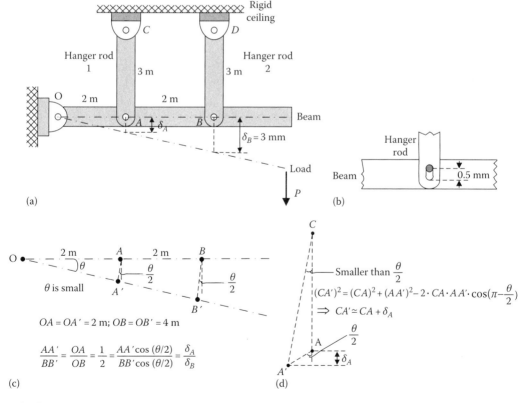

FIGURE 4.3 (a) A rigid beam supported by hanger rods; (b) clearance at a joint; (c) geometry of beam movement; (d) extension of rod A.

Hence, $\dfrac{AA'}{BB'} = \dfrac{OA}{OB} = \dfrac{AA'\cos(\theta/2)}{BB'\cos(\theta/2)} = \dfrac{\delta_A}{\delta_B} = \dfrac{2\,(m)}{4\,(m)} = \dfrac{1}{2}$ (from similar triangles)

$$\Rightarrow \delta_A = \dfrac{1}{2}\delta_B = \dfrac{1}{2} \times 3\,\text{mm} = 1.5\,\text{mm}$$

To determine the true extension of the hanger rod at A, see Figure 4.3d. Apply the cosine rule to triangle CAA′

$$(CA')^2 = (CA)^2 + (AA')^2 - 2 \cdot CA \cdot AA' \cdot \cos\left(\pi - \dfrac{\theta}{2}\right)$$

$$= (CA)^2 + (AA')^2 + 2 \cdot CA \cdot AA' \cdot \cos\left(\dfrac{\theta}{2}\right)$$

Since θ is very small, $\cos(\theta/2) \simeq 1$

$$\Rightarrow (CA')^2 = (CA)^2 + (AA')^2 + 2 \cdot CA \cdot AA' = (CA + AA')^2$$

$$\Rightarrow CA' \simeq CA + AA'$$

$\Rightarrow$ Extension in hanger rod $1 = CA' - CA \simeq AA'$

But $\delta_A = AA'\cos(\theta/2) \simeq AA'$ (since θ is small)

$\Rightarrow$ Extension in hanger rod $1 \simeq \delta_A = 1.5$ mm

Similarly,
Extension in hanger rod $2 \simeq \delta_B = 3.0$ mm

Note: In this manner, we showed that for small angular movements, the extension in the rod may be approximated as the vertical movement of the beam at the point to which the rod is attached.

Hence,

$$\text{Strain in hanger rod 1}: \varepsilon_1 = \dfrac{\delta_A}{3\,(m)} = \dfrac{1.5 \times 10^{-3}}{3}\,\text{m/m} = 0.5 \times 10^{-3}\,\text{m/m} = 500\,\mu\varepsilon$$

$$\text{Strain in hanger rod 2}: \varepsilon_2 = \dfrac{\delta_B}{3\,(m)} = \dfrac{3 \times 10^{-3}}{3}\,\text{m/m} = 1.0 \times 10^{-3}\,\text{m/m} = 1000\,\mu\varepsilon$$

b. In view of the 0.5 mm clearance at A,
Extension of hanger rod $1 = \delta_A - 0.5$ mm $= 1.5 - 0.5$ mm $= 1.0$ mm

$$\text{Hence, } \varepsilon_1 = \dfrac{1.0 \times 10^{-3}\,(m)}{3.0\,(m)} = 0.333 \times 10^{-3}\,\text{m/m} = 333\,\mu\varepsilon$$

Note: $\varepsilon_2 = 1000\,\mu\varepsilon$ as in part (a).

Strain

c. In view of the 0.5 mm clearance at B,
Extension of hanger rod $2 = \delta_B - 0.5$ mm $= 3.0 - 0.5$ mm $= 2.5$ mm

$$\text{Hence, } \varepsilon_2 = \frac{2.5 \times 10^{-3} \text{ (m)}}{3.0 \text{ (m)}} = 0.833 \text{ m/m} = 833 \, \mu\varepsilon$$

Note: $\varepsilon_1 = 500 \, \mu\varepsilon$ as in part (a).

Primary Learning Objectives
1. Proper use of geometric compatibility when solving complex problems of normal strain
2. Small angle approximation when solving problems of strain
3. Accommodation of joint clearance (loose joints) when solving problems of normal strain

■ **End of Solution**

Example 4.3

A rod is made of a slender segment AB of length 2 m and a thick segment BC of length 1 m, as shown in Figure 4.4. Under axial loading, the segment AB attains a normal strain of $\varepsilon_{AB} = 0.002 \, \varepsilon$. Also, the total elongation of the overall rod (AC) due to the loading is 0.005 m. Determine the elongation of AB; elongation of BC; strain in BC; and the average strain in the overall rod AC.

Solution

Length of AB: $L_{AB} = 2.0$ m
Hence, elongation of AB: $\delta_{AB} = \varepsilon_{AB} \times L_{AB} = 0.002 \times 2.0 = 0.004$ m
Given, total elongation of AC is $\delta_{AC} = 0.005$ m
Hence, elongation of BC: $\delta_{BC} = 0.005 - 0.004 = 0.001$ m

$$\text{Length of } BC: L_{BC} = 1.0 \text{ m}$$

$$\text{Hence, strain in } BC: \varepsilon_{BC} = \frac{\delta_{BC}}{L_{BC}} = \frac{0.001}{1.0} = 0.001 \varepsilon$$

$$\text{Average strain in } AC: \varepsilon_{AC} = \frac{\delta_{AC}}{L_{AC}} = \frac{0.005}{3.0} = 0.0017 \, \varepsilon$$

Note: This is not equal to the average of the strains in the two segments, which is 0.0015.

Primary Learning Objectives
1. Normal strain in a stepped axial member with multiple dissimilar segments
2. Deformation compatibility of a stepped rod with multiple dissimilar segments
3. Further considerations of average normal strain (in particular, individual strains should not be averaged to get the overall average strain)

■ **End of Solution**

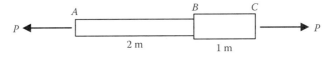

FIGURE 4.4 A rod made of two segments of unequal dimensions.

Example 4.4

A metal bar *OA* of unstrained length $a = 2$ m is hung from a rigid ceiling at point *O*. A cable carrying a vertical load is attached to the other end *A*, as shown in Figure 4.5a. The unstrained length of the cable is $AB = L = 1$ m.

As the system attains static equilibrium, under loading, point *A* moves down by $\delta_A = 1$ mm from the original unstrained position and point *B* moves down by $\delta_B = 3$ mm from the original unstrained position.

a. Assuming that the metal bar is light, determine the normal strain in the bar and the normal strain in the cable.
b. Now assume that the bar is heavy. Due to its own weight and the load from the cable, suppose that the normal strain varies along the bar according to the relation

$$\varepsilon = k(a - z) + \varepsilon_A$$

where $k = 0.2 \times 10^{-3}$ m^{-1}. *Note:* ε_A is the strain at *A*, where $z = a$, with *z* measured from the fixed end *O* as shown in Figure 4.5b.

Determine ε_A; maximum strain in the bar; and the distribution of the axial deflection of the bar from *O* to *A*.

Sketch the strain distribution and the deflection distribution of the bar.

Solution

a. Length of the bar, $a = 2.0$ m

Extension in the bar, $\delta_A = 1.0$ mm

Normal strain (average) in the bar, $\varepsilon_{OA} = \dfrac{1.0 \times 10^{-3} \text{ (m)}}{2.0 \text{ (m)}} = 0.5 \times 10^{-3}$ m/m $= 500$ µε

Length of the cable, $L = 1.0$ m

Extension in the cable $= \delta_B - \delta_A = 3.0 - 1.0 = 2.0$ mm

Normal strain (average) in the cable, $\varepsilon_{AB} = \dfrac{2.0 \times 10^{-3} \text{ (m)}}{1.0 \text{ (m)}} = 2.0 \times 10^{-3}$ ε $= 2000$ µε

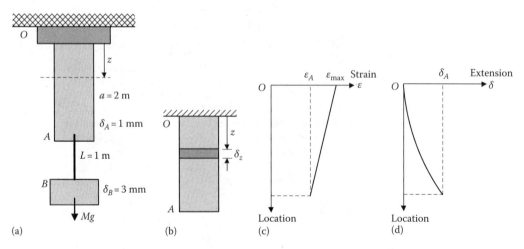

FIGURE 4.5 (a) A vertical bar with cable carrying a weight; (b) an elemental segment of the bar; (c) strain distribution in the bar; (d) extension profile of the bar.

Strain

b. Strain at location z from the fixed point O along the bar

$$\varepsilon = k(a - z) + \varepsilon_A \quad \text{(i)}$$

Consider an element δz at z (Figure 4.5b).
Its extension $= \varepsilon \delta z$

Total extension at $z = \int \varepsilon \cdot dz = \delta$ (say)

$$\rightarrow \delta = \int [k(a-z) + \varepsilon_A] dz = -\frac{k}{2}(a-z)^2 + \varepsilon_A z + C \quad \text{(ii)}$$

where C is the constant of integration
Point O is fixed $\Rightarrow \delta = 0$ at $z = 0$

$$\Rightarrow 0 = -\frac{k}{2}a^2 + C \Rightarrow C = \frac{1}{2}ka^2 \quad \text{(From ii)}$$

Also, $\delta = \delta_A$ at $z = a \Rightarrow \delta_A = \varepsilon_A a + C$ \quad (From ii)

$$\Rightarrow \varepsilon_A = \frac{\delta_A}{a} - \frac{C}{a} = \frac{\delta_A}{a} - \frac{1}{2}ka$$

Substitute $\delta_A = 1.0$ mm, $a = 2.0$ m, $k = 0.2 \times 10^{-6}$ m^{-1}

$$\varepsilon_A = \frac{1.0 \times 10^{-3} \text{ (m)}}{2.0 \text{ (m)}} - \frac{1}{2} \times 0.2 \times 10^{-3} (\text{m}^{-1}) \times 2.0 \text{ (m)}$$

$$\varepsilon_A = 0.5 \times 10^{-3} \text{ m/m} - 0.2 \times 10^{-3} \text{ m/m} \Rightarrow$$

$$\varepsilon_A = 0.3 \times 10^{-3} \text{ m/m} = 300 \, \mu\varepsilon$$

ε_{max} occurs at $z = 0$ (This is clear from Equation i)

$$\Rightarrow \varepsilon_{max} = ka + \varepsilon_A = 0.2 \times 10^{-3} \text{ (m}^{-1}) \times 2.0 \text{ (m)} + 0.3 \times 10^{-3} \text{ (m/m)}$$

$$\Rightarrow \varepsilon_{max} = 0.7 \times 10^{-3} \text{ m/m} = 700 \, \mu\varepsilon$$

Note: Average of ε_{max} and ε_A (strains at the two ends of the bar)

$$= \frac{1}{2}(0.3 + 0.7) \times 10^{-3} = 0.5 \times 10^{-3} \text{ m/m}$$

This average strain is the same as the strain we obtained for the case of light rod. This result holds because the strain of the rod is distributed linearly in the case of heavy rod.
  The strain in the heavy rod is given by (i): $\varepsilon = k(a - z) + \varepsilon_A$.
  This linear deflection strain profile is sketched in Figure 4.5c.

The deflection of the heavy rod is distributed quadratically, as seen from Equation ii:

$$\delta = -\frac{k}{2}(a-z)^2 + \varepsilon_A z + C$$

Substitute $C = (1/2)ka^2$ and $\varepsilon_A = (\delta_A/a) - (1/2)ka$ as obtained earlier. We get

$$\delta = \frac{k}{2}(a-z)z + \frac{\delta_A z}{a} \qquad \text{(iii)}$$

Note: Equation iii satisfies the boundary conditions: $\delta = 0$ at $z = 0$ and $\delta = \delta_A$ at $z = a$.
This quadratic deflection profile is sketched in Figure 4.5d.

Primary Learning Objectives
1. Analysis of the deflection and normal strain distribution of an axial member consisting of a bar segment and a cable segment
2. Analysis of the deflection of an axial member (rod) when its normal strain varies along the length
3. Deformation compatibility of axial members having multiple segments

■ **End of Solution**

4.3 SHEAR STRAIN

4.3.1 LOCAL SHEAR STRAIN AND AVERAGE SHEAR STRAIN

Since shear strain concerns change in angle between two intersecting line segments, we need to consider at least a 2-D (i.e., planar) situation to define it.

Consider two orthogonal (i.e., perpendicular) and infinitesimal line segments OP of length Δx and OQ of length Δy along the local x-axis and the local y-axis, respectively, as shown in Figure 4.6. When subjected to stresses, suppose that OP deforms to $O'P'$ and OQ deforms to $O'Q'$ as shown in the figure. The figure shows the general situation where all points in the body have moved due to its deformation.

Note: In general, a straight-line segment may not deform into a straight-line segment. For an infinitesimal (i.e., very minute) line segment, however, the deformed line segment may be assumed a straight line. Specifically, the line joining the two end points after deformation will approximate the deformed line segment.

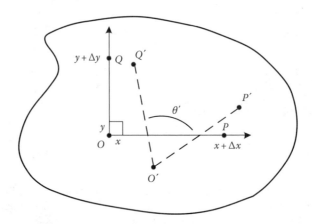

FIGURE 4.6 Definition of shear strain.

Strain

The angle between $O'P'$ and $O'Q'$ is denoted by θ'. The (local) shear strain at corner O is given by the "decrease" in the corner angle (in radians), in the limit

$$\gamma_{xy} = \frac{\pi}{2} - \lim_{\Delta x, \Delta y \to 0} \theta' \tag{4.3}$$

Note: Shear strain is associated with any shearing-like deformation in a body. Clearly, for a shear strain to exist, it is not necessary to have a right-angled corner in the body. A right-angled corner is used only as a reference in the formal definition of shear strain. In a body with shear deformation, a corner with any other angle also may experience a shear strain. This is rather analogous to using a straight-line segment of unity length as the reference when defining a normal strength even though a line of any length can experience a normal strain.

4.3.1.1 Average Shear Strain

Typically, the shear strain varies from point to point in the body. Then, an average shear strain may be used to represent the state of shear strain in the body. Consider a rectangular body (or part of it) of interest with a corner at O and the sides falling along x- and y-axes (see Figure 4.7). Due to shear stresses, suppose that one side slides through distance δ with respect to the opposite side, in the x direction.

Length of the side in the y direction = L

Then the average shear strain in the member is given by

$$\gamma_{avg} = \frac{\delta}{L} \tag{4.4}$$

Notes:

- Average normal strain is defined with respect to a line segment of finite length in an object
- Average shear strain is defined with respect to two perpendicular sides intersecting at a corner of a rectangular segment of finite dimensions in an object
- Averaging is done over the entire segment under consideration

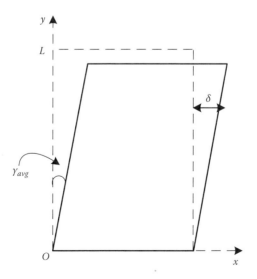

FIGURE 4.7 Definition of an average shear strain.

Example 4.5

A component of an engine suspension system is schematically shown in Figure 4.8a. The engine post AB is relatively deformable, which is welded to the rigid leg at B. The rigid leg is glued to two identical pieces of rubber, which in turn are glued to the inside wall of a fixed rigid well, as shown. Initially, the entire suspension unit is unstrained (neglect the weight).

When the engine load (P) is exerted on the post at the top (A), the point of application moves down through $\delta_A = 1.0$ mm. Also, as a result, the post is compressed axially with a normal strain of $\varepsilon_{AB} = -600$ με. (*Note:* The negative sign signifies the compressive strain.) If the length of the engine post is 50 cm and the thickness of each rubber lug is 7 cm, determine the shear stain at a corner of the rubber lug.

Solution

Let δ_B = downward movement of B on the application of the load.
Axial extension of the post = $\delta_B - \delta_A$
Length of the post = 50 cm = 0.5 m
Then, normal strain (average) in the post

$$\varepsilon_{AB} = \frac{\delta_B - \delta_A \text{ (m)}}{0.5 \text{ (m)}} = -600 \times 10^{-6}$$

$\rightarrow \delta_B = \delta_A - 0.5 \times 600 \times 10^{-6}$ m $= 1.0 \times 10^{-3} - 0.3 \times 10^{-3}$ m $= 0.7 \times 10^{-3}$ m $= 0.7$ mm

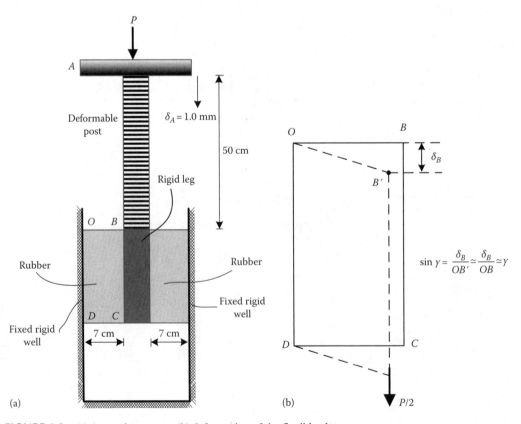

FIGURE 4.8 (a) An engine mount; (b) deformation of the flexible element.

Strain

The shear deformation of a rubber lug is sketched in Figure 4.8b, where the shrinking angle γ of the corner O gives the shear strain at O.

From geometry, we have

$$\sin\gamma = \frac{\delta_B}{OB'} \simeq \frac{\delta_B}{OB} = \frac{0.7 \times 10^{-3} \text{ (m)}}{7 \times 10^{-2} \text{ (m)}} = 0.01$$

Since γ is small, we get $\gamma = \sin^{-1} 0.01 \simeq 0.01$ rad

Note 1: Shear strain at corner $B = -0.01$ rad = shear strain at corner D (−ve sign is used because the corner angle there has increased). Shear strain at corner $C = +0.01$ rad = shear strain at corner O.

Note 2: $\tan\gamma = \delta_B/OB$. So, we could have used "tan" instead of "sin" in the present case of small angles.

Note 3: Strictly speaking, γ in this example is an "average" strain (as determined from the deformation of the entire lug, and not by considering an infinitesimal element of it) rather than a local strain. But, we will not make this distinction strictly, in the present example.

Primary Learning Objectives

1. Determination of average shear strain
2. Proper sign of a shear strain in a member
3. Deformation compatibility of a multicomponent device
4. Handling of small angles in the shear strain determination

■ **End of Solution**

Example 4.6

A triangular metal plate ABC is firmly attached to a rigid ceiling along the side AB and is maintained vertically in an unstrained state. The dimensions of the plate are

$$AC = b = 0.3 \text{ m}, \quad CB = a = 0.4 \text{ m}, \quad AB = c = 0.5 \text{ m}$$

Now a vertical load is applied to the corner C of the plate (see Figure 4.9a). As a result, C moves vertically to a new location C' with a movement of $\delta = 2$ mm.

Determine the resulting shear strain at the corner C of the plate.

Note: In this example, it is assumed that C moves vertically to C' (i.e., CC' is vertical) even though this may not be true in general.

Solution

In view of the fact that $c^2 = a^2 + b^2$, ACB is a right-angled triangle with $\angle ACB = \pi/2$.
See Figure 4.9b. Let $\angle BAC = \alpha$ and $\angle ABC = \beta$
Then,

$$\beta = \frac{\pi}{2} - \alpha$$

Also,

$$\cos\alpha = \frac{AC}{AB} = \frac{0.3}{0.5} = 0.6 = \sin\beta$$

$$\sin\alpha = \frac{CB}{AB} = \frac{0.4}{0.5} = 0.8 = \cos\beta$$

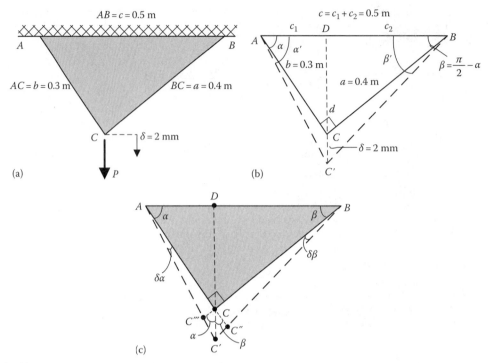

FIGURE 4.9 (a) A hung triangular plate carrying a load. (b) Plate deformation due to the load. (c) Plate deformation geometry for the approximate method.

$$\tan\alpha = \frac{CB}{AC} = \frac{0.4}{0.3} = \frac{4}{3} = \cot\beta$$

The vertical line through C (and C') meets AB at D. We have

$$AD = c_1(\text{say}) = AC\cos\alpha = 0.3 \times 0.6 = 0.18 \text{ m}$$

$$DB = c_2(\text{say}) = CB\cos\beta = 0.4 \times 0.8 = 0.32 \text{ m}$$

Check: $c_1 + c_2 = 0.18 + 0.32 \text{ m} = 0.5 \text{ m} = c$ (Checked)

$DC = d$ (say) $= AC\sin\alpha = 0.3 \times 0.8 = 0.24$ m

$$DC' \simeq DC + CC' \simeq DC + \delta = 0.24 + 2 \times 10^{-3} \text{ m}$$

Let $\angle BAC' = \alpha'$ and $\angle ABC' = \beta'$

We have

$$\tan\alpha' = \frac{DC'}{AD} = \frac{0.24 + 2 \times 10^{-3} \text{ (m)}}{0.18 \text{ (m)}} = 1.344 \Rightarrow \alpha' = 0.931274 \text{ rad}$$

$$\tan\beta' = \frac{DC'}{DB} = \frac{0.24 + 2 \times 10^{-3} \text{ (m)}}{0.32 \text{ (m)}} = 0.75625 \Rightarrow \beta' = 0.647489 \text{ rad}$$

Strain

Now $\angle AC'B = \pi - (\alpha' + \beta')$

Shear strain at corner A:
$$\gamma = \frac{\pi}{2} - \angle AC'B = \frac{\pi}{2} - [\pi - (\alpha' + \beta')] = \alpha' + \beta' - \frac{\pi}{2}$$

$$= 0.931274 + 0.647489 - \frac{\pi}{2} \text{ rad}$$

or $\gamma = 8.0 \times 10^{-3}$ rad

Note: Strictly speaking, γ in this example is an "average" strain (as determined from the deformation of the entire plate, and not by considering an infinitesimal element of it) rather than a local strain. But, we will not make this distinction strictly, in the present example.

Approximate Method

See Figure 4.9c. Line AC is extended until it meets line BC' at C''. Also, line BC is extended until it meets line AC' at C'''.

We have

$$CC''' = CC' \cos \alpha \text{ and } CC'' = CC' \cos \beta$$

Let $\delta\alpha$ be the increase in angle at corner A and $\delta\beta$ be the increase in angle at corner B. From right-angled triangles, we have

$$\tan \delta\alpha = \frac{CC'''}{AC} \approx \delta\alpha \text{ and } \tan \delta\beta = \frac{CC''}{BC} \approx \delta\beta \text{ (for small angles } \delta\alpha \text{ and } \delta\beta\text{)}$$

Hence,

$$\delta\alpha + \delta\beta = \frac{CC' \cos \alpha}{AC} + \frac{CC' \cos \beta}{BC} = \frac{\delta \cos \alpha}{b} + \frac{\delta \cos \beta}{a}$$

Now note that: Increase in angle A + Increase in angle B = Decrease in angle C = Shear strain at corner C.

Hence, $\gamma = \dfrac{\delta \cos \alpha}{b} + \dfrac{\delta \cos \beta}{a}$

Substitute numerical values: $\gamma = \dfrac{2 \times 10^{-3} \times 0.6}{0.3} + \dfrac{2 \times 10^{-3} \times 0.8}{0.4} = 8.0 \times 10^{-3}$ rad

This result is the same as what we obtained earlier.

Primary Learning Objectives
1. Determination of average shear strain
2. Proper use of geometric relations in the determination of shear strains
3. Proper sign of shear strain in a member
4. Approximations when handling small angles in the computation of shear strain

■ **End of Solution**

4.4 THERMAL STRAIN

Deformations in a body due to thermal expansions and contractions, which are caused by temperature changes, may be represented as thermal strains.

4.4.1 COEFFICIENT OF THERMAL EXPANSION (α)

Coefficient of thermal expansion α is defined as the expansion in a line segment of unit length of a body due to a temperature rise of 1°.

Hence, the elongation of a line segment of unit length due to a temperature rise of ΔT degrees $= \alpha \Delta T$.

It follows that, by definition of normal strain (elongation per unit length), the thermal strain due to a temperature increase of ΔT degrees is given by

$$\varepsilon_T = \alpha \Delta T \tag{4.5}$$

Note: For homogeneous (uniform; properties do not change from point to point) and isotropic (nondirectional; properties do not change according to the direction) material, α is the same in every location and in every direction of the body. Then "shear" thermal strains are not generated in the body. Only the normal thermal strains are considered in these problems. However, when a body is constrained in some complex manner, a temperature change may generate shear stresses and associated shear strains. These shear strains are analyzed by first determining the shear stresses.

Example 4.7

A polished bronze rod of length 1 m is placed inside a steel casing of internal length 1.002 m (i.e., there is a clearance of 0.002 m) as shown in Figure 4.10. Determine the maximum temperature rise ΔT that will not result in normal stresses in the rod or in the casing.

The coefficients of thermal expansion of bronze and steel are $\alpha_b = 20 \times 10^{-6}/°C$ and $\alpha_s = 10 \times 10^{-6}/°C$, respectively.

Solution

From deformation compatibility, for the expanded rod to just touch the expanded casing without stressing it, we require

Expansion of rod = 0.002 + Expansion of casing

$$\rightarrow 1.0 \alpha_b \cdot \Delta T = 0.002 + 1.002 \alpha_s \cdot \Delta T$$

$$\rightarrow \Delta T (1.0 \times 20 \times 10^{-6} - 1.002 \times 10 \times 10^{-6}) = 0.002$$

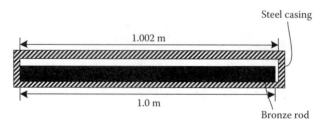

FIGURE 4.10 A bronze rod placed inside a steel casing.

$$\rightarrow \Delta T = \frac{2 \times 10^3}{20 - 10.02} \, °C \simeq 200°C$$

Primary Learning Objectives
1. Determination of thermal strain
2. Relationship between thermal strain and thermal deformation
3. Proper application of the deformation compatibility in thermal problems consisting of multiple components of different material and dimension

■ **End of Solution**

4.5 MEASUREMENT OF STRAIN

Strain is measured using strain gauges. Common are the resistance-type strain gauges. The change of electrical resistance in a material, when mechanically deformed, is the property that is used in these sensors. Modern strain gauges are manufactured primarily as metallic foil (e.g., using the copper–nickel alloy known as constantan) or semiconductor elements (e.g., silicon with trace impurity boron). Some examples are shown in Figure 4.11.

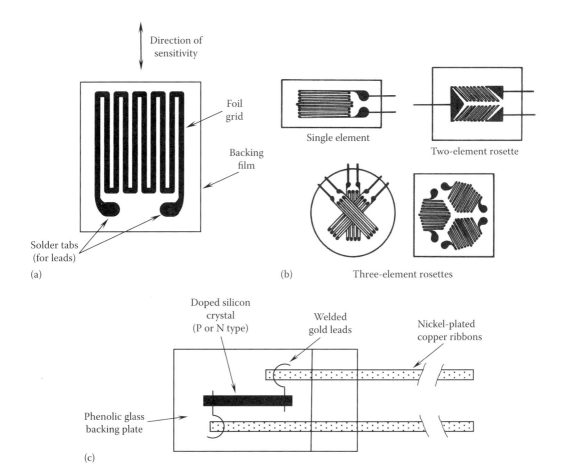

FIGURE 4.11 Typical strain-gauge elements: (a) a foil-type strain gauge; (b) different arrangements of strain gauges; (c) a semiconductor strain gauge.

TABLE 4.1
Properties of Common Strain Gauge Materials

Material	Composition	Gauge Factor (Sensitivity)	Temperature Coefficient of Resistance (10^{-6}/°C)
Constantan	45% Ni, 55% Cu	2.0	15
Isoelastic	36% Ni, 52% Fe, 8% Cr, 4% (Mn, Si, Mo)	3.5	200
Karma	74% Ni, 20% Cr, 3% Fe, 3% Al	2.3	20
Monel	67% Ni, 33% Cu	1.9	2000
Silicon (semiconductor)	p-type	100 to 170	70 to 700
Silicon (semiconductor)	n-type	−140 to −100	70 to 700

The following relationship can be written for a strain gauge element:

$$\frac{\delta R}{R} = S_s \varepsilon \tag{4.6}$$

where
  R is the strain gauge resistance
  δR is the change in strain gauge resistance due to strain ε
  S_s is the *gauge factor* (or *sensitivity*) of the strain gauge element

The gauge factor ranges from 2 to 6 for most *metallic strain gauge* elements and from 40 to 200 for *semiconductor strain gauges*. The values of gauge factors for several strain gauge materials are given in Table 4.1.

To measure strains in more than one direction, and to measure shear strain, multiple strain gauges (e.g., various rosette configurations) are available as single units, as shown in Figure 4.11b. These units have more than one direction of sensitivity. Principal strains (which exist in a plane where shear strains are absent; see Chapter 9) can be determined as well by using these multiple strain gauge units. Some analytical results that may be used in the strain measurement using multi-element rosettes are given in Chapter 9.

4.5.1 Bridge Circuit

The change in resistance δR of a strain gauge element, which determines the associated strain, is measured using a suitable electrical circuit. Typically, a direct-current (dc) bridge (a Wheatstone bridge) is used for this purpose. This is a resistance bridge with a constant dc voltage supply.

Consider the Wheatstone bridge circuit shown in Figure 4.12. One or more of the four resistors R_1, R_2, R_3, and R_4 in the bridge may represent strain gauges. Assuming that the bridge output is in open circuit (i.e., very high load resistance is present there), the output v_o may be expressed as

$$v_o = v_A - v_B = \frac{R_1 v_{ref}}{(R_1 + R_2)} - \frac{R_3 v_{ref}}{(R_3 + R_4)} = \frac{(R_1 R_4 - R_2 R_3)}{(R_1 + R_2)(R_3 + R_4)} v_{ref} \tag{4.7}$$

For a balanced bridge, the numerator of the RHS expression of (4.43) must be zero. Hence, the condition for bridge balance is

$$\frac{R_1}{R_2} = \frac{R_3}{R_4} \tag{4.8}$$

Strain

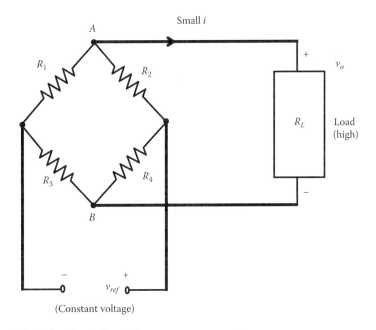

FIGURE 4.12 A dc bridge circuit for strain-gauge measurement.

Starting from the balanced condition (where the output voltage is zero), for small independent changes δR_1, δR_2, δR_3, and δR_4 in the four resistances of the bridge circuit (these may come from active strain gauges), using straightforward calculus in Equation 4.7, we can determine the bridge output voltage δv_o as

$$\frac{\delta v_o}{v_{ref}} = \frac{(R_2 \delta R_1 - R_1 \delta R_2)}{(R_1 + R_2)^2} - \frac{(R_4 \delta R_3 - R_3 \delta R_4)}{(R_3 + R_4)^2} \quad (4.9)$$

Suppose that at first $R_1 = R_2 = R_3 = R_4 = R$. Then, according to Equation 4.8, the bridge is balanced. Now increase R_1 by δR. For example, R_1 may represent the only active strain gauge while the remaining three elements in the bridge are identical dummy elements. In view of Equation 4.9, the change in the bridge output due to the change δR is given by

$$\delta v_o = \frac{\left[(R+\delta R)R - R^2\right]}{(R+\delta R + R)(R+R)} v_{ref} - 0$$

which can be written as

$$\frac{\delta v_o}{v_{ref}} = \frac{\delta R/R}{(4 + 2\delta R/R)} \quad (4.10)$$

Note: The output is nonlinear in $\delta R/R$. If, however, $\delta R/R$ is assumed small in comparison to 2, we have the linearized relationship:

$$\frac{\delta v_o}{v_{ref}} = \frac{\delta R}{4R} \quad (4.11)$$

The factor ¼ on the RHS of Equation 4.11 represents the *sensitivity* in the bridge, as it gives the change in the bridge output (nondimensional) for a given change in active resistance (nondimensional), while the other parameters are kept fixed.

4.5.2 BRIDGE CONSTANT

Any of the four resistors in a bridge circuit may represent active strain gauges; for example, tension in R_1 and compression in R_2, as in the case of two strain gauges mounted symmetrically at 45° about the axis of a shaft in torsion. In this manner, the overall sensitivity of a strain gauge bridge can be increased. It is clear from Equation 4.7 that if all four resistors in the bridge are active, the best sensitivity is obtained if, for example, R_1 and R_4 are in tension and R_2 and R_3 are in compression, so that all four differential terms have the same sign. Generally then, if more than one strain gauge is active, the bridge output may be expressed as

$$\frac{\delta v_o}{v_{ref}} = \frac{k}{4R} \delta R \qquad (4.12)$$

where $k = \dfrac{\text{Bridge output in the general case}}{\text{Bridge output if only one strain gauge is active}}$

This constant is known as the *bridge constant*. The larger the bridge constant, the better the bridge sensitivity.

Example 4.8

A strain-gauge load cell (a force sensor) consists of four identical strain gauges, forming a Wheatstone bridge, which are mounted on a rod that has a square cross section. One opposite pair of strain gauges is mounted axially and the other pair is mounted in the transverse direction, as shown in Figure 4.13a. To maximize the bridge sensitivity, the strain gauges are connected to the bridge as shown in Figure 4.13b. Determine the bridge constant k in terms of *Poisson's ratio v* of the rod material.

Note: When a rod is stretched, it contracts in the lateral direction. The corresponding lateral strain is related to the longitudinal (normal) strain according to (see Chapter 5)

$$\text{Transverse strain} = (-v) \times \text{Longitudinal strain} \qquad (4.13)$$

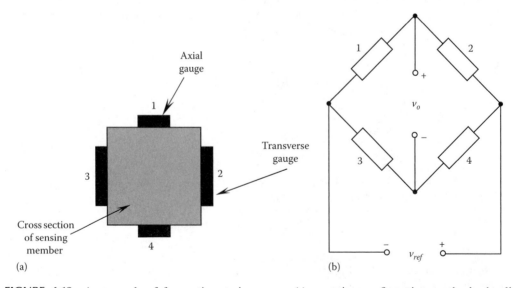

FIGURE 4.13 An example of four active strain gauges: (a) mounting configuration on the load cell; (b) bridge circuit.

Solution

Suppose that $\delta R_1 = \delta R$. Then, for the given configuration, using the definition (4.13), we have $\delta R_2 = -\nu \delta R$; $\delta R_3 = -\nu \delta R$; and $\delta R_4 = \delta R$.

Now, it follows from Equation 4.9 that

$$\frac{\delta v_o}{v_{ref}} = 2(1+\nu)\frac{\delta R}{4R}$$

According to Equation 4.12, the bridge constant is $k = 2(1+\nu)$.

Primary Learning Objectives
1. A practical application of strain gauges (force measurement)
2. Configuration of a bridge circuit to maximize its sensitivity

■ End of Solution

4.5.3 CALIBRATION CONSTANT

By combining Equations 4.6 and 4.12, we get the basic equation that is used in strain-gauge sensing:

$$\frac{\delta v_o}{v_{ref}} = \frac{1}{4}kS_s\varepsilon = C\varepsilon \qquad (4.14)$$

In Equation 4.12, C is called the *calibration constant* of the bridge. It determines the strain measurement once the bridge output δv_o and the bridge reference voltage v_{ref} are known.

SUMMARY SHEET

Displacement and deformation: A body can move (displace) without undergoing deformation. Then, the displacements are due to *rigid body motions* of the body components. No strains result from such displacements.

Strain:

- Caused by deformations in a body, as a result of stresses and temperature changes.
- It is an "intensity" of deformation at a location in the body.
- Deformations include change in length of a straight-line segment (normal strain) and change in the angle between two perpendicular line segments (i.e., change in the direction of the two line segments) intersecting at a corner.
- It is a ratio (is nondimensional), because it is measured either per unit length or as a change in angle.

Normal strain (ε): Deformation per unit length of a straight-line segment in a specified direction. Tensile is positive, compressive is negative.

Unit: m/m or 1 unit of strain or 1 ε (dimensionless).

$$1 \text{ microstrain} = 1 \text{ } \mu\varepsilon = 1 \text{ } \mu m/m = 1 \times 10^{-6} \text{ m/m} = 1 \times 10^{-6} \text{ } \varepsilon$$

To convert m/m into µε, multiply by 10^6.

$$1\% \text{ strain} = 1/100 \text{ strain} = 0.01 \text{ m/m} = 0.01 \text{ ε}$$

To convert m/m into % strain, multiply by 100.

Local normal strain: $\varepsilon_x = \lim_{\Delta x \to 0} (\Delta x' - \Delta x)/\Delta x$ along x.

Average normal strain: $\varepsilon_{avg} = \delta/L$; L is the original length in the specific direction, δ is the extension in that direction.

Variable normal strain: Total extension from $z=z_0$ to $z=z_1$ along z axis of a member: $\delta = \int_{z_0}^{z_1} \varepsilon_z dz$; ε_z is the normal strain at z.

Shear strain (γ): Change (decrease) in angle of a specified corner of angle $\pi/2$. Angle reduction is positive; angle increase is negative.

Unit: 1 rad (dimensionless).

$$1 \text{ µrad} = 1 \times 10^{-6} \text{ rad}$$

Local shear strain: $\gamma_{xy} = (\pi/2) - \lim_{\Delta x, \Delta y \to 0} \theta'$; θ' is the deformed angle of originally right-angled corner made of Δx and Δy.

Average shear strain: $\gamma_{avg} = \delta/L$; L is the length of the side in the y direction, δ is the deflection (relative movement of one end with respect to the other) of the same side in an orthogonal (x) direction (assuming that the direction of the other side of the corner remains as x).

Thermal strain (ε_T): Caused by deformation due to temperature change ΔT; thermal expansion is +ve, thermal contraction is −ve.

$\varepsilon_T = \alpha \Delta T$; α = coefficient of thermal expansion = expansion per unit length per unit increase in temperature.

Shear thermal strains: For an un-restrained homogeneous (uniform; properties do not change from point to point) and isotropic (nondirectional; properties do not change according to the direction) material, α is the same in every location and in every direction of the body; → "shear" thermal strains are not generated in the body.

Resistance change due to strain (in a strain gauge): $\delta R/R = S_s \varepsilon$; R is the original resistance, δR is the change in resistance due to strain ε, S_s is the *gauge factor* (or *sensitivity*) of strain gauge.

DC bridge (Wheatstone bridge): $\delta v_o / v_{ref} = (k/4R) \delta R$; δv_o is the bridge output, v_{ref} is the bridge reference voltage.

Bridge constant:

$$k = \frac{\text{Bridge output in the general case}}{\text{Bridge output if only one strain gauge is active}}$$

Strain sensing equation: $\delta v_o / v_{ref} = (1/4) k S_s \varepsilon = C \varepsilon$; C is the calibration constant of the bridge.

PROBLEMS

4.1 A soccer ball (spherical) was found to have a radius $r_0 = 11.0$ cm before inflating, when the internal pressure was equal to the atmospheric pressure. After inflating the ball, the new radius was $r = 11.5$ cm (see Figure P4.1). Estimate the average normal strain ε_θ of the leather skin of the ball.

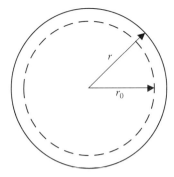

FIGURE P4.1 An inflated leather soccer ball.

4.2 A cable is tied to two points A and C and maintained horizontal under unstrained conditions. Then a weight is hung at point B (see Figure P4.2). As a result, the point B of the cable moves to B'. Given: $AB = a = 1.0$ m, $BC = b = 2.0$ m, and $BB' = \delta = 2.0$ mm.

Determine
  Strain in the cable segment AB'
  Strain in the cable segment $B'C$
  Average strain in the cable

4.3 A belt drive has two pulleys of radius r_1 and r_2 with a rubber belt passing around them (Figure P4.3). For tightening the belt, one pulley can be moved slightly, center-to-center, using a slider mechanism with tightening screws. Suppose that $2r_1 = 0.25$ m and $2r_2 = 0.35$ m. Initially, when the belt was not stretched (i.e., it was unstrained), the center-to-center distance between the pulleys was $a_0 = 1.0$ m. Next, the belt was tightened by increasing this distance to $a = 1.02$ m. Estimate the average normal strain of the belt in this configuration.

4.4 A rigid platform of length L (= 10 m) is hinged at one end and held horizontally using a vertical cable of length b (= 5 m), which is attached to the platform at a distance a (= 6 m) from

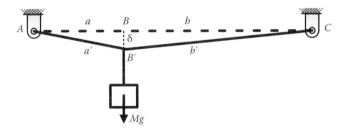

FIGURE P4.2 A horizontal cable carrying a load.

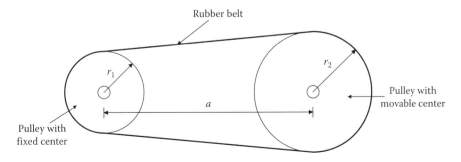

FIGURE P4.3 A belt drive.

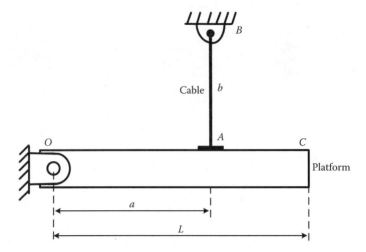

FIGURE P4.4 A rigid platform held by a cable.

the hinged end (see Figure P4.4). Due to a load on the platform, its free end moves through a vertical distance of $y_L = 0.03$ m. Assuming that the cable was unstrained in the beginning, determine the extension of the cable and the resulting average normal strain in it due to the platform movement.

4.5 A frame structure is made of two steel members BD and CD, which are pin-jointed at D, with the ends B and C hinged to a rigid vertical wall, as shown in Figure P4.5. The undeformed lengths of the members are $BD = 2.5$ m and $CD = 2.0$ m. A load P is hung at D. As a result, the member BD attains a tensile strain of 624.0 µε and the member CD attains a compressive strain of 500 µε. Determine the horizontal and vertical movements δ_h and δ_v, respectively, of joint D due to the deformation caused by the load.

4.6 A frame structure is made of two steel members BD and CD, which are pin-jointed at D, with the ends B and C hinged to a rigid vertical wall, as shown in Figure P4.6. The undeformed lengths of the members are $BD = 2.5$ m and $CD = 2.0$ m. A load P is hung at D. Due to the load, suppose that the member BD attains a compressive strain of 624.0 µε and the member CD attains a tensile strain of 500 µε. Determine the horizontal and vertical movements δ_h and δ_v, respectively, of joint D due to the deformation caused by the load.

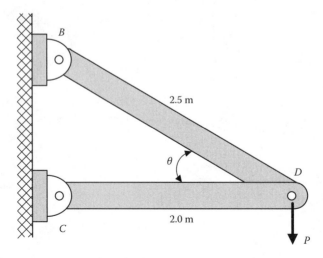

FIGURE P4.5 A frame structure carrying a load.

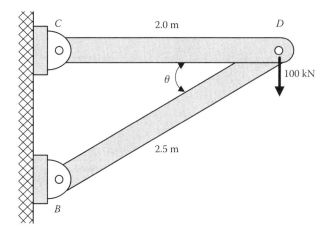

FIGURE P4.6 A frame structure carrying a load.

4.7 A uniform bar is hung from a rigid ceiling, as shown in Figure P4.7. Due to its own weight, the bar experiences a normal strain, which has a linear variation along its length, as given by $\varepsilon_z = k(L-z)$, where

   L is the length of the bar
   z is a general location along the bar, from the ceiling
   k is a constant that depends on the material properties (density and Young's modulus; see Chapter 5)

Determine
   a. The extension δ_A of the free end A of the rod due to its weight
   b. The average normal strain

Also, show that the average normal strain is equal to the average of the maximum normal strain and the minimum normal strain. Does this property hold for any bar, in general?

4.8 The normal strain along a nonuniform rod (Figure P4.8) varies according to $\varepsilon = kx^p$, where
   x is the axial location of the rod measured from one end
   k is a positive constant
   p is a constant, which can be negative

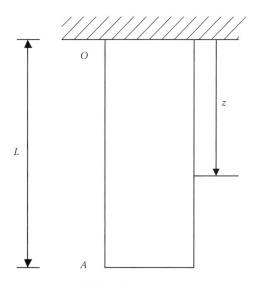

FIGURE P4.7 A uniform bar hung from a rigid ceiling.

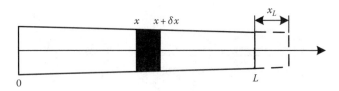

FIGURE P4.8 A nonuniform rod subjected to a variable strain.

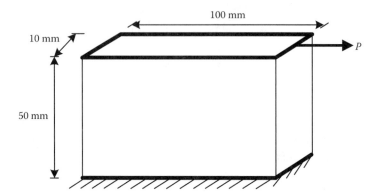

FIGURE P4.9 A rubber slab with fixed base and subjected to a shear force.

Note: In this problem, the variation in strain is attributed to both nonuniformity of the cross section and the material properties of the rod (e.g., inhomogeneity and anisotropy).

If the initial unstrained length of the rod is L, determine the overall extension (of one end with respect to the other) of the rod due to the strain distribution; and the average normal strain in the rod.

4.9 A solid slab of rubber is sandwiched between two thin horizontal rigid metal plates, with the bottom plate glued and fixed at the base, as shown in Figure P4.9. A horizontal force P is applied centrally to the top plate. The resulting horizontal deflection of the metal plate in the direction of P is $\delta = 5.0$ mm.
  a. Sketch the deformed shape of the object.
  b. Determine the average shear strain in the slab of rubber.

4.10 A uniform, rectangular plastic plate $ABCD$ of height 1.6 m and width 0.8 m is rigidly fixed at the bottom edge (AD) and anchored horizontally to a steel post at the corner C, using a steel bolt and nut (see Figure P4.10). The bolt and nut unit has a pitch of 2 mm. In the beginning, the plastic plate is unstrained and held vertically. Then the nut is tightened by turning it through two turns.
  a. Determine the average shear strain in the plastic plate with respect to the two perpendicular sides AB and AD (i.e., at corner A).
  b. Determine the average shear strain in the plate with respect to the perpendicular sides AD and DC (i.e., at corner D).

4.11 A rectangular fiberglass plate of size 1 m × 2 m is hinged to a rigid wall at corner O and supported on a vertical metal post of height 3 m at corner B, where OB is horizontal (see Figure P4.11). Initially, the system is unstrained (neglect the weight).

When a horizontal force is applied to corner C of the plate, point C moves through a horizontal distance of $\delta_C = 4$ mm. Also, the post assumes a normal strain $\varepsilon_B = 500$ µε (tensile) as a result. Determine the "average" shear strain in the plate with respect to corner O.

4.12 A uniform rectangular plate $OABC$ of length $L = 1.0$ m and width $h = 0.8$ m, made of a polymer material, is reinforced with a uniform metal bar along its edge OA. The reinforced plate is firmly attached to a rigid floor along the side CB and to a rigid wall along the side CO, as

Strain

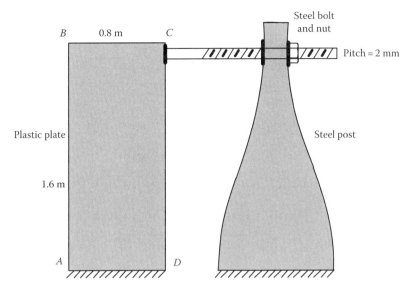

FIGURE P4.10 Plastic plate sheared using a bolt and nut.

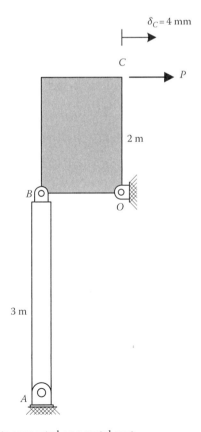

FIGURE P4.11 A fiberglass plate supported on a metal post.

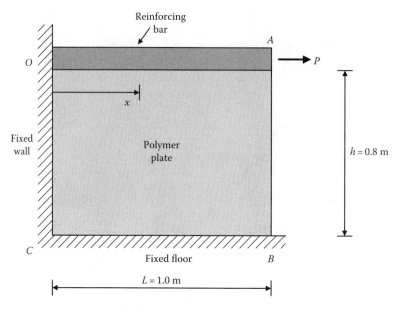

FIGURE P4.12 A polymer plate with edge reinforcement.

shown in Figure P4.12. A horizontal force P is applied to the bar at A. As a result, the plate deforms together with its reinforcement bar.

The resulting normal strain in the metal bar is given by

$$\varepsilon_x = \varepsilon_0 + kx$$

where x is a general location along the bar, from its fixed end O. The following parameters are given: $\varepsilon_0 = 1.0 \times 10^{-3}$ m/m, $k = 2.0 \times 10^{-3}$ m^{-1}.

Determine the average shear strains γ_A, γ_B, γ_C, and γ_O with respect to the corners A, B, C, and O, respectively, of the plate.

4.13 An elevated guideway of an urban transit system consists of identical multiple spans of length 40 m placed on support piers (Figure P4.13). There are expansion slots in the anchoring between the guideway ends and the support piers in order to accommodate change in length (longitudinal strain) of the guideway due to temperature changes and load variations. Considering the temperature extremes of −20°C and +40°C, estimate a suitable gap between two adjacent ends of guideway spans (or, length of an expansion slot) that can accommodate the variation of the longitudinal thermal strain of the guideway. The coefficient of thermal expansion of a guideway span is $\alpha = 11.7 \times 10^{-6}$/°C.

4.14 A steel tube of length 1 m is horizontally attached to a rigid wall. A bronze tube of length 1.5 m is horizontally attached to the opposite wall, which is also rigid. The two tubes face each other, and have a gap of 2 mm (see Figure P4.14). Determine the necessary temperature change ΔT so that the two tubes barely touch each other. The coefficients of thermal expansion of steel and bronze are $\alpha_s = 10 \times 10^{-6}$/°C and $\alpha_b = 20 \times 10^{-6}$/°C, respectively.

4.15 A frame structure is made of two aluminum members BD and CD (coefficient of thermal expansion $\alpha = 25.0 \times 10^{-6}$/°C), which are pin-jointed at D, with the ends B and C hinged to a rigid vertical wall, as shown in Figure P4.15. The undeformed lengths of the members are $BD = 2.5$ m and $CD = 2.0$ m.

The structure undergoes a temperature rise of $\Delta T = 50$°C.
  a. Show that the there are no axial forces in the members BD and CD.
  b. Determine the horizontal and vertical movements δ_h and δ_v, respectively, of joint D due to the deformation caused by the temperature change.

Strain

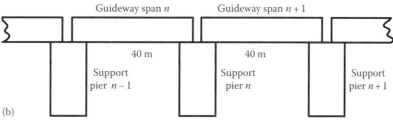

FIGURE P4.13 (a) A multispan guideway with thermal expansion slots (photo by C.W. de Silva); (b) a schematic diagram.

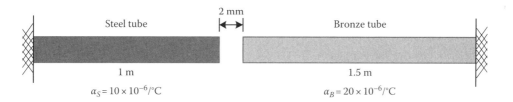

FIGURE P4.14 Horizontally facing tubes of steel and bronze.

Further Review Problems

4.16 A vertical tie rod is rigidly mounted in a foundation. The upper segment of the rod is threaded and passes through the hole of a rigid plate of the structure on which the tie rod would provide a holding force (see Figure P4.16). Initially, the tie rod is unstrained (relaxed) and a holding nut is screwed snugly against the plate. Next, the rod is tensioned by rotating the nut through two full turns using a wrench. Determine the normal strain in the segment of the tie rod below the nut. Given

Length of the tie rod below the nut $L = 5.0$ m
Pitch of the nut $= 1.0$ mm

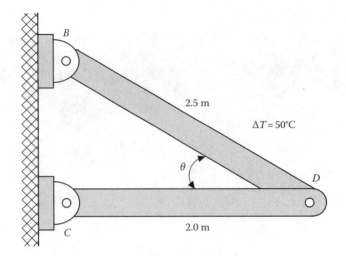

FIGURE P4.15 A frame structure subjected to thermal strains.

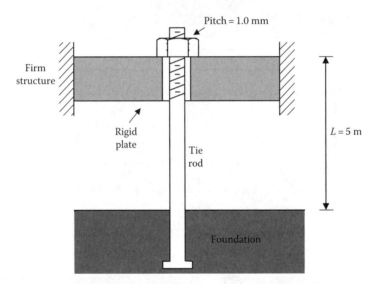

FIGURE P4.16 Tensioning rod of a rigid structure.

4.17 A slender metal shaft is firmly attached to two opposing walls at its two ends O and C. There is a rigid tube into which the shaft segment OA is inserted, and tightened using a nut at location A through the threading of the shaft. Also, there is a collar at point B of the shaft (see Figure P4.17). Initially, the shaft is relaxed, with the nut snugly placed against the tube, and without any holding force on the collar at B. Next, the nut is tightened through one full turn and simultaneously a force P is applied pushing the collar to the right. Then, the collar (B) moves to the right through $\delta_B = 1.75$ mm from the original relaxed position. Pitch of the nut = 1.0 mm. The initial lengths of the shaft segments are $OA = L_1 = 2.0$ m, $AB = L_2 = 3.0$ m, and $BC = L_3 = 3.5$ m.

Determine the average normal strains in the three segments of the shaft.

4.18 A two-bar truss is shown in Figure P4.18. The two-force members are pin-jointed at A, B, and C, as shown. Joint A has a roller, which is constrained to move along a horizontal guideway, when a horizontal force P is applied there. If this movement is $\delta_A = 1.0$ mm to the right, from

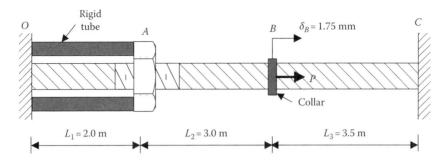

FIGURE P4.17 A slender shaft with a threaded segment through a rigid tube and a loading collar.

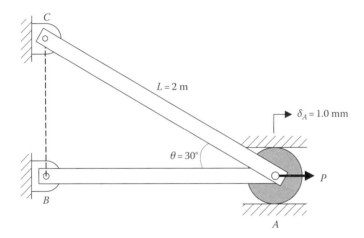

FIGURE P4.18 A two-bar truss with a constrained roller support at one joint.

the original unstrained position, determine the normal strains in the members AB and AC. Given $AC = L = 2$ m, $\angle CAB = \theta = 30°$. Hinge C is vertically above hinge B.

4.19 A nonuniform axial member AB is firmly hung from a rigid ceiling at A as shown in Figure P4.19. The axial deformation (downward vertical, due to the weight of the member) location z, as measured from A, was found to satisfy the equation $u = (k/3)[L^3 - (L-z)^3]$, where L is the length of the member and k is a constant.
  a. Show that the boundary condition at A is satisfied by this expression. What is the downward movement of the free end B of the member due to its own weight? Determine the average normal strain in the member.
  b. What is the normal strain at location z of the member? Determine the maximum and minimum normal strains in the member. What is the average of these two values? Compare the result with the average normal strain obtained in part (a).

4.20 A rectangular plate of metal $ABCD$ is deformed in plane into the quadrilateral $A'BCD'$ as shown in Figure P4.20. Note that the side BC is fixed. The deformed corner A' is 4 mm to the right of and 2 mm below the original corner A. If $AB = 0.4$ m and $BC = 0.5$ m, determine the average normal strain along the diagonal line $A'C$ and the average shear strain at the corner B, after the deformation.

4.21 A rigid circular concrete platform is firmly fixed on three symmetric steel posts of height $h = 10$ m and square cross section. The base of each steel post is firmly mounted in a rigid concrete foundation (see Figure P4.21). The central axis of a steel post is located at a radius $a = 1.0$ m from the center of the platform. Due to some structural loading, suppose that the

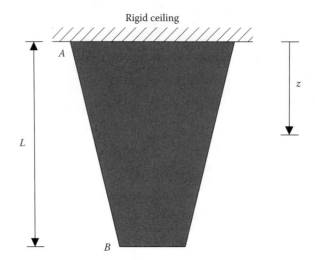

FIGURE P4.19 A uniform axial member suspended from a rigid ceiling.

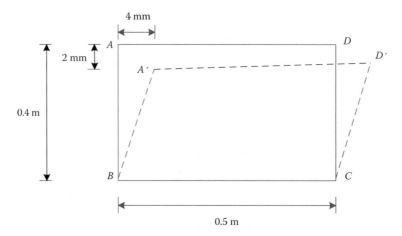

FIGURE P4.20 Planar deformation of a rectangular plate.

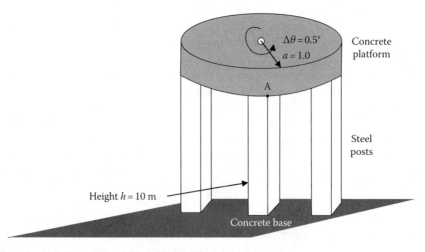

FIGURE P4.21 A rigid concrete platform firmly fixed on three steel posts.

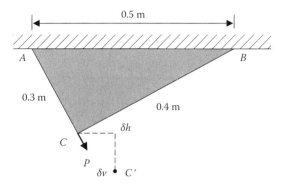

FIGURE P4.22 A steel plate hung vertically from a rigid ceiling and loaded at the bottom corner.

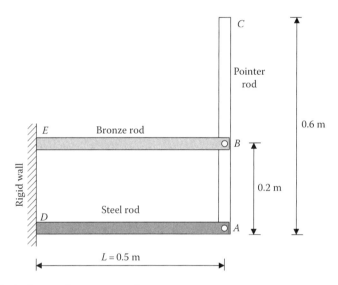

FIGURE P4.23 An indicator of temperature change.

platform rotates through an angle $\Delta\theta = 0.5°$ about the center (without any lateral motion at its center). Estimate the shear strain at the top corner A of a steel post.

4.22 A triangular steel plate ABC is firmly hung from a rigid ceiling along its edge AB with its plane kept vertically. A force is applied to the corner C (Figure P4.22). Due to this force, the corner C moves horizontally to the right through $\delta_h = 2$ mm and also vertically downward through $\delta_v = 2$ mm. Determine the resulting shear strain at corner C.

Given $AB = 0.5$ m, $AC = 0.3$ m, and $BC = 0.4$ m.

4.23 A steel rod AD of length $L = 0.5$ m and a bronze rod BE of equal length are fixed horizontally to a firm vertical wall at their ends D and E, and maintained horizontally at spacing 0.2 m. A vertical pointer rod ABC of length 0.6 m is pinned to the ends A and B (see Figure P4.23). If the temperature rises through $\Delta T = 60°C$, determine the resulting horizontal movement δ_C of point C of the pointer rod.

The coefficients of thermal expansion are, for steel, $\alpha_s = 12 \times 10^{-6}/°C$; for bronze, $\alpha_b = 18 \times 10^{-6}/°C$.

5 Mechanical Properties of Materials

CHAPTER OBJECTIVES

- Constitutive relation which represents the stress–strain behavior of a material
- Tensile test which experimentally determines the stress–strain behavior (stress–strain curve)
- Hooke's law, which gives the linear stress–strain relationship. Associated material parameters: Young's modulus E, Shear modulus G, and Poisson's ratio ν
- Material behavior: linear, nonlinear, elastic, plastic, work hardening, creep, fatigue, fracture
- Material properties: homogeneity, isotropy, elasticity, rigidity, strength, elongation, resilience, toughness, hardness
- Material types: ductile, brittle, soft, hard, strong, tough
- Strain energy: modulus of resilience, modulus of toughness

5.1 INTRODUCTION

Stress in a body is a result of internal loading, which is caused by external loading (forces and moments) acting on the body. Equations of equilibrium are used to determine internal loading (in statically determinate systems). We reviewed the associate subject of statics in Chapter 2. Strain is directly related to deformation or deflection (through geometry and kinematics of the body). We studied stress and strain separately, in Chapter 3 and Chapter 4, respectively. It is important to know how these two quantities are related in an object. For example, with the knowledge of stresses, we can determine strains (and deformations).

Deflections or deformations in a body are caused by loading in the body. The force–deflection relationship depends on the physics of the process. Correspondingly, the *stress–strain relation* is a mechanical property of the material. It depends of the "physics" of the material and is called a *constitutive relation*. These interrelations are summarized in Figure 5.1.

Compare: In a mechanical dynamic system, we are interested in loads (e.g., forces and moments), motions (e.g., linear and angular velocities), and their relationship (e.g., through Newton's second law). Relations among forces or moments are called *continuity equations* or *node equations* or *equilibrium equations* (i.e., force balance at a point; moment balance about an axis through a point); relations among linear or angular velocities are called *compatibility equations* or *loop equations* (velocity at a point is the same at a given instant, regardless of from which path we approach that point; i.e., there is no disintegration or rupture at the point). Newton's second law is a *constitutive equation* or a *physical law* (it determines the motion of a system in response to an applied force/moment). Also, Hooke's law is also a *constitutive equation* or a *physical law* (it determines the deformation of a system in response to an applied force/moment).

5.1.1 Problem of Mechanics of Materials

"Compatibility" is assumed (or, automatically satisfied) in the problem. "Equilibrium," which is governed by statics, must be satisfied. Stress–strain relationship is the "constitutive relation." This

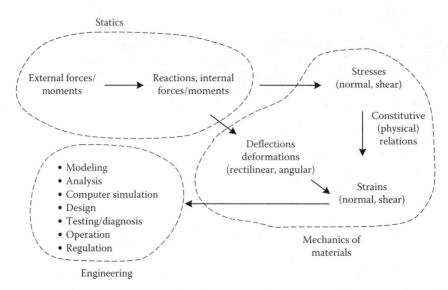

FIGURE 5.1 Interrelations in mechanics of materials.

"physical" relationship, which is governed by the material properties of body, is applied to solve the complete problem (see Figure 5.1).

5.1.2 HOMOGENEITY AND ISOTROPY

A material is *homogeneous* if the material properties (including composition and stress–strain relations) are the same at every point in the material. This does not imply that the material properties are the same in any direction at a given point. A material is *isotropic* if the material properties are the same in every direction, at a given point. In the present treatment, we assume homogeneity and isotropy of material.

5.2 STRESS–STRAIN BEHAVIOR

In theory, the stress–strain relation of a material should be derivable from the physics of the material. However, it will be a very complex (and often impossible) endeavor. A more realistic and practical approach would be to obtain this relation experimentally, as a stress–strain diagram (stress–strain curve). Normally, it is obtained by carrying out a *tensile test* (or a compressive test). Typically, the experimental result is a force–deflection diagram, which is then converted into a stress–strain diagram in a straightforward manner.

5.2.1 TENSILE TEST

A test specimen of the material is prepared, typically as a circular rod of uniform cross section (Figure 5.2a). Two punch marks (gauge points) are made on the specimen. The distance L_0 between them is known. This initial length is called the *gauge length*. The two ends of the specimen are made larger and provided with means (e.g., threads) to mount the specimen in the testing machine (Figure 5.2b). For testing, the lower end of the specimen is mounted on a stationary gripper. The upper end is mounted on a gripper that moves with the upper crosshead of the testing machine.

In a typical test, the upper crosshead is moved in small steps and the distance between the punch marks is accurately measured (e.g., using a laser extensometer, caliper pair, or a strain gauge). The tension exerted on the specimen in each step is accurately measured as well (e.g., using a load cell, which is built into the machine). The experiment is controlled (e.g., movement steps; speed of

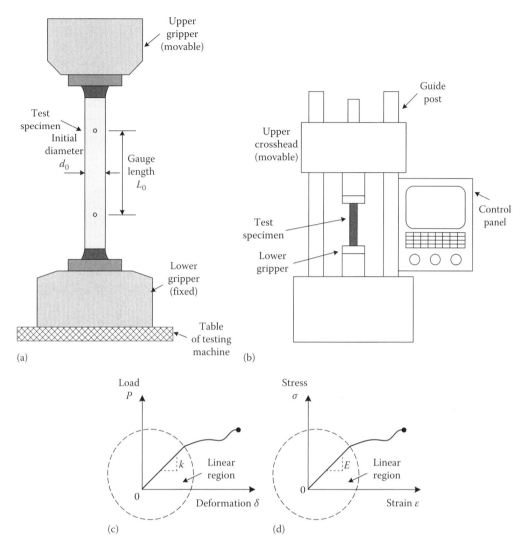

FIGURE 5.2 (a) Tensile test specimen; (b) tensile testing machine; (c) load versus deflection curve; (d) stress versus strain curve.

movement) by settings on the control panel, and the readings (force, deflection) are automatically recorded in the host computer of the machine for subsequent analysis (digital processing).

A plot of the force (vertical axis) versus deflection (horizontal axis) is called the *load–deflection diagram* (Figure 5.2c). This curve depends not only on the material properties but also on the dimensions of the specimen. The slope of the initial, linear region of this curve gives the *stiffness k* of the test specimen. A more general and useful relationship is provided by the stress–strain diagram (Figure 5.2d), which is indeed a normalized version of the load–deflection diagram. The slope of the initial, linear region of the stress versus strain curve gives the *modulus of elasticity E* of the material.

Test specimen: Only the central region, which includes the gauge length of the specimen, is machined to be uniform. The two end regions have gradually increasing dimensions in order to provide the necessary gripping strength and size for the test fixture. Professional standards (typically from ASTM—the American Society for Testing and Materials) specify the standard dimensions for test specimens. Specimens of circular *X*-section and flat specimens with rectangular *X*-section are commonly used. For metals, the recommended dimensions are $L_0 = 2.0$ in. (50 mm) and $d_0 = 0.5$ in. (13 mm). A miniature strain gauge (see Chapter 4), which directly measures the longitudinal strain, may be attached as well to the test specimen.

5.2.2 Stress–Strain Diagram

From the initial diameter d_0 of the test specimen, the initial area of cross section A_0 is computed:

$$A_0 = \frac{\pi}{4} d_0^2 \tag{5.1}$$

Then, for a given tensile load P on the specimen, the normal stress σ is computed using

$$\sigma = \frac{P}{A_0} \tag{5.2}$$

This is the "average" normal stress, because the stress profile across the section is not quite uniform in general. Furthermore, during the test, the specimen diameter (and hence the area of X-section) may decrease. This is called *necking*. Hence, what is computed using Equation 5.2 is not the true (actual) stress, but a "nominal stress." It is called the "engineering stress."

From the measured extension δ between the two punch marks of the test specimen, the strain ε in the specimen is computed using

$$\varepsilon = \frac{\delta}{L_0} \tag{5.3}$$

Note that the actual length L of the specimen after a load P is applied is greater than the original (gauge) length L_0. Furthermore, for various reasons (e.g., nonuniformity in material properties, geometry, and stress) the strain may vary from point to point in the test specimen. Hence, expression (5.3) does not represent the actual strain at a given instant, neither does it represent a local strain at a point along the specimen. Instead, it gives the "average" strain over the two gauge points (punch marks) over the deflection range of the test. The strain as given by (5.3) is called the "nominal strain" or "engineering strain."

Typically, the curve of engineering stress versus engineering strain is plotted at the end of the test. This gives us the "conventional stress–strain diagram" or "engineering stress–strain diagram." On the other hand, if the *true stress* given by $\sigma = P/A$ and the true strain given by $\varepsilon = \delta/L$ are used in the stress–strain diagram, we have the "true stress–strain diagram." Nevertheless, these "true" stresses and strains are "average" values, arranged over the effective segment of the test specimen. The engineering and true stress–strain diagrams are virtually the same when the strains are small (when necking does not occur). Typically, the host computer of the testing machine is programmed to perform the necessary computations and data plotting.

Example 5.1

A tensile test was carried out on a uniform cylindrical test specimen with the following dimensions: gauge length $L_o = 50$ mm; initial diameter $d_o = 10$ mm. The load (kN)–extension (mm) data obtained from the test are given in the first two columns of Table 5.1.

Compute the stress and strain values corresponding to the given data. Plot the curve of stress versus strain.

Solution

Area of cross section of the specimen $A = (\pi/4) \times 10^2$ mm^2 = 78.5 mm^2.
Divide the load values (N) by A (mm^2) to obtain the corresponding stress values (MPa).

TABLE 5.1
Stress versus Strain Data for a Test Specimen

Load (kN)	Extension (mm)	Stress (MPa)	Strain (%)
0	0	0	0
3.08	0.02	39.24	0.04
6.17	0.04	78.60	0.08
9.26	0.06	117.96	0.12
12.0	0.08	152.87	0.16
9.5	0.1	121.02	0.2
9.5	0.12	121.02	0.24
9.5	0.14	121.02	0.28
9.5	0.2	121.02	0.4
9.60	0.5	122.29	1
9.72	1	123.82	2
9.84	2	125.35	4
14.4	3.4	183.44	6.8
18.0	4	229.30	8
22.8	5	290.45	10
24.6	7	313.38	14
25.8	9	328.66	18
26.4	12	336.31	24
26.6	15	338.85	30
26.2	18	333.76	36
22.8	21	290.45	42
21.7	21.5 (Fracture)	276.43	43

Divide the extension values (mm) by length 50 mm and multiply by 100 to obtain the % strain values. These are tabulated in the third and the fourth columns of Table 5.1. The corresponding stress versus strain curve is plotted in Figure 5.3.

Primary Learning Objectives
1. Understanding the procedure of a tensile test
2. Conversion of load and deflection test data into stress and strain data
3. Experimental determination of a stress–strain curve

■ **End of Solution**

5.3 STRESS–STRAIN CHARACTERISTICS

Consider the engineering stress–strain diagram of a typical material that is ductile (e.g., low-carbon steel), as sketched in Figure 5.4.

Note: A ductile material is one that undergoes large changes in strain before fracture (failure).

Several important points and regions can be identified on this engineering stress–strain diagram.

Elastic region: This is the region from the initial undeformed state (point 0 in Figure 5.4) up to the *elastic limit* (point 2 in Figure 5.4). In this region (0–2), the material undergoes purely *elastic deformation* due to the loading. That means, when the load is removed, the material fully returns to the original undeformed state. In other words, there is no residual strain (permanent set).

Proportional limit: For many materials, initially, stress is proportional to strain (represented by a straight-line segment in the stress–strain curve). This is the *linear* region (from point 0 to point 1 in Figure 5.4). The remaining part of the elastic region is not linear (from point 1 to point 2). The

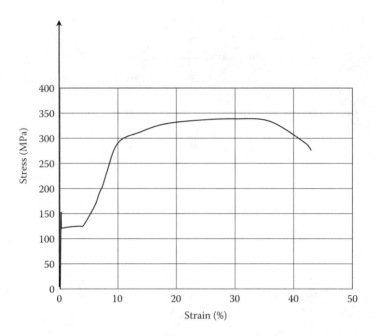

FIGURE 5.3 The stress versus strain curve corresponding to the data.

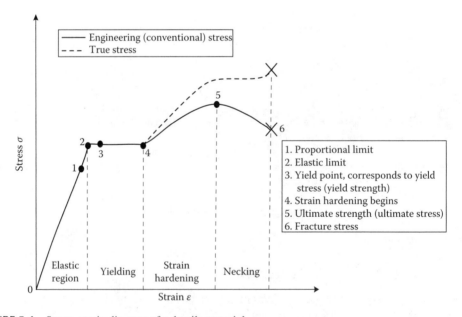

FIGURE 5.4 Stress–strain diagram of a ductile material.

maximum stress up to which the stress–strain relation is linear (point 1 in Figure 5.4) is called the proportional limit. *Note*: For many materials (e.g., steel, aluminum alloys), the proportional limit is very close to the elastic limit (point 2 in Figure 5.4).

Yielding: If loaded slightly beyond the elastic limit, the material will not fully recover once the loading is removed. A residual strain (a permanent set) will remain. This behavior is termed *yielding* or *plastic deformation*. The yielding region is from point 3 (slightly beyond point 2) up to point 4 in Figure 5.4. If this region is flat (i.e., if stress remains constant while strain increases), the material is in a *perfectly plastic* state. The stress at which yielding begins is called the *yield stress* or *yield strength*, and this is the stress at the *yield point* (point 3 in Figure 5.4). For some ductile materials, there exist two yield points called upper yield point and lower yield point. Yielding begins at the *upper yield point*, then the stress suddenly drops to the *lower yield point*, and yielding continues normally beyond that without an appreciable change in the stress level. Then, a recognizable "kink" will be seen at the yield point of the stress–strain curve, making the detection of the yield point easier. This is analogous to the behaviors of static friction and kinematic friction. Specifically, suppose that a force is applied in the direction of eventual movement, to a body that is resting on a surface. The body will remain stationary (in equilibrium) while the resisting frictional force equals the applied force. At some value of the applied force, the body will just begin to move. As it moves, the resisting frictional force will drop slightly (and may increase again). Analogously, if the tensile force that is applied to an axial member is gradually increased, at some point the elastic limit will be reached. Beyond that point, yielding will begin in the member. The stress will increase (similar to what happens in the elastic region when a rubber balloon is pressed) until it peaks. Just beyond that, the stress will drop slightly and rapidly (as when the balloon bursts) until the lower yield point is reached. On further increasing the load, the stress will continue to rise.

Offset method of determining yield strength: For many metals, stress will not remain constant during yielding, but will gradually (yet slightly) increase. Then the point at which the elastic region ends and the yielding begins is not easily distinguishable. Hence, a clear yield point will not be observable (no "kink" will be seen at the yield point) on the stress–strain curve. In this case, the *offset method* may be used to define the yield strength. In the 0.2% offset method, a line parallel to the linear part of the stress–strain curve is drawn from the 0.2% point of the strain axis (the offset). The point at which this straight line intersects the stress–strain curve is taken as the yield point (see Figure 5.5), which gives the yield strength.

Logic of the offset method: Beyond the yield point, the material will have a permanent set (i.e., residual strain) when the load is removed. The unloading path is approximately parallel to the linear segment of the original stress–strain curve (also, see strain hardening, as discussed later; Figure 5.6). The offset of this unloading curve on the strain axis is indeed the residual strain (permanent deformation). Through engineering experience, it has been found that for many materials, the minimum residual normal strain on releasing the load after the yield point is in the range 0.05%–0.3%.

Note: In practice, the 0.05% offset method is also used to determine the yield point, possible for less ductile material. The approach here is exactly the same as mentioned earlier, except that the offset point on the strain axis will be 0.05% (i.e., a strain of 0.0005 m/m) instead of 0.2%. If the

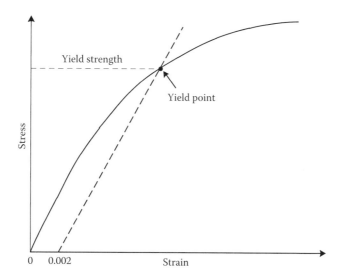

FIGURE 5.5 0.2% Offset method of yield strength determination.

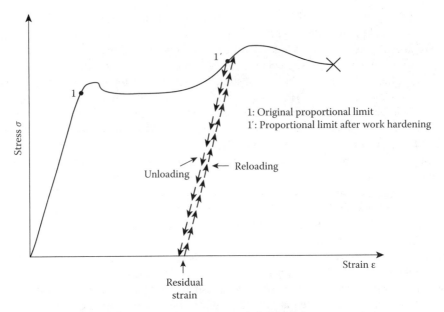

FIGURE 5.6 Increase of the proportional limit due to strain hardening.

stress–strain curve is relatively flat in the yielding region (i.e., almost perfectly plastic), the yield stress determined from the two methods will be almost the same.

5.3.1 Strain Hardening (Work Hardening)

After yielding ends (point 4 in Figure 5.4), a further change in strain is associated with an appreciable increase in stress. This is termed *strain hardening* (this occurs from point 4 to point 5 in Figure 5.4). Due to strain hardening, the material is able to support increased levels of stress (load) and hence increases its "strength." The maximum stress level up to which the strain hardening continues corresponds to the maximum load the material is able to support. This (point 5 in Figure 5.4) is called the *ultimate stress* or *ultimate strength*.

If we fully unload the member once it reaches the region of strain hardening, a permanent set (residual strain) will remain. Then if we reload the member, stress will increase elastically with strain until it reaches the original stress–strain curve (Figure 5.6), and will follow that curve on further loading. It is clear that the proportional limit has increased (from 1 to 1′ in Figure 5.6) as a result of the unloading and reloading cycle in the region of strain hardening. This phenomenon is called *work hardening* (hardening of the material as a result of a "work cycle" of loading and unloading).

Note: The slope (modulus of elasticity) of the new elastic region is roughly the same as that of the original elastic region. That means the stiffness of the material remains the same after work hardening, even though the yield strength has increased.

Ultimate stress (ultimate strength): This is the maximum stress that the material can withstand before failure (fracture).

Example 5.2

A rigid beam *ABC* of length 2*L* is hinged to a fixed rigid wall at end *A* and horizontally supported using a vertical hanger rod, which is hinged at the other end *C*. The hanger rod *DC* is uniform with area of cross section $A = 100$ mm², and length $l = 3$ m. The end *D* of the hanger rod is hinged to a fixed and rigid ceiling (see Figure 5.7a). A vertical downward load $P = 70$ kN is applied at the

Mechanical Properties of Materials

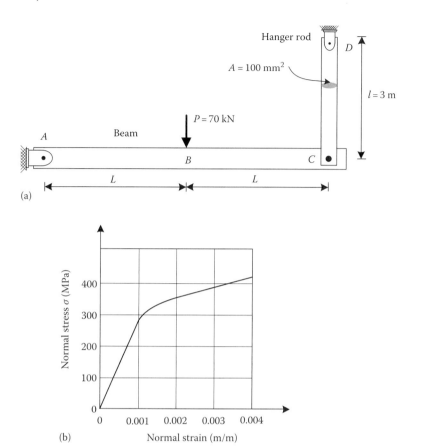

FIGURE 5.7 (a) A rigid beam supported by a hanger rod; (b) stress versus strain curve of the hanger rod.

midpoint B of the beam ($AB = L = BC$). The material of the hanger rod obeys the stress–strain curve shown in Figure 5.7b.

Determine
 a. Normal (tensile) stress in the hanger rod
 b. Downward movement at the point of load application (B)

Note 1: Assume that the beam is light (or, its weight is included in P).

Note 2: Assume that the movements due to P are small compared to the dimensions of the beam and the rod. (Hence, the circular arc of movement of the beam at C may be assumed to correspond to the extension of the rod DC. Similarly, the circular arc of movement of the beam at B may be assumed to be approximately vertical.)

Solution

 a. The free-body diagram (FDB) of the beam is shown in Figure 5.8.

$$P_C = \text{tensile force in the hanger rod}$$

Equilibrium conditions (take moments about A to exclude the reaction force R_A):

$$\curvearrowleft \sum M_A = 0: -P \times L + P_C \times 2L = 0$$

$$\rightarrow P_C = \frac{P}{2} = \frac{70}{2} \text{ kN} = 35 \text{ kN}$$

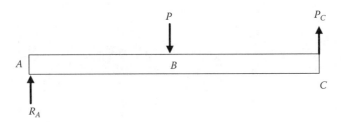

FIGURE 5.8 FBD of the beam.

Tensile stress in the hanger rod:

$$\sigma_{CD} = \frac{P_C}{A} = \frac{35 \times 10^3 \text{ (N)}}{100 \times 10^{-6} \text{ (m}^2\text{)}} = 350.0 \text{ MPa}$$

b. From the stress versus strain curve (Figure 5.7b), the corresponding normal strain in the hanger rod $\varepsilon_{CD} = 0.002$ m/m.

Note: The loading in the hanger rod is not within the linear region of the stress–strain curve.

Vertical movement at C = extension of the hanger rod = $\delta_C = \varepsilon_{CD} \times l = 0.002 \times 3.0$ m = 6.0 mm.
Vertical movement at B (since B is the midpoint and C is the end point of the beam):

$$\delta_B = \frac{1}{2}\delta_C = 3.0 \text{ mm}$$

Primary Learning Objectives
1. Solving problems using stress–strain relations
2. Solving nonelastic problems
3. Solving nonlinear stress–strain problems
4. Small movement approximation

■ **End of Solution**

5.3.2 NECKING

Beyond the ultimate stress, a material specimen typically undergoes a reduction in its cross-sectional area (from point 5 to point 6 in Figure 5.4) until it finally fails (fractures). The stress at which the specimen breaks (point 6 in Figure 5.4) is called the *fracture stress*.

Note: Reduction in the cross-sectional area is fairly uniform along the gauge length until the loading is close to fracture. Then a further sharp reduction in area occurs in a localized region in the midsection of the specimen where the failure will finally occur (see Figure 5.9a and b). This is called necking.

Fracture stress: This is the stress at the fracture point.

Percent elongation: This is the strain at the point of fracture, expressed as a percentage (i.e., percent strain at fracture).

5.3.3 TRUE STRESS–STRAIN DIAGRAM

Due to necking, the area of cross section of a specimen decreases considerably, and for a given load, the stress increases as a result. This increase in stress is not accounted for in the formula of engineering stress, as it uses the original area of cross section. However, this is accounted for in the true stress, as noted previously. Furthermore, the true strain that depends on the actual length of the specimen at a given instant (not the original length) is not accounted for in the

Mechanical Properties of Materials

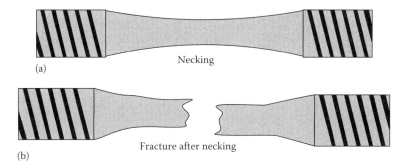

FIGURE 5.9 (a) Necking; (b) fracture after necking.

engineering strain. The true stress–strain diagram is shown as the broken line in Figure 5.4. In this diagram, the deviation from the engineering stress–strain diagram is significant (marked) only when the strain hardening begins, where the strain is considerably greater. The drop in stress in the necking region (point 5 to point 6 in Figure 5.4), which is a result of using the definition of engineering stress, is not present in the true stress–strain curve. Also, this clarifies why the fracture stress is less than the ultimate stress in the engineering stress–strain diagram, as it may not make sense at first glance.

5.4 HOOKE'S LAW

There are many different scenarios of load and deflection in practice, and associated relations of stress and strain. Four basic situations are shown in Figure 5.10, all of which are studied in this book.

In the linear elastic region of the constitutive relation (i.e., up to the proportional limit of the stress–strain curve), stress is proportional to strain. This behavior is expressed by Hooke's law (named after Robert Hooke, who observed this property in 1678). For the uniaxial (1-D, tensile) case of stress and strain (Figure 5.11a), Hooke's law is given by

$$\sigma = E\varepsilon \tag{5.4}$$

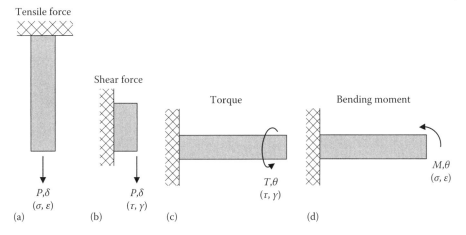

FIGURE 5.10 Four basic scenarios of load and deflection and associated stress and strain: (a) tensile; (b) shear; (c) torsion; (d) flexural.

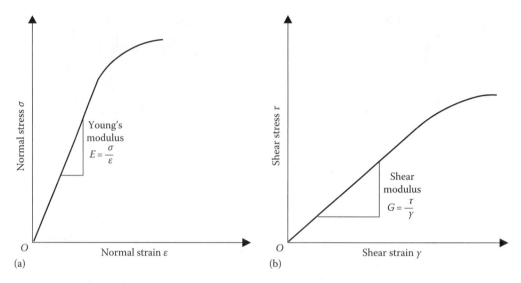

FIGURE 5.11 Stress versus strain relations: (a) tensile; (b) shear.

The constant of proportionality E is called the *modulus of elasticity* or *Young's modulus* (named after Thomas Young, who investigated the subject in 1807). It is a measure of *stiffness* of the material (the larger the E, the stiffer the material).

Hooke's law applies to the linear region of the shear stress (τ) versus shear strain (γ) relationship as well. In the purely shear, planar case (see Figure 5.10b), this relationship is expressed as (see Figure 5.11b)

$$\tau = G\gamma \tag{5.5}$$

The constant of proportionality G is called the *modulus of rigidity* or *shear modulus*. It can be determined from an experimental shear stress–shear strain diagram.

Example 5.3

In Example 5.1, determine the

  a. Proportional limit and the corresponding strain
  b. Young's modulus
  c. Yield stress (using the 0.2% offset method) and the corresponding strain
  d. Ultimate stress and the corresponding strain
  e. Fracture stress and the corresponding strain
  f. Percent elongation

Suppose that the member carries a tensile load of 10.0 kN. What is the corresponding factor of safety (FOS) with respect to (i) yield stress; (ii) ultimate stress?

Solution

  a. The initial segment of the stress versus strain curve of Figure 5.3 is shown in Figure 5.12a (with an enhanced scale).
     From this curve (or from the data in Table 5.1) it is seen that
       Stress at the proportional limit = 153.0 MPa
       Corresponding strain = 0.16%

Mechanical Properties of Materials

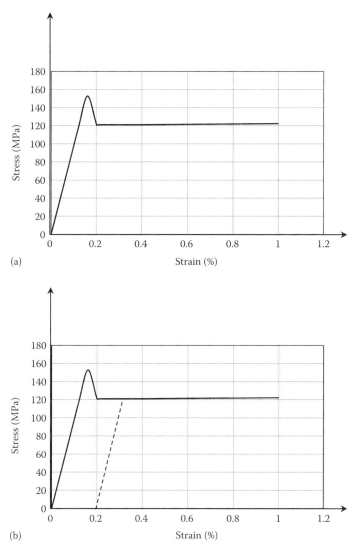

FIGURE 5.12 (a) The initial segment of the stress versus strain curve; (b) application of the offset method.

b. Slope of the linear segment = Young's modulus $E = 153.0/0.0016$ MPa = 95.63 GPa
c. Apply the 0.2% offset method (see Figure 5.12b)
   Yield stress = 121.0 MPa
   Corresponding strain = 0.32%
d. From Figure 5.3 (or from the data in Table 5.1): maximum stress value = ultimate strength = 340 MPa
   Corresponding strain = 34%
e. From Figure 5.3 (or from the data in Table 5.1): the stress at fracture (fracture stress) = 276.0 MPa
   Corresponding strain = 43%
f. From Figure 5.3 (or from the data in Table 5.1): % elongation = % strain at fracture = 43%
g. Stress corresponding to a load of 10.0 kN is $(10.0 \times 10^3 \text{ (N)})/((\pi/4) \times 10^2 \text{ (mm}^2)) = 127.4$ MPa
   Yield stress = 121.0 MPa
   FOS with respect to yield stress = 121.0/127.4 = 0.9

   Note: Typically, a FOS has to be greater than 1.0. Hence, 0.9 is not acceptable.
   Ultimate stress = 340 MPa
   FOS with respect to ultimate stress = 340/127.4 = 2.6

Primary Learning Objectives
1. Use of experimental data to estimate material properties (specifically, Young's modulus)
2. Determination of yield strength when the yield point is not clearly seen in the experimental data
3. Estimation of ultimate strength and fracture stress
4. Consideration of the FOS with respect to yield strength and with respect to ultimate strength

■ End of Solution

Example 5.4

A conical rod (peg) of rectangular cross section with base dimensions $a \times b$ and height h is hanging from a rigidly fixed ceiling, with the base firmly attached to the ceiling, as shown in Figure 5.13a. Material properties of the rod are

$$E = \text{Young's modulus}; \quad \rho = \text{mass density}$$

Determine the overall vertical extension at the bottom apex of the peg as a result of its weight.

Solution

Consider a small element of height δy at a height y from the bottom apex of the rod (Figure 5.13b). From similar triangles, width x of the element is related through

$$\frac{x}{a} = \frac{y}{h} \rightarrow x = \frac{a}{h} y \qquad \text{(i)}$$

Area of the triangular segment of the rod, from the apex up to y, is $\frac{1}{2} xy = \frac{1}{2}\frac{a}{h} y^2$ (on substituting (i))

Volume of the segment of the rod up to y is $\frac{1}{2} xy \times b = \frac{1}{2}\frac{ab}{h} y^2$.

Hence, weight of the segment of the rod up to y is $\frac{1}{2}\frac{ab}{h} y^2 \rho g$.

Area of X-section of the rod at y is $x \times b = \frac{a}{h} y \times b = \frac{ab}{h} y$.

Hence, normal, axial stress (force/area) at the X-section at y due to the weight below it

$$\sigma_y = \frac{(1/2)(aby^2 \rho g/h)}{(ab/h)y} = \frac{1}{2}\rho g y$$

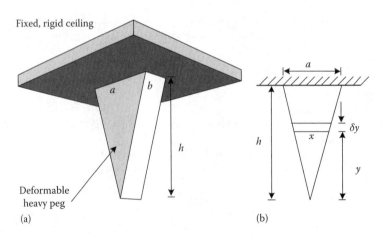

FIGURE 5.13 (a) A heavy peg hanging from a fixed ceiling; (b) geometry of analysis.

Mechanical Properties of Materials

Corresponding normal, axial strain at y is: $\varepsilon_y = \dfrac{\sigma_y}{E} = \dfrac{1}{2}\dfrac{\rho g}{E}y$

Extension in the element δy is $\varepsilon_y \delta y = \dfrac{1}{2}\dfrac{\rho g}{E}y \cdot \delta y$

The total extension of the rod at its apex is the sum of these elemental extensions. Hence, in the limit, the total extension at the apex is given by the integral

$$\int_0^h \frac{1}{2}\frac{\rho g}{E}y\,dy = \frac{1}{2}\frac{\rho g}{E}\int_0^h y\,dy = \frac{1}{4}\frac{\rho g h^2}{E}$$

Primary Learning Objectives

1. Solution of the stress–strain problems of nonuniform axial members with distributed loading
2. Determination of the extension of an axial member due to self-weight
3. Use of the integration formula of normal strain in determining the extension of an axial member

■ **End of Solution**

Example 5.5

A solid rubber slab is sandwiched between two thin, horizontal, rigid metal plates, with the bottom plate glued to a rigid, fixed base as shown in Figure 5.14a. A horizontal force $P = 200.0$ N is applied centrally to the top metal plate as shown. The horizontal deflection of the plate in the direction of P is $\delta = 5.0$ mm.

a. Sketch the deformed shape of the object.
b. Determine the average shear strain in the rubber slab.
c. Estimate the shear modulus G of the rubber.

Solution

a. The deformation of the slab is sketched in Figure 5.14b.

Note: Since the top surface is free, the inclined edge of the slab remains almost at the original length 55 mm.

b. Average shear strain is given by $\sin\gamma = \delta/50 = 5/50 = 0.1$.

$$\rightarrow \gamma = \sin^{-1} 0.1 = 0.1002$$

Note: Even though in this example it is more proper to use "sin" rather than "tan" or direct "radians" in the formula for shear strain (because the top shearing surface is free), the answer is almost identical in all three cases (i.e., 0.1002; 0.1000; and 0.0997) since the deformation is very small.

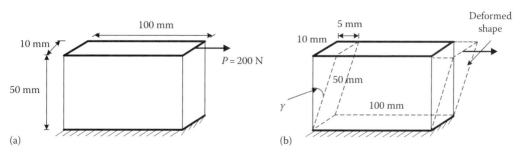

FIGURE 5.14 (a) A slab of rubber subjected to shear loading. (b) Shear deformation of the rubber slab.

c. Shear area $A = 100 \times 10$ mm²

$$\text{Average shear stress } \tau = \frac{P}{A} = \frac{200.0}{100 \times 10} \text{MPa} = 0.2 \text{ MPa}$$

$$\text{Modulus of rigidity } G = \frac{\tau}{\gamma} = \frac{0.2}{0.1} \text{MPa} = 2.0 \text{ MPa}$$

Primary Learning Objectives
1. Nature of shear deformation
2. Solution of problems involving shear stress and shear strain
3. Proper use of the geometric formula of shear strain
4. Estimation of shear modulus

■ **End of Solution**

Example 5.6

An engine mount should provide specified levels of flexibility and damping in order to mitigate vibration (i.e., to provide vibration isolation). A particular design uses two rubber-like flexible shear elements sandwiched between a rigid leg post, which supports the engine, and two rigid foot pieces that are anchored to the ground, as shown in Figure 5.15a. The two flexible elements are identical, having the following dimensions: height $L = 80$ mm; thickness $a = 20$ mm; width $b = 50$ mm. Due to a load of $P = 2.0$ kN on it, the leg post moves down through a distance of $\delta = 1.0$ mm.

a. Determine the average shear stress in a flexible element.
b. Determine the average shear strain in a flexible element.
c. Estimate the shear modulus (modulus of rigidity) of a flexible element.

Solution

a. Free-body diagram of the leg post is shown in Figure 5.15b.
From equilibrium, we have $2V - P = 0$
where V is the vertical shear force reaction from a flexible element on the leg post
Hence, $V = P/2 = 1.0$ kN.
Surface area of the flexible element on which the shear force is applied $= L \times b = 80.0 \times 50.0$ mm² $= 4.0 \times 10^3$ mm².
Average shear stress

$$\tau_{av} = \frac{V}{L \times b} = \frac{1 \times 10^3 \text{ (N)}}{4.0 \times 10^3 \text{ (mm}^2\text{)}} = 0.25 \text{ MPa}$$

b. The deformed shape of a rubber piece is shown in Figure 5.15c.
Average shear strain

$$\gamma_{av} = \tan^{-1} \frac{\delta}{a} = \tan^{-1} \frac{1.0}{20.0} = 0.05 = 5\%$$

Note: In this example, it is more proper to use "tan" rather than "sin" or direct "radians" in the formula for shear strain, because the two shearing surfaces are firmly attached to rigid planes that are separated by a constant distance of $a = 20$ mm. However, the answer is almost identical in all three cases since the deformation is very small.

Mechanical Properties of Materials

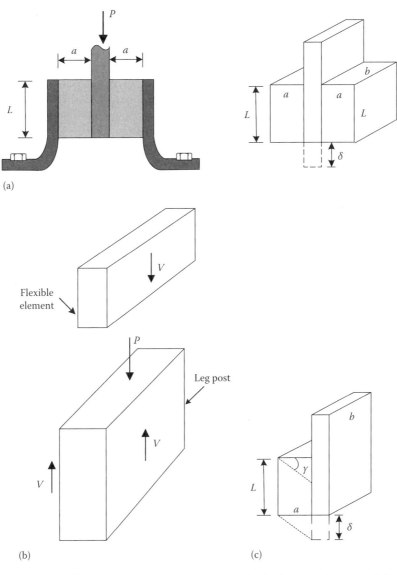

FIGURE 5.15 (a) A flexible mount of an engine; (b) free-body diagram; (c) shear deformation of a rubber piece.

c. Shear modulus

$$G = \frac{\tau_{av}}{\gamma_{av}} = \frac{0.25}{0.05} \text{ MPa} = 5.0 \text{ MPa}$$

Primary Learning Objectives
1. Solution of problems involving shear stress and shear strain
2. Proper use of the geometric formula of shear strain
3. Estimation of shear modulus

■ **End of Solution**

Example 5.7

A vertical load $P = 20$ kN is supported by a rigid structure using a structural device as shown in Figure 5.16a. In the device, a flexible rectangular piece (synthetic rubber) of height $h = 100$ mm, width $b = 50$ mm, and thickness t is firmly glued between two rigid metal plates in order to absorb shock and vibration. The device is attached to the structure using steel bolts and nuts of shank diameter $d = 10$ mm and allowable shear stress $\tau_{allw} = 25$ MPa.

   a. If the vertical movements δ at the point of application of the load P is to be limited to 5 mm, determine a suitable thickness t for the flexible piece. Shear modulus of the flexible piece, $G = 10$ MPa.
   b. Estimate how many bolts would be needed.

Note: Assume that the load is equally distributed among all the bolts.

Solution

   a. Shear area of the flexible element $= b \times h = 0.05 \times 0.1$ m²
   Shear force on the flexible element $= P = 20$ kN
   Shear stress (average) of the flexible element $\tau = P/bh$

$$\tau = \frac{20 \times 10^3 \text{ (N)}}{0.05 \times 0.1 \text{ (m}^2\text{)}} = 4.0 \times 10^6 \text{ Pa} = 4.0 \text{ MPa}$$

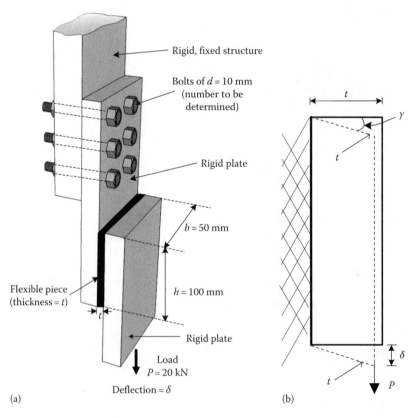

FIGURE 5.16 (a) A load supported on a rigid structure through a shock absorber; (b) deformation of the flexible element.

Mechanical Properties of Materials

Corresponding shear strain $\gamma = \tau/G$
Also, from Figure 5.16b, $\gamma = \sin^{-1}(\delta/t)$
where δ is the movement of the outer surface of the flexible element = movement of the rigid plate on which P is applied.

$$\rightarrow \frac{\delta}{t} = \sin\left(\frac{\tau}{G}\right) \rightarrow t = \frac{\delta}{\sin(\tau/G)} = \frac{5 \times 10^{-3} \text{ (m)}}{\sin(4.0 \times 10^6 \text{(Pa)}/\times 10.0 \times 10^6 \text{ (Pa)})}$$

$$\rightarrow t = 12.8 \times 10^{-3} \text{ m} = 12.8 \text{ mm}$$

Note 1: Since the shearing material is quite flexible (i.e., G is rather low) and since the two parallel shearing surfaces are not separated by a constant distance in this problem, we must use "sin" rather than "tan" or "radians" in the deformed geometry (Figure 5.16b).

Note 2: If we use "tan," we get $t = 11.8$ mm; if we use "radians," we get $t = 12.5$ mm.

b. Area of each bolt = $\frac{\pi}{4}d^2 = \frac{\pi}{4}(10 \times 10^{-3})^2$ m^2

Load on all the bolts = $P = 20$ kN

Note: The bolts are in single shear in this problem.

Suppose that we need n bolts to support this load, each experiencing the allowable shear stress $\tau_{allw} = 25$ MPa.

We have $n \times 25 \times 10^6$ (Pa) $\times \frac{\pi}{4}(10 \times 10^{-3})^2$ (m^2) = 20×10^3 (N)

$$\rightarrow n = \frac{20 \times 10^3}{25 \times 10^6 \times (\pi/4) \times 10^{-4}} = 10.2$$

We will need 11 bolts.

Primary Learning Objectives
1. Solution of problems involving shear stress and shear strain
2. Proper use of the geometric formula of shear strain
3. Design of bolted joints in single shear

■ **End of Solution**

5.5 POISSON'S RATIO

A body subjected to a longitudinal load will exhibit longitudinal extension (i.e., normal tensile strain) and accompanied with it a lateral contraction (i.e., a negative lateral strain). Conversely, when compressed longitudinally, it will expand laterally (see Figure 5.17). Poisson's ratio is defined as

$$\text{Poisson's ratio } \nu = -\frac{\text{Lateral strain } \varepsilon_{lat}}{\text{Longitudinal strain } \varepsilon_{long}} \quad (5.6)$$

Note: The negative sign accounts for the fact that, for most material, the lateral strain is negative when the longitudinal strain is positive, giving a positive value for ν. There are exceptions of some composite material whose Poisson's ratio is negative.

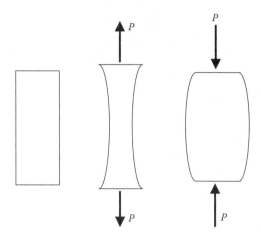

FIGURE 5.17 Poisson effect.

The dimensionless positive parameter ν cannot exceed the value 0.5, in the elastic range. A typical value for nonporous engineering materials is between ¼ and ⅓. For construction steel, its value is about 0.3.

In the linear elastic case, the modulus of elasticity (Young's modulus) E and the shear modulus of elasticity (modulus of rigidity) G are related through Poisson's ratio ν. The relationship is

$$E = 2(1+\nu)G \tag{5.7}$$

Note: Equation 5.7 is applicable only if the material is within its linear elastic range. However, Poisson's ratio may be defined for any region of the stress–strain curve of a material. The value of this parameter will not be the same throughout the range, however. In particular, when necking begins prior to failure of a ductile material (see Figure 5.9), the value of the Poisson's ratio can be different from that in the elastic range. This does not necessarily mean that the value will be larger in the region of necking (because the longitudinal strain is also relatively large in this region).

Example 5.8

A tensile test was carried out on a material specimen as follows. Two strain gauges were mounted in the middle region of the specimen, one axially and the other laterally. Their strains (ε_1 and ε_2, respectively, as indicated in Figure 5.18), on application of a tensile load of $P=1.5$ kN, were measured using a strain gauge bridge (i.e., a Wheatstone bridge or dc voltage bridge), as discussed in Chapter 4. The strain readings were found to be $\varepsilon_1 = 450$ με and $\varepsilon_2 = -150$ με. The specimen dimensions are

Gauge length = 30 mm; area of cross section (initial) = 40 mm²

Determine

    a. Poisson's ratio ν of the material
    b. Young's modulus E of the material
    c. Shear modulus G of the material
    d. Extension of the gauge length on application of P

Solution

    a. By definition, $\nu = -\dfrac{\varepsilon_2}{\varepsilon_1} = \dfrac{150\,(\mu\varepsilon)}{450\,(\mu\varepsilon)} \approx 0.33$
    b. Average stress in the specimen, $\sigma = \dfrac{P}{A}$

Mechanical Properties of Materials

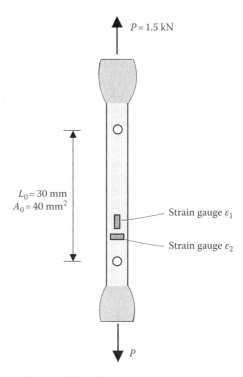

FIGURE 5.18 A tensile test specimen with strain gauges.

$$= \frac{1.5 \times 10^3 \text{ (N)}}{40 \times 10^{-6} \text{ (m}^2\text{)}} = \frac{3}{8} \times 10^8 \text{ Pa} = 37.5 \text{ MPa}$$

Hence, Young's modulus

$$E = \frac{\sigma}{\varepsilon_1} = \frac{37.5 \times 10^6 \text{ (Pa)}}{4.5 \times 10^{-4} \text{ (m/m)}} = 8.33 \times 10^{10} \text{ Pa} = 83.3 \text{ GPa}$$

c. Shear modulus

$$G = \frac{E}{2(1+\nu)} = \frac{83.3}{2 \times (1+0.33)} \text{ GPa} = 31.3 \text{ GPa}$$

d. Extension $\delta = \varepsilon_1 \times$ gauge length

$$= 4.5 \times 10^{-4} \text{ (m/m)} \times 30 \times 10^{-3} \text{ (m)} = 0.0135 \text{ mm}$$

Primary Learning Objectives
1. Interpretation of Poisson's ratio
2. Solution of problems involving Poisson's ratio
3. Relation between material properties in normal strain and shear strain, within the elastic limit
4. Concepts of material testing

■ **End of Solution**

Example 5.9

A flexible sleeve of inside radius r and thickness t is fitted snugly inside a uniform cylindrical bore of a rigid engine block (bore radius is very slightly smaller than the outer radius of the sleeve). A rigid shaft passes tightly through the sleeve (see Figure 5.19a). When a torque T is applied to the shaft, determine the resulting angle of rotation θ, assuming that there is no slip (between the shaft and the sleeve, and the sleeve and the engine block) and that the sleeve remains within its proportional limit. L = length of the sleeve; G = shear modulus of the sleeve.
Express your answer in terms of t, r, L, G, and T.

Solution

Consider an elemental (miniscule) sector of angle $\delta\beta$ in the sleeve (see Figure 5.19b).
Arc length of the sector = $r \cdot \delta\beta$
Inside surface area of the elemental segment = $L \cdot r \cdot \delta\beta$.
Tangential force on the elemental segment: $\delta F = \tau \cdot L \cdot r \cdot \delta\beta$
where τ is the tangential shear stress on the inside surface of the sleeve.
Corresponding torque $\delta T = r \cdot \delta F = \tau \cdot L \cdot r^2 \cdot \delta\beta$

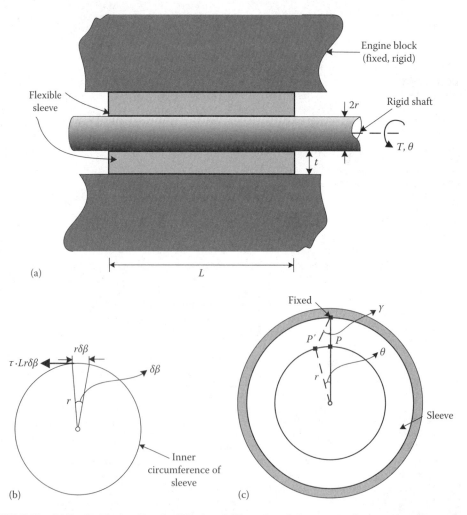

FIGURE 5.19 (a) Engine shaft with a flexible sleeve; (b) analyzed element; (c) shear strain geometry.

Overall torque applied to the sleeve by the shaft is the summation of these elemental torques, which (in the limit) is given by the integral

$$T = \int_0^{2\pi} \tau \cdot L \cdot r^2 \cdot d\beta = \tau \cdot L \cdot r^2 \cdot 2\pi \quad \text{(i)}$$

Also,

$$\tau = G\gamma \quad \text{(ii)}$$

where γ is the shear strain (average) in the sleeve.

Let θ = angle of shaft rotation due to T.

Consider a point P on the inside surface of the sleeve, which moves to P' on the application of T (see Figure 5.19c).

$$\text{Arc } PP' = r \cdot \theta \approx t \cdot \tan^{-1}\gamma \approx t \cdot \gamma \text{ (since } \gamma \text{ is small)} \rightarrow$$

$$\theta = \frac{t \cdot \gamma}{r} \quad \text{(iii)}$$

Note: In this example, it is more proper to use "tan" rather than direct "rad" or "sin" in the formula for shear strain (because the two shearing surfaces are firmly attached to rigid surfaces that are separated by a constant distance of t). However, the answer is almost identical in all three cases since the deformation is very small.

Substitute (iii) into (ii) for γ, and substitute the result into (i) for τ. We get

$$T = G \cdot \gamma \cdot L \cdot r^2 \cdot 2\pi = G \cdot \frac{r\theta}{t} \cdot L \cdot r^2 \cdot 2\pi$$

or

$$\theta = \frac{t}{2\pi r^3 LG} T$$

Primary Learning Objectives
1. Solution of problems involving shear stress and shear strain, within the proportional limit
2. Understanding how torque is related to shear stress and angle of rotation is related to shear strain
3. Use of proper approximations and geometric formulas under small deformations in shear

■ **End of Solution**

5.6 MATERIAL TYPES AND BEHAVIOR

Materials may be classified according to many considerations. Some classifications and characteristics that specifically depend on the stress–strain behavior of a material are indicated now.

5.6.1 DUCTILE MATERIALS

Materials that undergo large deformations (strains) before fracture (failure) are called ductile (see Figure 5.20). They are capable of absorbing relatively large amounts of energy due to their large deformations. Hence, they are particularly useful in applications that need buffering against high impacts and shock (e.g., automobiles).

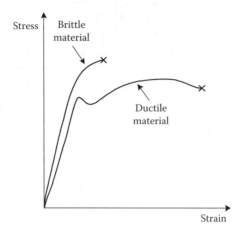

FIGURE 5.20 Characteristics of ductile and brittle materials.

Examples: Low-carbon steel (mild steel), copper alloys, molybdenum, zinc.

Ductility: This is the amount of strain at fracture. It also represents the capacity for plastic deformation.

5.6.2 Ductility Measures

A common measure of ductility is the percentage elongation at fracture of a specimen in tensile testing:

$$\text{Percent elongation} = \frac{L_f - L_0}{L_0} \times 100\% = \varepsilon_f \times 100\% \quad (5.8)$$

where
 L_0 is the initial length of the specimen
 L_f is the final length of the specimen (at fracture)
 ε_f is the final strain of the specimen (at fracture)

Note: For mild steel, percent elongation is about 40%.

Since a material shrinks laterally as it extends longitudinally, another measure of ductility is the percentage reduction of the cross-sectional area at fracture of a specimen in tensile testing. This is expressed (for a circular specimen) as

$$\text{Percent reduction of area} = \frac{A_0 - A_f}{A_0} \times 100\% = \frac{(d_0)^2 - (d_f)^2}{(d_0)^2} \times 100\% \quad (5.9)$$

where
 A_0 is the initial area of X-section of the specimen
 A_f is the final area of X-section of the specimen (at the neck of fracture)
 d_0 is the initial diameter of X-section of the circular specimen
 A_f is the final diameter of X-section of the circular specimen (at the neck of fracture)

Note: Necking is a common manifestation of ductility.

For mild steel, percent reduction of area is about 60%.

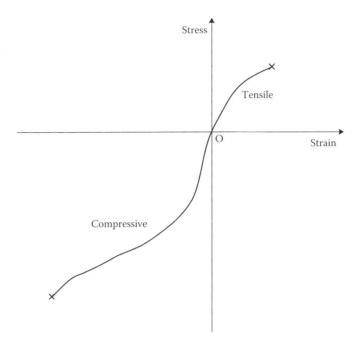

FIGURE 5.21 Typical stress–strain behavior of a brittle material.

5.6.3 Brittle Materials

Brittle materials are those that do not exhibit large deformations (yielding) before fracture (see Figure 5.20). These are the opposite of ductile materials. They fail abruptly. Also, they typically have a greater compressive strength than tensile strength (Figure 5.21).

Examples: Cast iron, concrete. In concrete, the tensile strength is negligible compared to its compressive strength. Reinforcing steel rods are used to support tensile loads in concrete members.

5.6.4 Hardness

Hardness is a measure of the resistance of a material to surface indentation. A typical test for hardness (e.g., Rockwell test, Brinell test) involves applying a specified load to the material surface using an indentation head of specific shape, and measuring the depth of the resulting dent. As noted earlier, hardness of a material can be improved through strain hardening (work hardening), and this is commonly used in the manufacturing processes. Since strain hardening increases the yield strength, the remaining capacity to yield (plastic deformation) will decrease, making the material less ductile (more brittle).

5.6.5 Creep

When a load is applied to a material, it may continue to deform (i.e., increase its strain) over a very long period of time even under constant loading (constant stress) conditions. This is called creep.

Creep strength: This is the highest strength (stress) a material can withstand over a specified period of time (say, 1000 h), without exceeding a specified strain (say, 1%) called the creep strain. Hence, the creep strain associated with creep strength may be specified as a strain rate (e.g., 0.1% per year).

Note: Creep strain typically increases with the material temperature and the magnitude of the load (stress).

5.6.6 Fatigue

A material may abruptly fail when a loading cycle (and associated deformation) is repeated many times even if the corresponding average stress levels are well below the ultimate strength of the material. This is called *fatigue failure*.

Reason: Microscopic imperfections produce high local stress (typically on the surface of the member). As a result, small cracks are generated and hence the load-bearing X-section gradually decreases. In this manner, the average stress at the X-section gradually increases, until failure.

Note: Fatigue failure of a ductile material may appear as a brittle failure (abrupt failure without changing the strain significantly).

Endurance limit (fatigue limit): This is the stress below which no failure occurs if the corresponding load is repeated over a specified number of times (e.g., 500×10^6 cycles).

Parameter values of some important mechanical properties of engineering material are given in Table 5.2.

5.7 STRAIN ENERGY

A load will cause a deformation (deflection) in a deformable member, thereby doing "work" on the body. Analogously, stresses (created by the loading) do "work" in deforming the member and generating strains. In the elastic region of a material, the work done is stored as a form of elastic potential energy called "strain energy," which may be fully recovered when the load is removed from the member. Once the elastic region is exceeded, however, some of the work done on the member is used for permanent deformation of the material and is not stored as recoverable strain energy. In that case, only part of the absorbed energy is recoverable when the applied load is removed.

To obtain a relation for *strain energy density* (i.e., strain energy per unit volume of body), consider an elemental (extremely small) cuboid of sides $\delta x \times \delta y \times \delta z$ located in the orthogonal (Cartesian) directions x, y, and z (see Figure 5.22). Suppose that two equal and opposite tensile stresses σ are acting in the z direction of the cuboid. Corresponding tensile force $F = \sigma \times \delta x \times \delta y$.

In the current state of stress (σ), the normal tensile strain in the z direction is denoted by ε.

Now, if the strain is increased by $\delta \varepsilon$, the corresponding extension in the z direction of the element is $\delta q = \delta \varepsilon \times \delta z$.

Note: The stress hardly changes during this infinitesimal change in strain.

The elemental work done by force F due to this deflection (δq in the z direction of the element) is

$$\delta W = F \times \delta q = (\sigma \times \delta x \times \delta y) \times (\delta \varepsilon \times \delta z) = \sigma \times \delta \varepsilon \times (\delta x \times \delta y \times \delta z) = \sigma \times \delta \varepsilon \times \delta V$$

= strain energy absorbed by the element

Dividing this expression by the elemental volume, $\delta V = \delta x \times \delta y \times \delta z$ gives the corresponding "increase" in "strain energy density" that is absorbed by the material at the location of the element.

Hence, increase in strain energy density

$$\delta u = \frac{\delta U}{\delta V} = \sigma \times \delta \varepsilon \tag{i}$$

This is in fact the shaded elemental area under the stress–strain curve, as shown in Figure 5.23.

TABLE 5.2
Mechanical Properties of Some Engineering Materials

Material	Density (kg/m³)	Yield Strength Tensile; Compr.; Shear (MPa)	Ultimate Strength Tensile; Compr.; Shear (MPa)	Modulus of Elasticity (GPa)	Shear Modulus (GPa)	Poisson's Ratio	Coef. of Therm. Exp. (×10⁻⁶/°C)	Ductility (% Elongation)
Aluminum alloy 2014-T6	2790	414; 414; 172	469; 469; 290	73	27	0.35	23	12
Brass (Red) C83400	8740	70; 70; —	241; 241; —	101	37	0.35	18	35
Bronze C86100	8830	345; 345; —	655; 655; —	105	40	0.34	17	20
Cast iron ASTM A48	7190	—; —; —	140; 600; —	84	34	0.22	9	0.6
Concrete	2380	—; —; 12	—; —; —	22	—	0.15	11	
Plastic: Kevlar 49	1450	—; —; —	717; 483; 20	130	—	0.34	—	3
Steel, structural A36	78550	248; 248; —	400; 400; —	200	79	0.30	12	30
Titanium alloy (6% Ti; 4% Al by V)	4430	924; 924; —	1000; 1000; —	120	43	0.35	9.5	16
Wood: Douglas Fir	470	—; —; —	2.1; 26; 6.2	13	—	0.29	—	

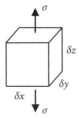

FIGURE 5.22 A cuboidal element of material.

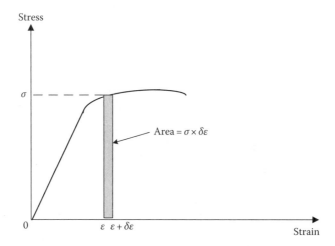

FIGURE 5.23 Incremental strain energy density.

From the initial state of zero stress and zero strain, up to the present state (σ, ε), the total strain energy density is given by the summation of these incremental energies (i). This sum, in the limit, is the integral (or, equivalently, the area under the stress–strain curve). Hence, for normal strain

$$\text{Strain energy density } u = \int_0^\sigma \sigma \, d\varepsilon = \text{area under the stress–strain curve} \qquad (5.10)$$

Note: Equation 5.10 is applicable regardless of whether the stress–strain relationship is linear.

If the stress–strain relationship is linear (i.e., for linear elastic material), Hooke's law (5.4), as given by $\sigma = E\varepsilon$, holds. Substituting this into Equation 5.10 and carrying out the integration gives

$$u = \frac{1}{2}\frac{\sigma^2}{E} \qquad (5.11)$$

In this linear case, in view of (5.4), Equation 5.11 may be written as

$$u = \frac{1}{2}\sigma\varepsilon \qquad (5.12)$$

Note: Beyond the elastic limit of the material, the strain energy density as given by (5.10) is not fully recoverable because part of the energy is used in the permanent deformation of the material.

Mechanical Properties of Materials

5.7.1 Strain Energy in Shear

Equations relating shear strain energy to shear stress and shear strain, analogous to (5.10) through (5.12), may be written in a straightforward manner. We have

$$\text{Shear strain energy density } u = \int_0^\tau \tau \, d\gamma = \text{Area under shear stress – Shear strain curve} \quad (5.13)$$

Note: Equation 5.13 is applicable regardless of whether the stress–strain relationship is linear.

If the stress–strain relationship is linear (i.e., for linear elastic material), Hooke's law (5.5), as given by $\tau = G\gamma$, holds. Substituting this into Equation 5.13 and carrying out the integration gives

$$u = \frac{1}{2} \frac{\tau^2}{G} \quad (5.14)$$

In this linear case, in view of (5.5), Equation 5.14 may be written as

$$u = \frac{1}{2} \tau \gamma \quad (5.15)$$

Note: Energy is a scalar. To obtain the overall strain energy, various strain energy terms (corresponding to normal strain and shear strain) may be simply added together.

5.7.2 Modulus of Resilience

This is the maximum strain energy per unit volume (strain energy density) that can be stored in the "elastic region" of the body. It is given by the area under the elastic region of the stress–strain curve. This energy is fully recoverable when the load is removed.

For the linear case (see Figure 5.24), from the result (5.11), the modulus of resilience may be expressed as

$$u_r = \frac{1}{2} \frac{\sigma_p^2}{E} \quad (5.16)$$

where σ_p is the stress at the proportional limit.

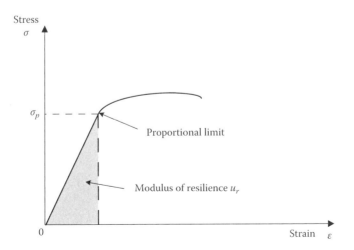

FIGURE 5.24 Modulus of resilience.

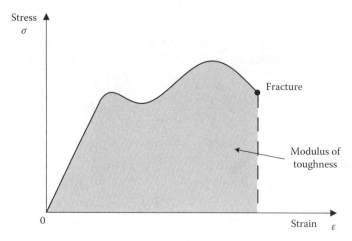

FIGURE 5.25 Modulus of toughness.

5.7.3 Modulus of Toughness

Modulus of toughness of a member is the maximum strain energy per unit volume (strain energy density) that can be absorbed by the member, up to the fracture point, when work is done by straining the member using an external load. It is given by the area under the entire stress–strain curve up to the point of fracture (see Figure 5.25).

Note: Since a member undergoes permanent deformation once its elastic limit is exceeded, some of the work is used to cause that permanent deformation, which is not recoverable. Hence, only part of the strain energy included in the modulus of toughness is recoverable.

Strength and toughness: A stronger material has a higher ultimate stress. It can withstand a greater stress (i.e., a greater load) before failure. A tougher material has a greater capacity to absorb energy before failure. Clearly, strength and toughness are not the same, yet both are important properties of engineering materials. In particular, a brittle material can be stronger than a ductile material, but typically a ductile material is tougher.

Example 5.10

Estimate the modulus of resilience and the modulus of toughness of the material that is approximately represented by the engineering stress–strain curve in Figure 5.26.

Solution

Modulus of resilience = area of the stress–strain curve up to the proportional limit

$$u_r = \frac{1}{2}\sigma_p \varepsilon_p = \frac{1}{2} \times 400.0 \times 10^6 \text{ (N/m}^2\text{)} \times 0.2 \times 10^{-2} \text{ (m/m)} = 400.0 \text{ kN} \cdot \text{m/m}^3 = 400.0 \text{ kJ/m}^3$$

Note: The energy unit 1 J (joule) = 1 N·m.

Modulus of toughness = area under the entire stress–strain curve

$$u_t = 400.0 + \frac{1}{2}(400.0 + 500.0) \times 10^3 \times (1.4 - 0.2) \times 10^{-2} \text{ kN} \cdot \text{m/m}^3$$

$$= 5800.0 \text{ kN} \cdot \text{m/m}^3 = 5800.0 \text{ kJ/m}^3$$

Mechanical Properties of Materials

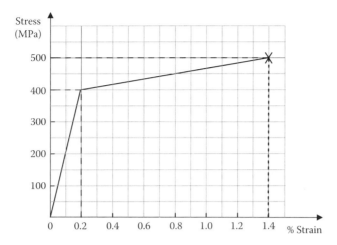

FIGURE 5.26 An idealized stress–strain behavior.

Primary Learning Objectives
1. Computation of the modulus of resilience graphically, using the stress–strain curve
2. Computation of the modulus of toughness graphically, using the stress–strain curve
3. Appreciation of the difference in the magnitudes of modulus of resilience and modulus of toughness
4. Recognition of the proper units of strain energy density

■ **End of Solution**

SUMMARY SHEET

Constitutive relation: Stress–strain curve; Hooke's law (linear case):

$$\sigma = E\varepsilon = G\gamma$$

where
  E is the modulus of elasticity or Young's modulus
  G is the modulus of rigidity or shear modulus

$$\text{Poisson's ratio } \nu = -\frac{\text{Lateral strain } \varepsilon_{lat}}{\text{Longitudinal strain } \varepsilon_{long}}$$

$E = 2(1+\nu)G$ (for the linear elastic case)
"Nominal stress" or "Engineering stress"

$$\sigma = \frac{P}{A_0}$$

where A_0 is the initial area of X-section
True stress

$$\sigma = \frac{P}{A}$$

where A is the actual area of X-section

Nominal strain or engineering strain

$$\varepsilon = \frac{\delta}{L_0}$$

Original length (gauge length) = L_0
True strain

$$\varepsilon = \frac{\delta}{L}$$

Actual length = L

Elastic: Deformation is completely reversible when the load is removed.

Proportional limit: Maximum stress up to which the stress–strain relation is linear (obeys Hooke's law).

Yield point: Point at which the material ceases to be elastic (then, material will not fully recover once loading is removed).

Elastic limit: Yield point.

Yield stress (yield strength): Stress at yield point.

0.2% Offset method: Determines yield stress. Draw a line parallel to linear part of stress–strain curve from 0.2% point on strain axis (offset). Point of intersection with *stress–strain* curve → yield point.

0.05% Offset method: Similar to the 0.2% offset method, except the strain offset used is 0.05%

Perfectly plastic state: Yielding region is flat (stress is constant as the strain increases).

Ultimate stress (ultimate strength): Maximum stress that the material can withstand before failure.

Necking: Beyond ultimate stress, abrupt and significant reduction in the cross-sectional area until the specimen fails (fracture).

Fracture stress: The stress at the fracture point

Percent elongation: Strain at the point of fracture, expressed as a percentage (i.e., percent strain at fracture).

Strain hardening (work hardening): Beyond yielding, further change in strain is associated with an appreciable "increase" in stress. → Material can support increased levels of stress (load) → Increases its yield point (on reloading after unloading); makes the material less ductile (more brittle).

Hardness: Resistance of a material to surface indentation.

Residual strain: Plastic strain; remaining strain when the stress is removed.

Ductile materials: Undergo large deformations (strains) before fracture; capable of absorbing relatively large amounts of energy (i.e., they are tough).

Ductility: Level of strain at fracture. Capacity for plastic deformation.

Ductility measures (in tensile testing):

$$\text{Percent elongation} = \frac{L_f - L_0}{L_0} \times 100\% = \varepsilon_f \times 100\%$$

$$\text{Percent reduction of area} = \frac{A_0 - A_f}{A_0} \times 100\% = \frac{(d_0)^2 - (d_f)^2}{(d_0)^2} \times 100\%$$

Brittle materials: Do not exhibit large deformations (yielding) before fracture; they fail abruptly; typically they have a greater compressive strength than tensile strength.

Strain energy density: In the linear case, strain energy per unit volume of body = area under the stress–strain curve

$$u = \int_0^\sigma \sigma d\varepsilon \quad \text{In the linear case,} \quad u = \frac{1}{2}\sigma\varepsilon = \frac{1}{2}\frac{\sigma^2}{E}$$

Modulus of resilience: Strain energy density up to the elastic limit; area under the linear region of stress–strain curve (this energy is fully recoverable when the load is removed).

Modulus of toughness: Energy density up to the fracture point; area under entire stress–strain curve up to fracture point (only part of this energy is recoverable when the load is removed, because part of it is used for permanent deformation of the material).

Creep: Material continues to deform (strain increases) over a very long period, at constant loading (constant stress).

Creep strength: Highest strength (stress) a material can withstand over a specified period of time (1000 h) without exceeding creep strain (a specified strain; e.g., 1%).

Note: Creep strain (associated with a creep strength) is specified as a strain rate (e.g., 0.1% per year).

Note: Creep strain typically increases with material temperature and size of loading (stress).

Fatigue failure: Abrupt failure when a loading cycle (and associated deformation) is repeated many times even if the average stress levels are well below the ultimate strength.

Endurance limit (fatigue limit): A stress below which no failure occurs on repeated application of the corresponding load over a specified number of times (e.g., 500×10^6 cycles).

Homogeneous: Material properties (including composition and stress–strain relations) are the same at every point in the material.

Isotropic: Material properties are the same in every direction, at a given point.

Factor of safety: [Harmful limit (value at yielding or fracture)] / [Allowable value].

PROBLEMS

5.1 A uniform metal rod of original (unstrained) length L and area of X-section A is welded to a rigid and fixed structure at one end. A spring of stiffness k is attached to the other end of the rod. When a tensile force P is applied to the free end of the spring, that end moves through δ (see Figure P5.1).

Express Young's modulus E of the rod in terms of P, δ, k, L, and A.

Note: Assume that the rod is within its proportional limit:

$$\text{Spring stiffness} = \frac{\text{Tension}}{\text{Extension}}$$

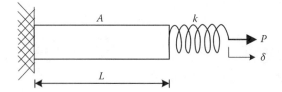

FIGURE P5.1 A metal rod with end spring subjected to tensile load.

5.2 A uniform rubber band holds three identical and rigid plastic rods of circular X-section in the symmetric configuration shown in Figure P5.2a. The outer surface of the tubes is smooth and the rubber band is unstretched in this configuration. Then the top tube is carefully pressed into the configuration shown in Figure P5.2b. The tension in the rubber band in this configuration was found to be $T = 150$ N. If the area of X-section of the rubber band is $A = 2$ mm^2, determine the Young's modulus of the rubber band. Assume that the rubber band remains within its proportional limit.

5.3 A uniform, heavy cable is hanging from a fixed point of a ceiling (Figure P5.3). The following parameters are given:

L: length of the cable
A: area of X-section of the cable
ρ: mass density of the cable material
E: Young's modulus of the cable material
g: acceleration due to gravity

  a. Determine an expression for the extension at the free (bottom) end of the cable due to its own weight. *Note*: Use only the parameters given in the question.
  b. Show that this is equal to the extension of a light cable of same dimensions and E, with a weight of half the cable mass hung at the free end.
  c. Is the normal strain distribution the same in the two cases?

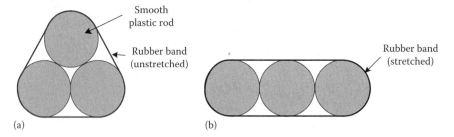

FIGURE P5.2 (a) Original unstrained symmetric configuration; (b) final strained longitudinal configuration.

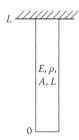

FIGURE P5.3 Heavy cable hanging from a rigid ceiling.

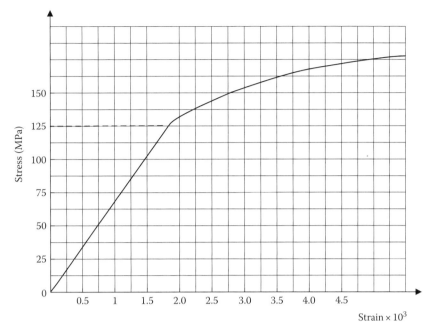

FIGURE P5.4 Stress versus strain curve of the material.

5.4 Consider the stress versus strain curve shown in Figure P5.4.
  a. Determine the proportional stress.
  b. Determine the Young's modulus.
  c. Using the 0.2% offset method, estimate the yield stress (yield strength) of the material.
  d. If the test specimen is stressed up to 175.0 MPa, and then the stress is released, estimate the residual strain in the specimen. What is the new proportional limit?

5.5 Consider the idealized stress versus strain curve of a material as shown in Figure P5.5. Suppose that a circular rod (i.e., with circular X-section) made of this material has a cross-sectional diameter of 25 mm. Determine
  a. The proportional limit, corresponding strain, and the Young's modulus
  b. The axial load that will take the rod to its proportional limit
  c. The yield stress and the corresponding strain
  d. The fracture stress (fracture strength), load at fracture, and the corresponding strain
  e. The ultimate stress (ultimate strength) and the corresponding strain
  f. The residual strain that remains in the rod, if an axial load of 220 kN is applied to the rod and then released
  g. The percentage elongation (at fracture). If the length of the rod is 200 mm, what is the elongation?

5.6 Determine the total extension in the rod shown in Figure P5.6a, which is made of material that obeys the stress–strain curve of Figure P5.6b.

5.7 Consider the stress versus strain curve shown in Figure P5.7. *Note*: The lower scale of strain is 10 times the upper scale (i.e., the initial segment of the curve is zoomed in 10 times). For this material, determine
  a. The proportional stress
  b. The Young's modulus
  c. The yield stress (yield strength), using the 0.2% offset method
  d. The fracture stress (fracture strength) and the corresponding strain
  e. The ultimate stress (ultimate strength) and the corresponding strain

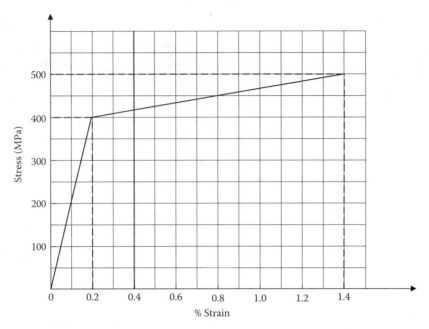

FIGURE P5.5 Stress–strain curve of the material.

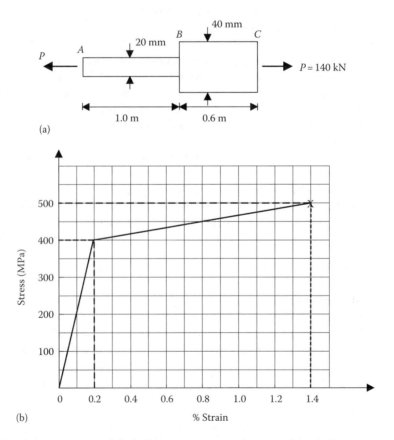

FIGURE P5.6 (a) A two-segmented shaft; (b) stress versus strain curve of the shaft.

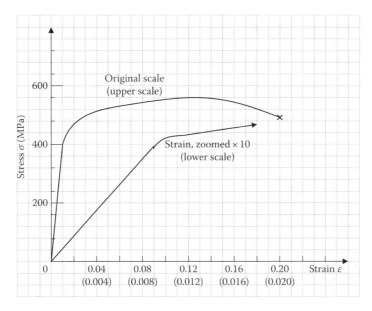

FIGURE P5.7 Stress versus strain curve of the material.

5.8 A tensile test was carried out on a uniform cylindrical test specimen with the following dimensions: gauge length $L_o = 50$ mm; initial diameter $d_o = 10$ mm. The experimental load (kN) – extension (mm) data obtained from the test are given in Table P5.8.
Determine
   a. The stress versus strain curve
   b. The proportional limit and the corresponding strain
   c. The Young's modulus
   d. The yield stress (by the 0.2% offset method) and the corresponding strain
   e. The ultimate stress and the corresponding strain
   f. The fracture stress and the corresponding strain
   g. The percent elongation
   h. If the actual tensile load on the member is 8.0 kN, what is the corresponding FOS with respect to (a) yield stress; (b) ultimate stress

5.9 Struts are used to anchor a bridge deck to a bridge tower (see Figure P5.9). To monitor the load-deformation condition of the strut AB, an extensometer is mounted over a small axial segment CD of the strut. The following data are known: $AB = 3$ m, $CD = 100$ mm, extensometer reading (for CD) = 0.01 mm, area of X-section of the strut = 5.0 cm², Young's modulus of the strut material (structural steel) = 200 GPa.
Determine
   a. The normal strain in CD
   b. The normal strain in AB
   c. The extension of the strut
   d. The normal stress at a strut X-section
   e. The axial load in the strut

5.10 Consider a uniform cylindrical test specimen used in a standard tensile test, as shown in Figure P5.10. Gauge length $L_o = 50$ mm; initial diameter $d_o = 10$ mm. When a tensile load of 5.0 kN is applied (in the elastic region), the extension was found to be 0.04 mm. Also, the contraction in diameter (in the midsection of the specimen) was measured to be 0.003 mm.
   Estimate the Young's modulus of elasticity, Poisson's ratio, and shear modulus of the material.

TABLE P5.8
Experimental Load–Deflection Data for a Test Specimen

Load (kN)	Extension (mm)
0	0
2.57	0.02
5.14	0.04
7.72	0.06
10	0.08
10.5	0.1
12.2	0.12
12.5	0.14
13.2	0.2
14.7	0.5
16.4	1
17.34	2
18.69	3.4
19.19	4
19.91	5
20.97	7
21.5	9
21.3	12
18.8	15
18.5	15.2 (Fracture)

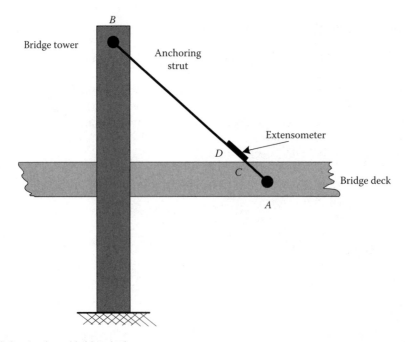

FIGURE P5.9 Anchored bridge deck.

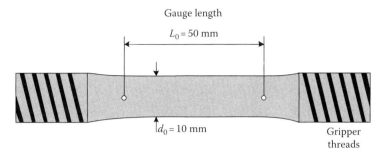

FIGURE P5.10 A tensile test specimen.

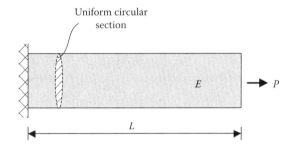

FIGURE P5.11 A uniform rod under tension.

5.11 A tensile force P is applied to a uniform rod of length L, circular X-section, Young's modulus E, and Poisson's ratio ν (Figure P5.11). Assuming that square terms of strain can be neglected in comparison to linear terms of strain, show that the increase in volume of the rod may be expressed as

$$\Delta V = \frac{PL}{E}(1-2\nu)$$

For what value of ν will the rod not increase its volume due to extension?

5.12 A rectangular wooden pole (assumed rigid) of width 5 cm is glued to a rigid structural opening of length 10 cm as shown in Figure P5.12. When set, the top layer of glue is 4 mm thick

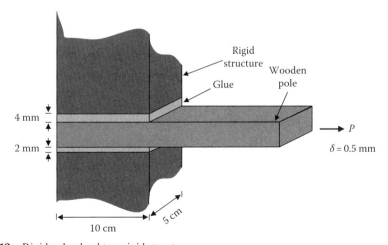

FIGURE P5.12 Rigid pole glued to a rigid structure.

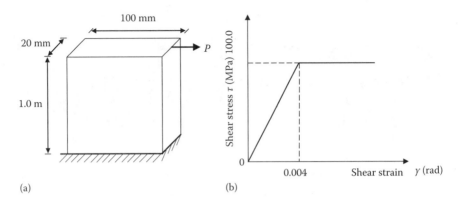

FIGURE P5.13 (a) A shear loaded rectangular plate; (b) shear stress versus shear strain of the plate material.

and the bottom layer is 2 mm thick. When the pole is pulled with force P, it moves through 0.5 mm. If the shear modulus of glue is $G = 10$ MPa, determine P.

5.13 A rectangular plate of material is rigidly fixed at its bottom edge and a lateral force P is applied at its top edge, as shown in Figure P5.13a. The shear stress versus shear strain of the plate material is approximated as in Figure P5.13b.
  a. What is the shear modulus G of the material?
  b. Determine the force P that will begin yielding due to shear in the plate.
  c. What is the corresponding deflection of the top edge in the direction of P?

5.14 A tensile test was carried on a cylindrical alloy specimen of gauge length $L_0 = 30$ mm and initial diameter $d_0 = 10$ mm. When tensile load $P = 6.0$ kN was applied, the gauge length was extended (as measured using an extensometer, see Figure P5.14) by $\delta L = 0.025$ mm and the diameter (in the midsegment) was contracted by $-\delta d = 0.0025$ mm.
Determine
  a. The Poisson's ratio
  b. The Young's modulus
  c. The shear modulus
of the specimen

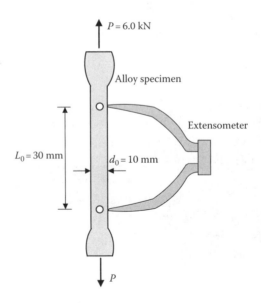

FIGURE P5.14 A tensile test arrangement.

Mechanical Properties of Materials

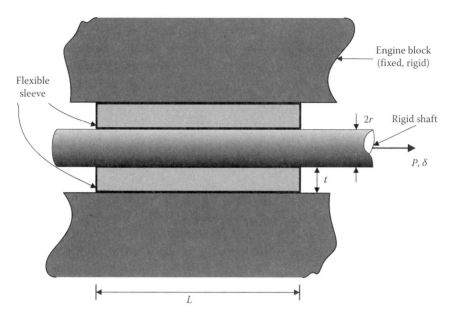

FIGURE P5.15 A shaft fitted to engine block through a flexible sleeve.

5.15 A flexible sleeve of inside radius r, thickness t, and length L is fitted snugly inside a uniform cylindrical bore of a rigid engine block (bore radius is very slightly smaller than the outer radius of the sleeve). A rigid shaft passes tightly through the sleeve (see Figure P5.15). When an axial force P is applied to the shaft, determine the resulting movement δ assuming that there is no slip (between the shaft and the sleeve, and also the sleeve and the engine block):

G = shear modulus of the sleeve

Express your result in terms of t, r, L, G, and P.

5.16 The carriage arm of a machine is made of rigid metal parts and a synthetic rubber attachment piece, as shown in Figure P5.16. *Note*: All the components except the rubber attachment piece are assumed rigid. One side of the rubber piece is firmly glued to a rigid wall of the machine. The other side is firmly glued to the top end of the carriage arm. The metal arm is made of two pieces joined using a single bolt and nut, in double shear arrangement, as shown.

When a vertical load P is applied to the free end of the carriage arm, it moves vertically through 2 mm. The dimensions of the rubber attachment piece are 50 mm × 40 mm × 20 mm as shown. The shear modulus of the rubber attachment piece is $G = 0.5$ MPa.
  a. Determine the value of the load P in newtons (N).
  b. If the diameter of the bolt is $d = 10$ mm, determine the average shear stress in the X-section of the bolt.

5.17 A uniform rod of area of X-section A and Young's modulus E is subjected to a tensile force P. Within the proportional limit, what is the strain energy density of the strained rod?

5.18 A metal specimen was subjected to a tensile test. From the load–deflection data, the stress–strain curve shown in Figure P5.18 was computed.

Note: The bottom curve gives the details of the initial part of the original curve where the strain axis is zoomed in by a factor of 10.

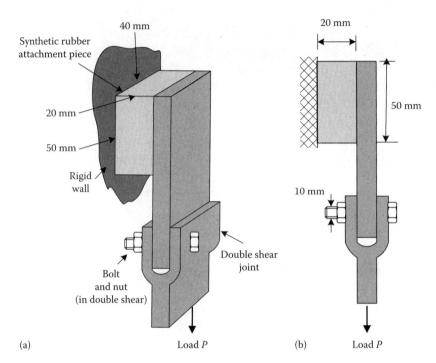

FIGURE P5.16 (a) Perspective view. (b) Front view.

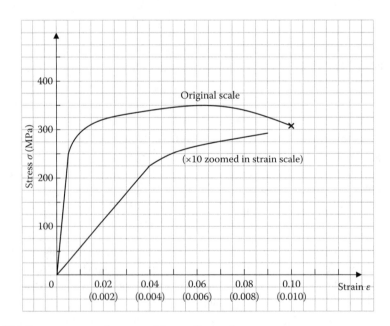

FIGURE P5.18 Stress versus strain curve of the tensile specimen.

Using the given curves, determine the following parameters and indicate the proper units for them:
a. The proportional limit
b. The Young's modulus
c. The yield strength (by the 0.2% offset method)

d. The ultimate strength
e. The fracture strength
f. The % elongation (at fracture)
g. The modulus of resilience
h. The modulus of toughness

5.19 The stress–strain relationship of a metal is given by

$$\sigma = 70\varepsilon \quad \text{for } 0 \leq \varepsilon \leq 2$$
$$= 70\varepsilon - 1.35(\varepsilon - 2)^2 \quad \text{for } 2 \leq \varepsilon \leq 52$$
$$= 265 \quad \text{for } 52 \leq \varepsilon \leq 250$$

where
σ is in MPa
ε is in 1×10^{-3} units of strain (i.e., 1000 $\mu\varepsilon$)

For this material, estimate the following parameters:
Young's modulus E; proportional limit σ_p; ultimate strength σ_u; and % extension

5.20 For the material in Problem 5.19, estimate
a. The modulus of resilience u_r
b. The modulus of toughness u_t

5.21 A frame structure is made of two steel members BD and CD, which are pin-jointed at D, with the ends B and C hinged to a rigid vertical wall, as shown in Figure P5.21. A load of weight 100 kN is hung at D. The lengths of the members are $BD = 2.5$ m and $CD = 2.0$ m. The area of X-section of each member is 13.33 cm². The Young's modulus of the members is $E = 200$ GPa.
Determine
a. The normal stresses σ_{BD} and σ_{CD} in the two members
b. The overall axial deformations δ_{BD} and δ_{CD} in the two members
c. The horizontal and vertical movements δ_h and δ_v, respectively, of joint D due to the deformation caused by the load

5.22 A frame structure is made of two steel members BD and CD which are pin-jointed at D, with the ends B and C hinged to a rigid vertical wall, as shown in Figure P5.22. A load of weight 100 kN is hung at D. The lengths of the members are $BD = 2.5$ m and $CD = 2.0$ m. The area of X-section of each member is 13.33 cm². The Young's modulus of the members is $E = 200$ GPa.

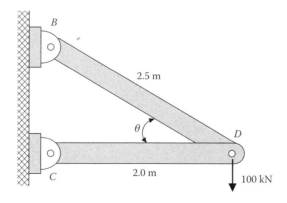

FIGURE P5.21 A frame structure attached to a wall and carrying an end load.

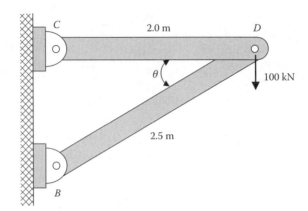

FIGURE P5.22 A frame structure attached to a wall and carrying an end load.

Determine
a. The normal stresses σ_{BD} and σ_{CD} in the two members
b. The overall axial deformations δ_{BD} and δ_{CD} in the two members
c. The horizontal and vertical movements δ_h and δ_v, respectively, of joint D due to the deformation caused by the load

Further Review Problems

5.23 A two-member truss is shown in Figure P5.23. The members have the same Young's modulus, $E = 200$ GPa, and the same area of cross section, $A = 400$ mm². The length of the horizontal member AC is $L = 4.0$ m. The member BC is inclined at $\theta = 30°$ to AC, as shown. All the joints are smooth (pin joints).

A horizontal force P_x and a vertical (downward) force P_y are applied at joint C, and as a result, C moves through $u_x = 1.0$ mm horizontally and $u_y = 2.0$ mm vertically (downward). Determine the values of the applied loads P_x and P_y.

5.24 Two slabs of polyurethane of identical rectangular cross section $L \times w = 50$ cm × 20 cm, and heights $a = 25$ cm and $b = 10$ cm are firmly glued on to the two sides of a rigid metal plate. The top and the bottom of the composite block are glued to a fixed rigid top, and a fixed, rigid base, respectively, as shown in Figure P5.24. The plate is constrained to move

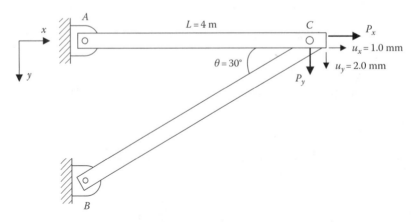

FIGURE P5.23 A truss with external loads.

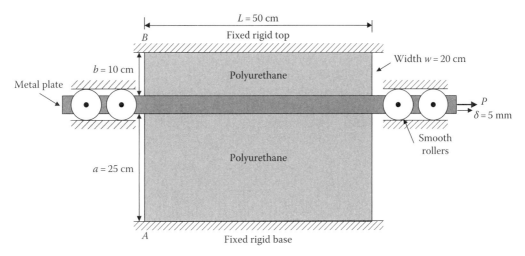

FIGURE P5.24 Two sandwiched polyurethane blocks with reinforced plate on smooth rollers.

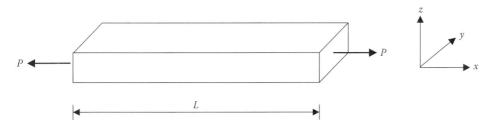

FIGURE P5.25 A rectangular rod under tensile loading.

only in the horizontal direction at a fixed level, by means of a channel mechanism with smooth rollers.

When a horizontal force P is applied to the metal plate, the associated horizontal movement was found to be $\delta = 5$ mm. Taking the shear modulus of polyurethane as $G = 2.0$ MPa, determine P.

5.25 A uniform rod of length L_0 and rectangular cross section is subjected to a tensile load P, as shown in Figure P5.25. Show that the increase in volume may be expressed as

$$\Delta V = (1 - 2\nu)\frac{L_0 P}{E}$$

where
   ν is the Poisson's ratio
   E is the Young's modulus of the material
Assume that the rod remains within its proportional limit.

5.26 Figure P5.26 gives the engineering stress–strain diagram of an aluminum alloy. Note that the lower scale presents the initial segment of the curve, with a zooming-in factor of 20 for strain (i.e., on the horizontal axis).
   a. Estimate the Young's modulus E of the material.
   b. Determine the yield strength σ_Y using the 0.05% offset method and also the 0.20% offset method. Compare the two values.

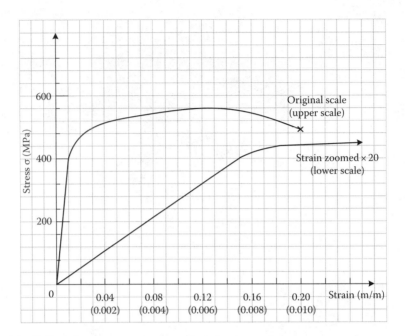

FIGURE P5.26 Engineering stress–strain curve of an aluminum alloy.

    c. What is the ultimate strength σ_U of the material?

    d. What is the fracture stress σ_F of the material? If the Poisson's ratio of the material, when close to the fracture point, is 0.4, estimate the true stress at fracture. What is the percentage elongation?

5.27 For the aluminum alloy whose stress–strain curve is given in Figure P5.26, estimate
    a. The modulus of resilience
    b. The modulus of toughness

6 Axial Loading

CHAPTER OBJECTIVES

- Axial loading; use of Saint-Venant's principle and equivalent loading to simplify the analysis
- Determination of axial force diagram
- Analysis of axially loaded members having: (a) uniform X-section; (b) continuously varying nonuniform section; (c) multiple segments of uniform X-section
- Application of the principle of superposition (valid when force–deflection relation is linear)
- Analysis of statically indeterminate structures (where statics—equilibrium equations—alone are not adequate to solve the problem)
- Thermal deformation due to temperature change, and associated strains and stresses; application in statically indeterminate problems
- Stress concentration due to geometric nonsmoothness and discontinuities

6.1 INTRODUCTION

The solution of problems in mechanics of materials involves three basic considerations in general (see Chapter 1):

1. Statics (equations of equilibrium). This determines reactions at supports, and internal loads, which determine stresses.
2. Nature of deformation (hence, nature of strains).
3. Stress–strain relations (i.e., constitutive relations or physical relations). These are needed to determine the strains (deformations) once the stresses are known (through the knowledge of internal loads).

In Chapter 2, we reviewed the subject of statics. In Chapter 3, we studied stress (both normal stress and shear stress) and its determination from the knowledge of internal loading. In Chapter 4, we studied strain (both normal strain and shear strain) and its representation using the geometry of deformation. In Chapter 5, we studied the physical relation between stress and strain (i.e., constitutive relation), as applicable to any type of object under any type of loading. As illustrative examples, we considered several types of members including axially loaded members with normal stresses and strains, and shear loaded members with shear stress and shear strain. In the present chapter, we will integrate all three considerations of mechanics of materials (stress, strain, and stress–strain relation), as listed earlier, and apply them in the solution of a specific class of problems—members with axial loading. A similar treatment for two other common types of members (those with torsion and bending, respectively) will be given in Chapters 7 and 8.

6.1.1 Basic Types of Loading

Consider a rectangular segment of a deformable member with a rectangular grid marked on its outside, as shown in Figure 6.1a. Four basic scenarios of loading and deformation of such a member are studied in this book:

1. *Axial loading*: The loads are forces (tensile or compressive) which are applied along the main axis of a member that is relatively long. The resulting deformations are primarily extensions or compressions, which occur along the loading axis even though deformations (strains) may occur perpendicular to this axis as well (Poisson effect). An example is a pin-jointed tension rod (strut), which is a two-force member (see Chapter 2), as shown in Figure 6.1b.

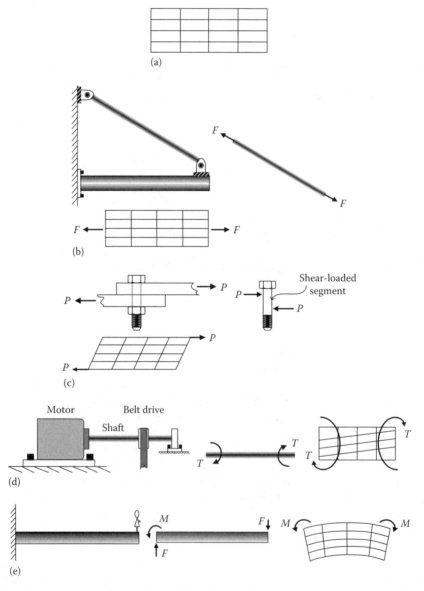

FIGURE 6.1 Representation of four basic loading scenarios: (a) original unstrained member; (b) axial loading; (c) shear loading; (d) torsional loading; (e) flexural loading.

2. *Shear loading*: In this case, two equal and opposite parallel forces exist along two closely spaced parallel sections of the member. The resulting deformation involves sliding (shearing) of one section with respect to the other along the direction of the loading. As we have noted in Chapter 4, in shear deformation, the angle of a corner will change (producing a shear strain) in the deformed element. An example is a "bolt of a lap joint," where one section of the bolt is pulled in one direction and a section next to it is pulled in the opposite direction, as shown in Figure 6.1c.
3. *Torsional loading*: Here the external loads are "torques," which tend to twist the member. An example is a shaft of a motor that is connected to a load, as shown in Figure 6.1d. The motor applies a torque at one end of the shaft. The driven load, at the other end of the shaft, exerts a resisting torque (in the opposite direction). Under steady conditions, the two torques are equal (and opposite). This subject is treated in Chapter 7.
4. *Bending (flexural) loading*: In this case, the external loads (forces and moments) cause bending deformations (i.e., flexure) of the member. An example is a diving board, which is bent by the force of the diver, as shown in Figure 6.1e. The external load applied by the diver is a force while the reaction at the clamped end of the board will include both a shear force and a bending moment. Hence, both normal stresses and shear stresses will be present in general, in conditions of bending (see Chapter 8).

In practical problems, two or more of these basic loading scenarios may exist in combination, in the same member. If the conditions are linear, then the *principle of superposition* (PoS) may be applied to combine the individual effects of them.

6.1.2 CHAPTER OBJECTIVES

In this chapter, we will specifically study in more detail the loading and the resulting deformation of axially loaded members. Examples of axial members are supporting cables and hangers of bridges, tie bars and tensioning struts of aircraft and other structures, tall structural columns, and truss members in building roofs. In these members (e.g., strut in Figure 6.1b), the load (force) is applied along a single axis (typically, the central axis). The resulting normal stresses and strains on sections normal to this axis (i.e., normal sections or cross sections) are of particular importance. In the present chapter, we will revisit the relations of stress, strain, and deformation in axial members. Particularly, we will learn how these results may be used in the design of axial members and also in solving statically indeterminate problems (where the equilibrium equations alone are not adequate to solve the problem) of axial members.

Temperature changes will result in deformations (thermal expansions and thermal contractions) and associated thermal strains. If the body is not restrained and is free to deform in all directions, then no thermal loading and thermal stresses will be generated. In the general case of a restrained body, however, there will be both thermal strains and thermal stresses. We will study the generation of thermal stresses and thermal strains in axial members due to temperature variation.

Using Saint-Venant's principle, complex and extreme types of loading may be approximated by statically equivalent, yet simplified, loading conditions. According to this principle, uniform stress and strain distributions (which are far simpler to handle) may be assumed at locations that are not very close to the loading points of an axial member. We will present Saint-Venant's principle in the context of an axially loaded member. Stress concentration and residual stresses will exist in axial members that have geometric nonsmoothness and discontinuities. We will study these concepts as well.

6.2 SAINT-VENANT'S PRINCIPLE

The basic idea of Saint-Venant's principle is as follows: In a body with the same restraints (supports), statically equivalent externals loads produce the same stress and strain conditions at sufficiently farther locations from the external load.

Note 1: A statically equivalent load has the same resultant load (magnitude and direction) at the location of the originally applied load.

Note 2: A statically equivalent load produces the same reactions at the supports as with the original load.

Note 3: The required minimum distance from the load where the Saint-Venant's property is valid is equal to the largest dimension of the cross section of the body in the neighborhood of loading.

Example 6.1

Consider an axial member with a concentrated axial force P applied at the center of one end. The other end is firmly attached to a rigid foundation (see Figure 6.2). A uniform grid is drawn on the unstrained member (before applying P) as in Figure 6.2a. On application of P, the member will deform due to the loading, as illustrated by the shape of the deformed grid in Figure 6.2b. Note the pronounced nonuniform deformations (and hence nonuniform strains) that occur near the load and also at the support location (base) of the member. The nonuniform deformation at the support location is primarily due to the Poisson effect of lateral shrinking at that end of the member, since the base is rigidly restrained (which restricts lateral shrinking and longitudinal extension).

The stress profile at the cross section A–A of the member (see Figure 6.2b) peaks sharply near the point of application of the load. The sharpness of the peak reduces, and the profile tends to level off, in cross sections farther and farther from the point of application of the load (e.g., X-section B–B). As this distance becomes larger than the largest cross-sectional dimension (e.g., X-section C–C), the stress (and strain) profile becomes rather uniform, whose value is given by the average stress

$$\sigma_{av} = \frac{P}{A} \tag{6.1}$$

where A is the area of X-section. This is so according to Saint-Venant's principle, because a uniformly distributed axial load having this value is indeed a load that is equivalent to the original point load P.

If the axial point force P at the center of the end section of the member is replaced by two point forces each of magnitude $P/2$, which are symmetrically applied about the center of the end section (see Figure 6.2d), this loading system is statically equivalent to the loading system of Figure 6.2b. It is seen that in the neighborhood of the loading, the strain distribution and the stress profile are quite different from those of the previous case (see Figure 6.2e). However, at a distance equal to at least the largest dimension of the X-section, the stress profile is uniform and equal to the value of the previous case ($\sigma_{av} = P/A$). Also, in this region, the strain distribution will be uniform and equal to the value of the previous case.

Primary Learning Objectives

1. Understanding of Saint-Venant's principle
2. The concept of statically equivalent loading
3. Nature of stress profiles in an axial member subjected to a concentrated axial force
4. Simplification of the analysis of axial members through Saint-Venant's principle

■ **End of Solution**

The exact stress–strain behavior in the neighborhood of an applied load (and near abrupt structural nonuniformities) may be analyzed using the theory of elasticity and continuum mechanics, the treatment of which is outside our scope. Related computational studies may be carried out using the finite element method.

Axial Loading

6.3 AXIALLY LOADED MEMBER

Mechanics of materials concerns stresses (as determined by the internal loading), strains (as determined by the geometry of deformation), and the relationship between stresses and strains (the constitutive relation, as determined by the physics of the material) of deformable bodies. We will now study the application of these concepts to axial members.

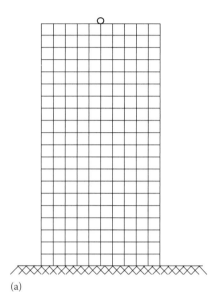

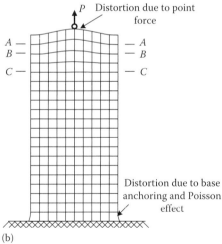

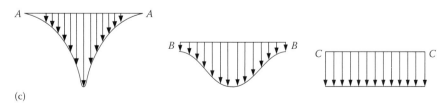

FIGURE 6.2 Illustration of the Saint-Venant's principle: (a) original axial member; (b) deformation due to an axial point load; (c) stress profiles at different cross sections of the member.

(*continued*)

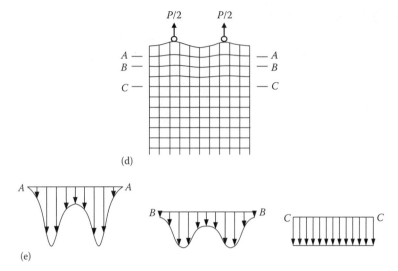

FIGURE 6.2 (continued) Illustration of the Saint-Venant's principle: (d) a loading system that is statically equivalent to what is shown in (b); (e) corresponding stress profiles at different X-sections.

The general case of an axial member is where its cross section varies continuously along the main axis. Special cases can be formed by joining two or more axial segments having either uniform or varying cross sections. We will first analyze the general case of an axial member, for normal stresses and normal strains. Then we will consider the special case of combined multiple segments of uniform X-section. In the analysis, we will assume that the material is homogeneous and isotropic in each segment. Concerning deformation of an axial member in particular, the following two assumptions are made:

1. The axis of the member remains straight.
2. Cross sections (which are planes perpendicular to the axis) remain planar and also perpendicular to the axis.

6.3.1 Continuously Varying Nonuniform Section

Let us now formally revisit a general, axially loaded member. In particular, consider the case where the member is not uniform, having a variable cross section $A(x)$, as shown in Figure 6.3a. Also suppose that a variable (distributed) axial force is applied externally along the member (or the member is restrained along its length), as indicated by the distributed left-hand arrows in the figure. The member

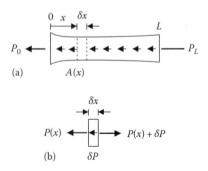

FIGURE 6.3 (a) A general axial member having a variable cross section and a variable external load; (b) a small element of the member.

Axial Loading

is maintained in equilibrium using the end forces P_0 and P_L. (*Note*: These two end forces are not equal, under equilibrium, due to the presence of an external axial load along the length.) If we make a virtual cut in the member at a distance x from its left end, the resulting cross section will have an "internal" force $P(x)$. Clearly, this internal axial force is also variable along the length of the member (again, due to the presence of an external axial load along the length). Consider a small element of length δx at distance x from the left end of the member, as shown in Figure 6.3b. Its area of cross section is $A(x)$. The end force on the cross section at x is $P(x)$, which is positive when acting to the left. The end force on the cross section at $x+\delta x$ is $P(x)+\delta P$, which is positive to the right. The equilibrium of this small element is maintained by the small external force of magnitude δP exerted on the element, as shown in Figure 6.3b.

The normal stress on the cross section at x is

$$\sigma(x) = \frac{P(x)}{A(x)} \tag{6.2}$$

The corresponding strain is

$$\varepsilon(x) = \frac{\sigma(x)}{E} = \frac{P(x)}{EA(x)} \tag{6.3}$$

where E is the Young's modulus of the material (constant for a homogeneous material)

Similarly, the normal stress on the cross section at $x+\delta x$ is

$$\sigma(x+\delta x) = \frac{P(x)+\delta P}{A(x)}$$

The corresponding strain is

$$\varepsilon(x+\delta x) = \frac{\sigma(x+\delta x)}{E} = \frac{P(x)+\delta P}{EA(x)} \approx \frac{P(x)}{EA(x)} = \varepsilon(x)$$

Note: This last result is true because the variation δP of the external force on the small element is very small compared to $P(x)$, since the element length δx itself is very small and $P(x)$ is finite.

Hence, the normal strain along the small element δx may be assumed constant (uniform) along its length.

By the definition of normal strain (see Chapter 4), extension in the small element δx is $\varepsilon(x)\delta x$. Substituting (6.3), we have $\varepsilon(x)\delta x = (P(x))/(EA(x))\delta x$.

The total extension in the member is given by the summation of these small extensions of all the elemental segments of the member. As $\delta x \to 0$, in the limit, the total extension in the member is given by the integral

$$\delta = \int_0^L \varepsilon(x)dx = \int_0^L \frac{\sigma(x)}{E}dx = \int_0^L \frac{P(x)}{EA(x)}dx \tag{6.4}$$

If P, A, and E are constant (i.e., a uniform, homogeneous, and isotropic member with no distributed external loading along its length), we get by direct integration of (6.4), which now has a constant integrand, the following result:

$$\delta = \frac{PL}{EA} \tag{6.5}$$

Example 6.2

A solid conical rod (frustum) of Young's modulus E is attached horizontally at its base A to a rigid vertical wall. A horizontal force P is applied centrally at the free end B as shown in Figure 6.4a. The following parameters are given:

L is the length of the rod
d_A is the base diameter of the rod
d_B is the end diameter of the rod

Determine an expression for the horizontal extension of the rod at end B due to the applied load P. Neglect the weight of the rod.

$$\text{Hint}: \int \frac{dx}{(a+bx)^2} = -\frac{1}{b(a+bx)}$$

Solution

Consider a cross section at horizontal distance x from the base A, as shown in Figure 6.4b, where the diameter is d_x.

From similar triangles we have $\dfrac{\frac{1}{2}(d_x - d_B)}{\frac{1}{2}(d_A - d_B)} = \dfrac{L-x}{L}$

Hence, $d_x = d_B + \left(1 - \dfrac{x}{L}\right)(d_A - d_B) = d_A - \dfrac{x}{L}(d_A - d_B)$

Area of cross section at x is $A_x = \dfrac{\pi}{4} d_x^2$

From the standard formula (6.4), the extension of the rod is

$$\delta = \int_0^L \frac{P}{A_x E} dx = \frac{P}{E} \int_0^L \frac{1}{(\pi/4)d_x^2} dx = \frac{4P}{\pi E} \int_0^L \frac{1}{d_x^2} dx = \frac{4P}{\pi E} \int_0^L \frac{1}{[d_A - (x/L)(d_A - d_B)]^2} dx$$

$$= \frac{4P}{\pi E} \cdot \frac{L}{(d_A - d_B)} \cdot \frac{1}{[d_A - (x/L)(d_A - d_B)]} \Bigg|_0^L \quad \text{(Using the integration formula)}$$

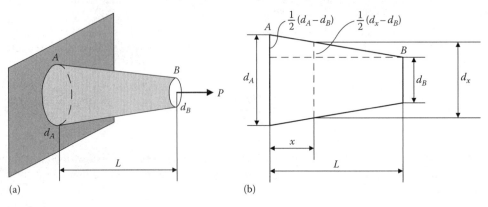

FIGURE 6.4 (a) A horizontally mounted frustum with an axial load. (b) Geometry and general axial location of the frustum.

$$= \frac{4P}{\pi E} \cdot \frac{L}{(d_A - d_B)} \left[\frac{1}{d_A - d_A + d_B} - \frac{1}{d_A} \right] = \frac{4P}{\pi E} \cdot \frac{L}{(d_A - d_B)} \left[\frac{1}{d_B} - \frac{1}{d_A} \right]$$

or

$$\delta = \frac{4PL}{\pi E d_A d_B}$$

Primary Learning Objectives
1. Determination of the extension of a nonuniform axial member
2. Use of proper geometric relations for a nonuniform axial member
3. Analytical solution of load–deflection problems involving axial members

■ **End of Solution**

Example 6.3

A water storage tank of internal diameter 4 m, depth 4 m, and wall thickness 0.25 m is made of steel. It is welded to a steel column of height 6 m and diameter 0.5 m, which is firmly fixed to a rigid foundation. Assume that the tank is completely filled with water (see Figure 6.5a).

Given: Young's modulus of steel $E_s = 200$ GPa; specific weight of steel $\rho_s g = 80$ kN/m³; specific weight of water $\rho_w g = 10$ kN/m³.

Determine the following:

a. Distribution of the average normal stress of the tank wall cross section, along the tank depth, and its maximum value
b. Distribution of the average normal stress of the column cross section, along its height, and its maximum value
c. The downward settlement at the tank top due to the gravitational loading on the system

Solution

a. Make a virtual X-section B–B in the tank at a depth of z and consider the segment above it (see Figure 6.5b).

$$\text{Area of tank wall } A_B = \frac{\pi}{4} \times (4.5^2 - 4^2) \text{ m}^2$$

Weight of the tank segment above B–B =

$$A_B \times z \times \rho_s g = \frac{\pi}{4}(4.5^2 - 4^2) \times z \times 80 \text{ kN} = 267.0 \, z \text{ kN}$$

Equilibrium → Compressive force on the tank wall X-section, $P_B = 267.0 \, z$ kN

Note: The weight of the water (P_w) is completely balanced by the upward water pressure at depth z (neglect friction between water and the interior wall of the tank). So, it need not be considered in the equilibrium equation of the free body of Figure 6.5b.

$$\text{Average normal stress on the tank wall } X\text{-section at } B - B : \sigma_B = \frac{P_B}{A_B} = 80z \text{ kPa}$$

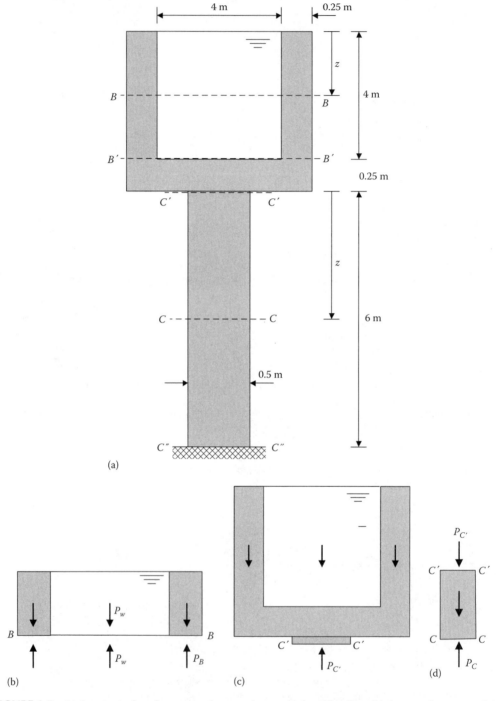

FIGURE 6.5 (a) A water tank and support column under gravitational loading. (b) A general segment of the tank with water. (c) Tank with water. (d) A general segment of the support column.

Axial Loading

The maximum normal stress in the tank X-section occurs at the bottom of the tank (i.e., at section $B' - B'$).

$$\sigma_{B'} = \sigma_B\big|_{z=4\,m} = 80 \times 4 \text{ kPa} = 0.32 \text{ MPa}$$

b. Make a virtual X-section $C' - C'$ in the column at the base of the tank, and consider the segment above it (Figure 6.5c).
$P_{C'}$ = compressive force on the column at $C' - C'$ = reaction force on the tank at $C' - C'$
= weight of the tank + weight of the water (from the equilibrium condition).

$$\text{Weight of the tank} = \left[\frac{\pi}{4}(4.5^2 - 4^2) \times 4 + \frac{\pi}{4} \times 4.5^2 \times 0.25\right] \times 80 \text{ kN} = 1386.2 \text{ kN}$$

$$\text{Weight of the water} = \frac{\pi}{4} \times 4^2 \times 4 \times 10 \text{ kN} = 502.7 \text{ kN}$$

$$\Rightarrow P_{C'} = 1386.2 + 502.7 = 1888.9 \text{ kN}$$

Area of X-section of the column

$$A_{C'} = \frac{\pi}{4} \times 0.5^2 \text{ m}^2$$

Normal stress in the column at $C' - C'$

$$\sigma_{C'} = \frac{P_{C'}}{A_{C'}} = \frac{1888.9}{(\pi/4) \times 0.5^2} \text{ kN/m}^2 = 9.62 \text{ MPa}$$

Make virtual cuts of the column at $C' - C'$ and $C-C$ at depth z below $C' - C'$, and consider the resulting column segment (Figure 6.5d).

$$\text{Weight of the column segment} = \frac{\pi}{4} \times 0.5^2 \times z \times 60 \text{ kN}$$

$$\left(\text{Note}: X\text{-section area}\left(A_C = A_{C'} = \frac{\pi}{4} \times 0.5^2 \text{ m}^2\right)\right)$$

$$\text{Equilibrium}: P_C = P_{C'} + \frac{\pi}{4} \times 0.5^2 \times z \times 80$$

where P_C is the compressive force in the column at $C-C$.

$$\Rightarrow P_C = 1888.9 + \frac{\pi}{4} \times 0.5^2 \times z \times 80 \text{ kN} = 1888.9 + 15.7z \text{ kN}$$

Average normal stress at $C-C$:

$$\sigma_C = \frac{P_C}{A_C} = \frac{1888.9 + (\pi/4) \times 0.5^2 \times z \times 80 \text{ kN}}{A_C} = \sigma_{C'} + 80z \text{ kPa}$$

$$\Rightarrow \sigma_C = 9.62 \times 10^3 + 80z \text{ kPa}$$

The maximum normal stress in the column X-section occurs at the base of the column $C''-C''$. We have

$$\sigma_{C''} = \sigma_C\big|_{z=6m} = 9.62 \times 10^3 + 80 \times 6 \text{ kPa} = 10.1 \text{ MPa}$$

c. Movement of the tank top
   δ = Shrinking of the tank height + shrinking of the column height
   (Neglect the shrinking of the tank base, which has a relatively small depth and larger area.)
   Apply Equation 6.4 to the two segments

$$\delta = \delta_{B'} + \delta_{C'C''} = \int_0^{z=4} \frac{\sigma_B}{E} dz + \int_0^{z=6} \frac{\sigma_C}{E} dz$$

$$= \frac{1}{E}\left[\int_0^4 80z \cdot dz + \int_0^6 (9.62 \times 10^3 + 80z) dz\right]$$

$$= \frac{1}{200 \times 10^6}\left[80 \times \frac{4^2}{2} + 9.62 \times 10^3 \times 6 + 80 \times \frac{6^2}{2}\right] \text{m}$$

(1/kPa)   (kPa·m)

$\Rightarrow \delta = 1.2$ mm

Primary Learning Objectives

1. Determination of the extension of an axial structure with several uniform segments of distributed loading
2. Choice of proper free-body diagrams
3. Handling of liquid weight and pressure in structural analysis

■ **End of Solution**

Example 6.4

A cylindrical punching tool of length $L = 10$ cm is used to punch holes in a metal block of thickness $h = 6$ cm. The maximum punching force $P_0 = 10$ kN.

a. Determine the minimum diameter d that is required for the punching tool so that the tool is safe with regard to yielding, with a factor of safety = 4. The yield strength of the tool material is $\sigma_Y = 250$ MPa. If the Young's modulus of the tool material is $E = 200$ GPa, what would be the maximum axial deformation in the tool of this diameter when operating at the maximum punching force?
b. When the tool completes a punching operation, suppose that it is in equilibrium against the restraining force of the punched hole, with the lower segment of the tool just clearing the thickness $h = 6$ cm of the block and is in tightly holding contact with the block (Figure 6.6a). Assume that this restraining force distribution (force per unit length) is uniform (i.e., constant) along the tool segment. A tool segment of length $L - h = 4$ cm remains just above the metal block, without any restraining force on it from the metal block. Determine the axial force distribution and the normal stress distribution in the tool under these conditions. What is the overall axial deflection in the tool in these conditions?

Axial Loading

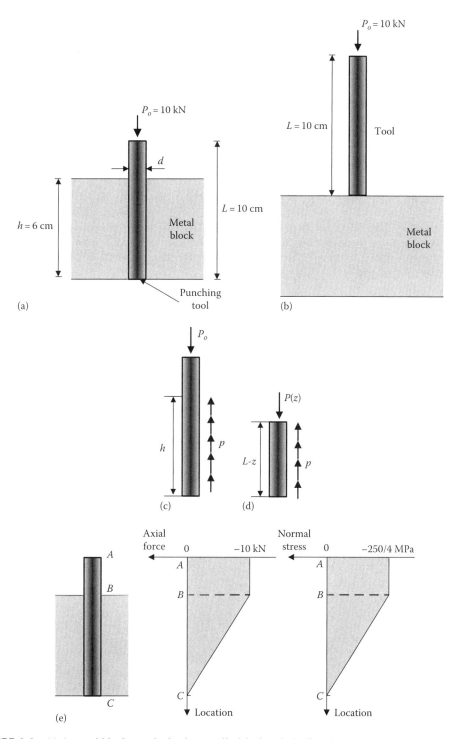

FIGURE 6.6 (a) A metal block punched using a cylindrical tool. (b) Tool in the beginning of the punching operation. (c) Free-body diagram of the punching tool. (d) FBD of a general segment of the tool inside the block. (e) Distributions of axial force and normal stress.

Solution

a. Maximum axial force in the tool = P_o = 10 kN (compressive)
Maximum allowable normal stress $\sigma_o = \sigma_Y/4$ (with a factor of safety = 4)

$$\rightarrow \frac{\sigma_Y}{4} = \frac{P_o}{(\pi/4)d^2} \rightarrow d = 4\sqrt{\frac{P_o}{\pi\sigma_Y}} \quad \text{(i)}$$

Substitute numerical values:

$$d = 4\sqrt{\frac{10 \text{ (kN)}}{\pi \times 250 \times 10^3 \text{ (kPa)}}} \text{ m} = 0.014 \text{ m} = 14.0 \text{ mm}$$

Maximum tool deflection occurs when the tool tip just hits the punched block. Then the axial force is constant at P_o along the tool axis, and the corresponding normal stress is also constant along the tool axis at $\sigma_o = P_o/((\pi/4)d^2)$. This should be clear from Figure 6.6b. For equilibrium of the tool, there has to be a reaction force equal to P_o from the metal block, on the tool tip, and there is no other external force along the length of the tool. Hence, Equation 6.5 applies:

$$\text{Tool deflection } \delta = \frac{\sigma_o}{E} \times L = \frac{\sigma_Y}{4E} \times L$$

$$= \frac{250 \text{ (MPa)}}{4 \times 200 \times 10^3 \text{ (MPa)}} \times 100 \text{ m} = 0.00313 \text{ mm (a shrinking)}$$

b. When the tool just completes a punching operation, its top segment of length $L - h$ will have a constant axial force P_o = 10 kN (see Figure 6.6a).

Note: This can be confirmed by making a virtual cut in the tool X-section at any location above the block and considering the equilibrium of the tool segment above the cut.

Let the restraining force per unit length from the punched hole, which is constant, be p. Consider the free-body diagram of the tool, as shown in Figure 6.6c.

$$\text{Equilibrium equation}: P_o - ph = 0 \rightarrow p = \frac{P_o}{h} \quad \text{(ii)}$$

Now make a virtual cut in the tool X-section at any general location (z) below the block and consider the equilibrium of the tool segment (length = $l - z$) below the cut, as shown in Figure 6.6d. Let the axial force at the cut X-section be $P(z)$.
Equilibrium equation: $P(z) - p \times (L - z) = 0$

$$\text{Substitute (ii)}: P(z) = P_o \frac{(L-z)}{h} \quad \text{(iii)}$$

It is seen from (iii) that the axial force (internal) of the tool segment will decrease linearly from P_o at the top to zero at the bottom (tool tip) within the punched block. This axial force distribution is shown in Figure 6.6e.

The corresponding stress distribution is proportional to the axial force distribution (because the X-section is uniform), decreasing for the maximum value of compressive stress, $\sigma_o = \sigma_Y/4$, to zero. This is given by

$$\sigma(z) = \sigma_o \frac{(h-z)}{h} = \frac{\sigma_Y}{4} \frac{(h-z)}{h} \text{ (compressive)} \quad \text{(iv)}$$

Axial Loading

The axial (normal) stress distribution is also shown in Figure 6.6e.

The corresponding tool deflection is obtained by applying Equation 6.5 to the top segment of the tool, and Equation 6.4 to the bottom segment:

$$\delta = \frac{\sigma_o}{E} \times (L-h) + \int_0^h \frac{\sigma(z)}{E} dz \text{ (a shrinking)}$$

Substitute (i) and (iv):

$$\delta = \frac{\sigma_Y}{4E}(L-h) + \int_0^h \frac{\sigma_Y}{4E} \frac{(h-z)}{h} dz = \frac{\sigma_Y}{4E}\left[(L-h) + \frac{1}{2h}(h-z)^2\right]_0^h = \frac{\sigma_Y}{4E}\left[L-h+\frac{h}{2}\right]$$

$$\rightarrow \delta = \frac{\sigma_Y}{4E}\left(L - \frac{h}{2}\right) \text{ (a shrinking)}$$

Substitute numerical values:

$$\delta = \frac{250 \text{ (MPa)}}{4 \times 200 \times 10^3 \text{ (MPa)}} \times \left(100 - \frac{60}{2}\right) \text{ mm} = 0.00219 \text{ mm (a shrinking)}$$

Primary Learning Objectives
1. Handling a uniform axial member with distributed loading on a segment of it
2. Determination of the axial force distribution (internal) of a member with distributed external loading
3. Determination of normal stress distribution of a member with distributed external loading
4. Determination of the extension of an axial member with distributed loading
5. Proper use of free-body diagrams

■ **End of Solution**

6.3.2 MULTIPLE SEGMENTS OF UNIFORM CROSS SECTION

Now we will consider the special case of a member consisting of several uniform segments. Suppose that each segment i has specific constant values of axial force P_i, length L_i, Young's modulus E_i, and area of cross section A_i.

Note: Here, we have allowed for different material properties in different segments, hence different E_i.

According to Equation 6.5, each segment will have an axial deflection of $\delta_i = P_i L_i / E_i A_i$.

Hence, the total extension of the composite member is

$$\delta = \sum_i \frac{P_i L_i}{E_i A_i} \tag{6.6}$$

Example 6.5

A shaft is made of two uniform segments of the same material, but different areas of cross section. It is attached to a fixed load-transmission point A and supported using thrust bearings at locations B and C, as shown in Figure 6.7a. An axial load P is applied at the free end D. The axial loads exerted by the thrust bearings are P_B and P_C, as shown.

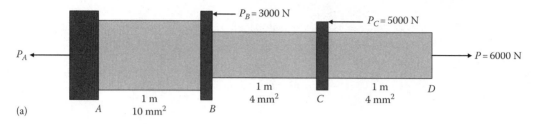

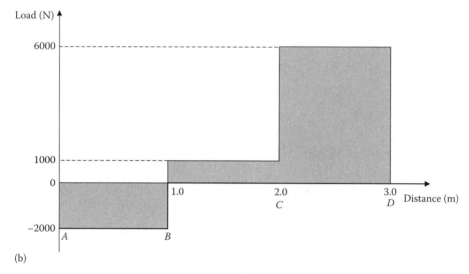

FIGURE 6.7 (a) A shaft with two uniform segments and two thrust bearings. (b) Axial force diagram of the shaft.

Given: Young's modulus of the material $E = 100$ GPa; forces $P = 6000$ N; $P_B = 3000$ N; $P_C = 5000$ N; lengths $AB = BC = CD = 1$ m; area of cross section in the uniform segment $AB = 10$ mm²; area of cross section in the uniform segment $BD = 4$ mm².

Sketch the axial force diagram for the shaft. Determine the axial movement of the shaft at location C with respect to location A, and the axial movement of the shaft at location D with respect to location A.

Solution

Let the transmitted axial force (tension) at A be P_A.

Equilibrium of the entire shaft gives: $-P_A - P_B - P_C + P = 0$

Hence, $P_A = P - P_B - P_C = 6000 - 3000 - 5000 = -2000$ N

The axial force diagram is sketched in Figure 6.7b.
Deflection at B with respect to A

$$\delta_{B/A} = \frac{P_A L_{AB}}{E A_{AB}} = \frac{-2000 \times 1.0}{100 \times 10^9 \times 10 \times 10^{-6}} \text{ m} = -2.0 \text{ mm (i.e., a compression, to the left)}$$

Similarly,

$$\delta_{C/B} = \frac{1000 \times 1.0}{100 \times 10^9 \times 4 \times 10^{-6}} \text{ m} = 2.5 \text{ mm (i.e., a tension, to the right)}$$

$$\delta_{D/C} = \frac{6000 \times 1.0}{100 \times 10^9 \times 4 \times 10^{-6}} \text{ m} = 15.0 \text{ mm (i.e., a tension, to the right)}$$

Hence,

$$\delta_{C/A} = \delta_{C/B} + \delta_{B/A} = 2.5 - 2.0 = 0.5 \text{ mm (i.e., a tension, to the right)}$$

$$\delta_{D/A} = \delta_{D/C} + \delta_{C/A} = 15.0 + 0.5 = 15.5 \text{ mm (i.e., a tension, to the right)}$$

Primary Learning Objectives
1. Determination of the force diagram of an axial member with multiple uniform segments
2. Determination of relative deflections of a multisegmented axial member
3. Establishment of the proper directions of movement at specific locations of an axial member of multiple segments with different loading conditions

■ **End of Solution**

6.4 PRINCIPLE OF SUPERPOSITION

The principle of superposition (PoS) is a powerful method that can be used to simplify the solution of engineering problems. It is valid only for linear systems. This principle may be expressed as follows: *When the cause–effect relationship is linear, the overall effect of the combined application of several causes is equal to the sum of the effects produced by applying the causes separately.* The PoS, in the context of Mechanics of Materials, is further explained later.

6.4.1 LINEAR ELASTIC SYSTEMS

In mechanics of materials, linearity means deflection is proportional to the applied load. As a result, the strain will be proportional to stress. In general, the load–deflection relation is nonlinear, as shown in Figure 6.8a. Typically, there will be a linear region in this relation, which corresponds to relatively smaller loads and deflections, as shown in the figure. The associated constant of proportionality is called the stiffness (k).

Similarly, as shown in Figure 6.8b, the stress–strain relation is also nonlinear in general (see Chapter 5). Again, there will be a linear region as shown (which exactly corresponds to the linear region of the load–deflection relation), where the constant of proportionality is called the modulus (Young's modulus E or the shear modulus G).

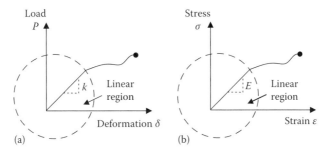

FIGURE 6.8 (a) Linear region of a load versus deflection relation; (b) linear region of a stress versus strain relation.

6.4.2 Load Reversal

A useful property that follows from linearity is the deflection reversal resulting from load reversal. This is illustrated in Figure 6.9. Suppose that an external load P is applied to a structure, as shown in Figure 6.9a. As a result, the internal load at some location B changes by ΔF. Also, a corresponding deflection δ results at B, as shown. Since P is a general load, its sign is arbitrary. Let us then replace P by $-P$ (in the same structure, with the same support conditions). If the structural deformation is in the linear region, the internal load at B will now change by $-\Delta F$, and also the resulting deflection at B will be $-\delta$. This is shown in Figure 6.9b. Since a negative sign means a change in direction, the result in Figure 6.9b is exactly the same as what is shown in Figure 6.9c.

In summary, for a linear structure, when an external load is reversed, the changes of the internal loads that are caused by the load also will reverse. Furthermore, the deflections caused by the load also will reverse.

6.4.3 Principle of Superposition

The PoS is illustrated in Figure 6.10. Suppose that a load P_1 applied to a structure causes a deflection δ_1 at some location A in the structure (Figure 6.10a). In the same structure with the same support conditions, suppose that an external load P_2 causes a deflection δ_2 at the same location A in the

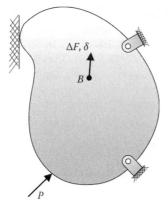

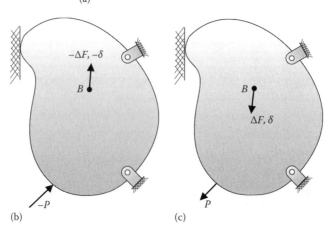

FIGURE 6.9 (a) Load and associated changes in internal loading and deflection; (b) sign change of the loading; (c) load reversal.

Axial Loading

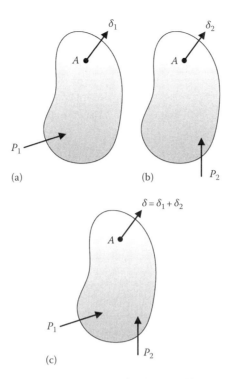

FIGURE 6.10 Principle of superposition: (a) a load–deflection condition; (b) another load–deflection condition; (c) combined load–deflection condition.

structure (Figure 6.10b). If the structure remains in the linear region of the load–deflection relation, according to the PoS, when both loads P_1 and P_2 are applied simultaneously, the deflection at A will be $\delta_1 + \delta_2$ (Figure 6.10c).

Example 6.6

As an application example of the PoS for loads, consider a uniform rod of length L and area of cross section A subjected to an axial tensile force P (Figure 6.11). We can show that PoS applies for the extension caused by this force, assuming that the rod remains in its linear elastic region.

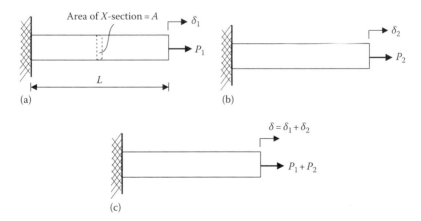

FIGURE 6.11 An example of the principle of superposition: (a) tensile force on a rod and resulting extension; (b) a different tensile force and extension; (c) combined tensile force and resulting extension.

According to Equation 6.5, the applied force P_1 and the resulting extension δ_1 are related through (Figure 6.11a)

$$\delta_1 = L \cdot \varepsilon_1 = L \cdot \frac{\sigma_1}{E} = L \cdot \frac{P_1}{EA} \qquad \text{(i)}$$

Similarly, the applied force P_2 and the resulting extension δ_2 in Figure 6.11b are related through

$$\delta_2 = L \cdot \frac{P_2}{EA} \qquad \text{(ii)}$$

Now suppose that the combined force P_1 and P_2 is applied to the rod, as shown in Figure 6.11c. As discussed earlier, the resulting extension δ is obtained using Equation 6.5. It is easy to show that this extension is equal to $\delta_1 + \delta_2$, as follows:

$$\delta = L \cdot \frac{(P_1 + P_2)}{EA} = \frac{P_1 L}{EA} + \frac{P_2 L}{EA} = \delta_1 + \delta_2 \qquad \text{(iii)}$$

Clearly, (iii) is equal to (i) + (ii). This verifies the PoS.

Note: Here, the PoS applies because the force–deflection relationship (6.5) is linear.

Primary Learning Objectives

1. Application of the PoS to an axial member
2. Verification of the PoS
3. Illustration that the PoS is applicable because the associated relationship is linear
4. The use of PoS to simplify the solution of a problem

■ **End of Solution**

Example 6.7

Now, we will consider an example where the PoS does not apply because the governing relationship is nonlinear. A uniform rod of length L, rectangular cross section of area A, and side dimension b is shown in Figure 6.12. Suppose that we are interested in the change in area of one side of the rod due to the applied force. We now show that PoS does not apply for this relation.

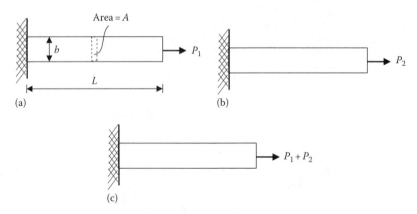

FIGURE 6.12 An example where PoS is not applicable: (a) application of one force; (b) application of another force; (c) application of the combined force.

Axial Loading

Given: E is Young's modulus; v is Poisson's ratio.

$$\text{Original area of a side of the rod}: A_s = L \times b \qquad \text{(i)}$$

On application of P_1 (Figure 6.12a), the new area of the side

$$A_s + \delta A_{s1} = \left(L + L \times \frac{P_1}{AE}\right) \times \left(b - bv\frac{P_1}{AE}\right) \qquad \text{(ii)}$$

Change in the side area, (ii) – (i)

$$\delta A_{s1} = \left(L + \frac{LP_1}{AE}\right)\left(b - bv\frac{P_1}{AE}\right) - Lb = \frac{Lb}{AE}(1-v)P_1 - \frac{Lbv}{(AE)^2}P_1^2 \qquad \text{(iii)}$$

Similarly, on application of P_2 (Figure 6.12b), the corresponding change in the side area

$$\delta A_{s2} = \frac{Lb}{AE}(1-v)P_2 - \frac{Lbv}{(AE)^2}P_2^2 \qquad \text{(iv)}$$

On application of the combined force $P_1 + P_2$ (Figure 6.12c), the corresponding change in the side area can be determined in the same manner. It is easy to show that this change in area is not equal to the sum of the previous two changes in area ((iii) and (iv)). Specifically,

$$\delta A_s = \frac{Lb}{AE}(1-v)(P_1 + P_2) - \frac{Lbv}{(AE)^2}(P_1 + P_2)^2 \neq \delta A_{s1} + \delta A_{s2}$$

Primary Learning Objectives

1. Consideration of nonlinear relationships of cause and effect
2. Illustration that the PoS is not applicable because the associated relationship is nonlinear

■ **End of Solution**

6.4.4 Summary of PoS

Summarizing, for linear systems, the PoS is applicable. In the present context, this principle may be applied as follows.

Suppose that a given set of external loads generates a certain stress distribution and a corresponding strain distribution in a structure; and another set of external loads, with the same restraints (supports, bearings, end conditions, etc.) on the structure as mentioned earlier, generates corresponding stress and strain distributions in the structure. Then, if both sets of external loads are applied simultaneously, the resulting stress and strain distributions will be equal to the sum of the individual distributions for the two separate cases of loading.

Note 1: A restraint may be replaced by a corresponding external load that exists at the restraint. Then, this restraint load may be treated as an external load in the application of the PoS to the structure, with the particular restraint considered to be absent.

Note 2: If the deformations are large (i.e., a case of *geometric nonlinearity*), the material will not remain in its linear elastic region (i.e., a case of *physical nonlinearity*), and the PoS does not hold. So, geometric nonlinearity and physical nonlinearity go hand in hand in problems of mechanics of materials.

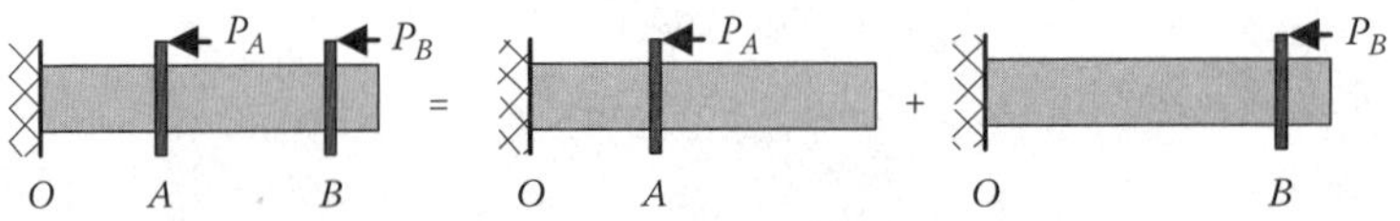

FIGURE 6.13 Representation of the PoS for load–deformation relationship.

Note 3: In general, if the relationship of cause and effect is not linear, then the PoS does not hold for that particular effect.

So, in order to apply the PoS for deformations (and strains), we require the following:

1. Deformation or strain is proportional to the applied load (physical linearity).
2. This also means that the applied load does not significantly change the geometry of the structure (geometric linearity).

This concept is graphically illustrated in Figure 6.13.

6.5 STATICALLY INDETERMINATE STRUCTURES

Structures whose unknown loads (at support points, etc.) and the corresponding stresses cannot be determined using the equations of equilibrium (principles of statics) alone are known as *statically indeterminate* structures. To solve these problems, "compatibility" in structural deformation has to be utilized in addition to the equations of equilibrium.

Compatibility practically means "structural integrity," which assures that joints are not broken, material is not ruptured or dislocated, supports and restraints are maintained, and so on, during structural deformation due to loading.

6.5.1 SOLUTION APPROACH

1. Write equations of equilibrium and establish that the number of unknowns is greater than the number of equations (i.e., the problem is statically indeterminate).
2. Select a suitable location (point) in the structure and write an equation for its deformation by considering the contribution to this deformation from various deformed members of the structure, assuring that the location remains integral (this guarantees *compatibility* or *structural integrity*). *Note*: If another unknown needs to be determined, another compatibility condition will be required.

Example 6.8

A metal sleeve is tightened using a bolt and a nut as shown in Figure 6.14a. Initially, the nut is snugly in contact with the sleeve, but without exerting any force. Subsequently, the nut is tightened through n turns. Determine

  a. The tension in the bolt
  b. The average normal stress and normal strain in the sleeve

The following parameters are given:
  L is the original (unstrained) length of the sleeve
  A_s is the solid area of cross-section of the sleeve
  E_s is the Young's modulus of the sleeve material
  A_b is the area of cross section of the bolt
  E_b is the Young's modulus of the bolt material
  δ is the pitch of the nut

Axial Loading

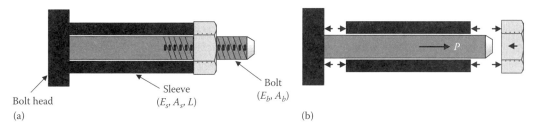

FIGURE 6.14 (a) A sleeve tightened using a bolt and a nut. (b) Free-body diagrams of a sleeve–bolt unit.

Note: Assume that the bolt head and the nut are rigid (i.e., they do not deform).
Compute the answers using the numerical values: $L=0.4$ m, $A_b=4$ mm², $A_s=8$ mm², $E_b=200$ GPa, $E_s=100$ GPa, $\delta=0.5$ mm, and $n=4$ turns.

Solution

a. For the equilibrium of the nut (see the free-body diagrams in Figure 6.14b), we require
Tensile force in the bolt = Compressive force in the sleeve = P (say)

$$\text{Extension in the bolt (from the bolt head up to the nut, using Equation 6.5)} = \frac{PL}{E_b A_b}$$

Also,

$$\text{Compression in the sleeve (using Equation 6.5)} = \frac{PL}{E_s A_s}$$

Movement of the nut = $n\delta$

The sum of the extension of the bolt and the compression of the sleeve is the clearance that has to be closed by turning the nut so that it will remain snugly in contact with the sleeve. Hence,

$$\text{Compatibility condition}: \frac{PL}{E_b A_b} + \frac{PL}{E_s A_s} = n\delta \qquad (i)$$

$$\rightarrow P = \frac{n\delta / L}{1/(E_b A_b) + 1/(E_s A_s)} \qquad (ii)$$

Substitute the given numerical values. We get

$$P = \frac{(0.5 \times 4 \times 10^3)/0.4}{\left[1/\left(200 \times 10^9 \times 4 \times 10^{-6}\right) + 1/\left(100 \times 10^9 \times 8 \times 10^{-6}\right)\right]} \text{N}$$

$$= 200 \times 10^3 \text{ N} = 2.0 \text{ kN}$$

b. Average normal stress in the sleeve:

$$\sigma_s = -\frac{P}{A_s} = -\frac{2.0 \times 10^3}{8 \times 10^{-6}} \text{ Pa} = -0.25 \text{ GPa (i.e., compressive)}$$

Average normal strain in the sleeve:

$$\varepsilon_s = \frac{\sigma_s}{E_s} = -\frac{0.25}{100} = -0.0025 \; \varepsilon = -0.25\%$$

Primary Learning Objectives
1. Solution of a statically indeterminate problem of an axial member
2. Proper use of the compatibility condition in deflection
3. Use of free-body diagrams

■ **End of Solution**

Example 6.9

A shaft is made of two uniform segments with parameters E_1, L_1, A_1 and E_2, L_2, A_2, where E_i is the Young's modulus, L_i is the length, and A_i is the area of cross section. One end of the shaft is rigidly fixed. There is a rigid wall with clearance δ_0 (under unloaded conditions) at the free end of the shaft. Also, there is a collar at the joint of the two segments on which an axial external force P is applied toward the free end of the shaft (Figure 6.15a).

a. Determine the limiting value of P, denoted by P_0, such that the system is statically determinate (i.e., when the clearance between the free end of the shaft and the wall just becomes zero, with zero reaction force at the wall).
b. Assuming that P exceeds the limiting value obtained in (a), determine the axial forces (internal) P_1 and P_2 in the two segments of the shaft (express them in terms of the given parameters P, E_i, L_i, A_i, and δ_0).
c. Given that $L_1 = L_2 = 1.0$ m, $A_1 = 5$ mm², $A_2 = 10$ mm², $E_1 = E_2 = 100$ GPa, and $\delta_0 = 3$ mm, compute the value of P_0 in part (a).

If $P = 2$ kN, compute the values of P_1 and P_2 in part (b).

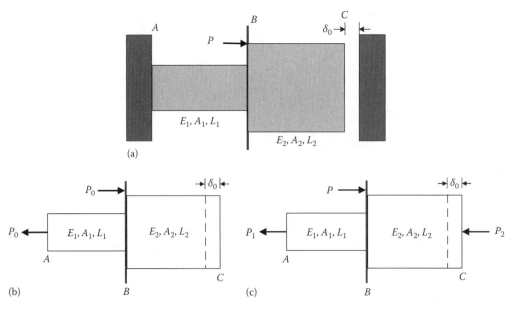

FIGURE 6.15 (a) Two-segmented shaft with a wall clearance at one end. (b) Free-body diagram in the absence of wall reaction. (c) Free-body diagram in the presence of wall reaction.

Axial Loading

Solution

a. Since in this case there is no reaction force from the wall, we have the following from the equilibrium conditions of the two segments of the shaft (see Figure 6.15b):
Axial force from A to B is P_0.
Axial force from B to C is zero.
From Equation 6.5, the extension in AB is $P_0 L_1/E_1 A_1$.
Also, the extension in BC is zero.

$$\rightarrow \frac{P_0 L_1}{E_1 A_1} = \delta_0$$

or

$$P_0 = \frac{E_1 A_1 \delta_0}{L_1}$$

b. **Method 1: Compatibility Method**

In the presence of wall reaction, the free-body diagram of the shaft unit is shown in Figure 6.15c.

$$\text{Shaft equilibrium}: -P_1 + P - P_2 = 0 \rightarrow P_1 + P_2 = P \tag{i}$$

Note: Two unknowns P_1 and P_2 and one equation $\Rightarrow$ statically indeterminate

Compatibility condition: (Extension in AB) − (Compression in BC) = δ_0

$$\rightarrow \frac{P_1 L_1}{E_1 A_1} - \frac{P_2 L_2}{E_2 A_2} = \delta_0 \tag{ii}$$

Substitute (i) in (ii) for P_2: $\dfrac{P_1 L_1}{E_1 A_1} - \dfrac{(P - P_1)L_2}{E_2 A_2} = \delta_0$

$$\rightarrow P_1 = \frac{\delta_0 + (PL_2/E_2 A_2)}{(L_1/E_1 A_1) + (L_2/E_2 A_2)} \tag{iii}$$

Substitute (i) in (ii) for P_1: $\dfrac{(P - P_2)L_1}{E_1 A_1} - \dfrac{P_2 L_2}{E_2 A_2} = \delta_0$

$$\rightarrow P_2 = \frac{(PL_1/E_1 A_1) - \delta_0}{(L_1/E_1 A_1) + (L_2/E_2 A_2)} \tag{iv}$$

Method 2: Superposition Method

Suppose that the wall is absent at the free end of the shaft. Hence, the internal axial load in BC is zero even when P exceeds P_0.

Deflection at the free end (C) due to P is $\delta_1 = \dfrac{PL_1}{E_1 A_1}$.

Now suppose that P is removed and a compressive axial force P_2 is applied at the free end (C). Consequently, there is a constant compressive axial force along the entire shaft ABC.

Deflection at the free end due to P_2 is $\delta_2 = -\left[\dfrac{P_2 L_1}{E_1 A_1} + \dfrac{P_2 L_2}{E_2 A_2}\right]$.

Note: The negative sign denotes a compression.

Combined deflection at the free end when both P and P_2 are present is $\delta_1 + \delta_2$. We must have (due to the deflection constraint at the wall) $\delta_1 + \delta_2 = \delta_0$.

$$\rightarrow \frac{PL_1}{E_1A_1} - \left[\frac{P_2L_1}{E_1A_1} + \frac{P_2L_2}{E_2A_2}\right] = \delta_0$$

$$\rightarrow \frac{(P-P_2)L_1}{E_1A_1} - \frac{P_2L_2}{E_2A_2} = \delta_0$$

Substitute (i) for P:

$$\rightarrow \frac{P_1L_1}{E_1A_1} - \frac{P_2L_2}{E_2A_2} = \delta_0$$

This result is identical to (ii).

c. Substitute numerical values: $\frac{P_0 \times 1.0}{100 \times 10^9 \times 5 \times 10^{-6}} = 3 \times 10^{-3} \rightarrow P_0 = 1.5 \times 10^3 \text{ N} = 1.5 \text{ kN}$
With $P = 2.0$ kN

$$\frac{PL_1}{E_1A_1} = \frac{2 \times 10^3 \times 1.0}{100 \times 10^9 \times 5 \times 10^{-6}} \text{ m} = 4 \times 10^{-3} \text{ m} = 4.0 \text{ mm}$$

$$\frac{PL_2}{E_2A_2} = \frac{2 \times 10^3 \times 1.0}{100 \times 10^9 \times 10 \times 10^{-6}} \text{ m} = 2 \times 10^{-3} \text{ m} = 2.0 \text{ mm}$$

Add: $\frac{L_1}{E_1A_1} + \frac{L_2}{E_2A_2} = \frac{1.0}{100 \times 10^9 \times 5 \times 10^{-6}} + \frac{1.0}{100 \times 10^9 \times 10 \times 10^{-6}}$ m/N $= 3.0 \times 10^{-6}$ m/N

Substitute in (iii) and (iv):

$$P_1 = \frac{(3+2) \times 10^{-3}}{3.0 \times 10^{-6}} \text{ N} = \frac{5}{3} \times 10^3 \text{ N} = \frac{5}{3} \text{ kN}$$

$$P_2 = \frac{(4-3) \times 10^{-3}}{3.0 \times 10^{-6}} \text{ N} = \frac{1}{3} \times 10^3 \text{ N} = \frac{1}{3} \text{ kN}$$

Check: $P = P_1 + P_2 = \frac{5}{3} + \frac{1}{3} = 2$ kN

Primary Learning Objectives
1. Solution of a statically indeterminate problem of an axial member
2. Proper use of the compatibility condition in deflection
3. Using the PoS as an alternative method to solve statically indeterminate problems

■ **End of Solution**

6.6 THERMAL EFFECTS

Deformations in a body due to thermal expansions (+ve) and contractions (−ve), which are caused by temperature changes, manifest as thermal strains. As discussed in Chapter 4, the thermal strain (which is a normal strain) in a member is given by

Axial Loading

$$\varepsilon_T = \alpha \Delta T \quad (6.7)$$

where
- α is the coefficient of thermal expansion (expansion in a line segment of unit length due to a temperature increase of 1°)
- ΔT is the temperature increase (°)

Hence, for a longitudinal member of length L, the extension due to a temperature increase of ΔT is given by

$$\delta_T = L\varepsilon_T = L\alpha\Delta T \quad (6.8)$$

The introduction of thermal strain into a problem is straightforward regardless of whether the problem is statically determinate or statically indeterminate. Specifically, we use
Total strain = Strain due to temperature change + Strain due to loading
The only difference in the solution process for a statically indeterminate structure is that, in addition to the equations of equilibrium, "compatibility conditions" must be used.
The following are useful points to remember.

1. For a given set of external loads, the internal loading can change as well due to temperature change. Hence, stress can change in various locations of the structure, as a result of temperature change.
2. The direction of the thermal strain depends on whether the temperature is increased (+ve) or decreased (−ve). This direction is independent of the direction of the strain due to external loading.

Example 6.10

In Example 6.8, suppose that there is a temperature rise of $\Delta T = 50°C$. Also,
α_s = coefficient of thermal expansion of the sleeve = $17 \times 10^{-6}/°C$
α_b = coefficient of thermal expansion of the bolt = $12 \times 10^{-6}/°C$
Determine the new tension in the bolt and the new average normal stress and normal strain in the sleeve.

Solution

In Figure 6.14b, force P is different now. Let us denote it by P^*.

$$\text{Total extension in the bolt} = \frac{P^* L}{E_b A_b} + L\alpha_b \Delta T$$

$$\text{Total compression in the sleeve} = \frac{P^* L}{E_s A_s} - L\alpha_s \Delta T$$

The compatibility condition (which is the same as mentioned earlier, except the values of deformation are different now) dictates that the nut is snugly in contact with the sleeve. In other words, the sum of the bolt extension and the sleeve compression must be taken up by the nut movement.

$$\frac{P^* L}{E_b A_b} + L\alpha_b \Delta T + \frac{P^* L}{E_s A_s} - L\alpha_s \Delta T = n\delta$$

$$\rightarrow \frac{P^* L}{E_b A_b} + \frac{P^* L}{E_s A_s} = n\delta + (\alpha_s - \alpha_s)L.\Delta T \quad (i)^*$$

Note: This is essentially the same as Equation (i) of Example 6.8, except that the RHS has been modified by the thermal term. Hence, P^* is expressed simply by inspection of (ii) in Example 6.8. Specifically, replace $n\delta$ in (ii) of Example 6.8 by $n\delta + (\alpha_s - \alpha_b)L \cdot \Delta T$.

$$\rightarrow P^* = \frac{[(n\delta/L) + (\alpha_s - \alpha_b)\Delta T]}{1/(E_b A_b) + 1/(E_s A_s)} \quad \text{(ii)*}$$

Substitute numerical values:

$$P^* = \frac{(0.5 \times 4 \times 10^{-3})/0.4 + (17 - 12) \times 10^{-6} \times 50}{1/(200 \times 10^9 \times 4 \times 10^{-6}) + 1/(100 \times 10^9 \times 8 \times 10^{-6})} \text{ N}$$

$$= \frac{(5 + 0.25) \times 10^{-3}}{\left((2/0.8) \times 10^{-6}\right)} \text{ N}$$

$$= 2.1 \times 10^3 \text{ N} = 2.1 \text{ kN}$$

Average normal stress in the sleeve: $\sigma_s^* = -\frac{P^*}{A_s} = -\frac{2.1 \times 10^3}{8 \times 10^{-6}}$ Pa $= -0.2625$ GPa

Average normal strain in the sleeve

$$= -\frac{\text{Total compression}}{\text{Original length}} = -\frac{((P^* L)/(E_s A_s)) - L\alpha_s \Delta T}{L} = -\frac{P^*}{E_s A_s} + \alpha_s \Delta T = -\frac{\sigma_s^*}{E_s} + \alpha_s \Delta T$$

$$= -\frac{0.2625}{100} + 17 \times 10^{-6} \times 50 = -0.001775 \text{ m/m} = -0.18\%$$

Primary Learning Objectives
1. Introduction of thermal strain into statically indeterminate problems
2. Proper use of the compatibility condition in statically indeterminate problems with temperature change
3. Simplification of the solution process in view of the similarity to the expressions in the solution of the nonthermal case
4. Illustration that both stress and strain may change due to a temperature change (*Note*: If the member is not restrained and is free to expand or contract, the stress will not change due to a temperature change)

■ **End of Solution**

Example 6.11

A beam of length 12 m is hinged at one end and supported by a cylindrical bronze post of diameter 5 cm at the other end, on rollers (see Figure 6.16a). Initially, the beam is maintained horizontally (*Note*: Weight of the beam is not neglected), and the external vertical force on the beam, $P = 0$.

   a. If the temperature increases by $\Delta T = 40°C$, determine the vertical force P that should be applied at the midspan of the beam such that the beam would remain horizontal.
   b. Also, determine the associated change in diameter of the post.

Axial Loading

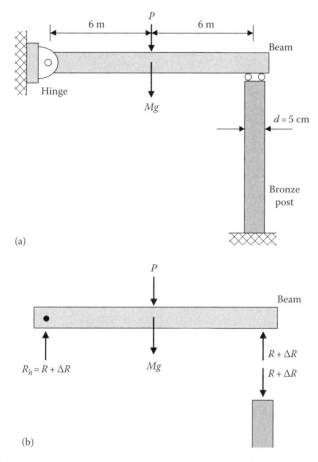

FIGURE 6.16 (a) A beam horizontally supported by a post. (b) Free-body diagram of the beam.

The following data are given:
  Coefficient of thermal expansion of the post $\alpha = 25 \times 10^{-6}/°C$
  Young's modulus of the post $E = 100$ GPa
  Shear modulus of the post $G = 38$ GPa

Solution

a. Thermal strain in the post due to temperature change:

$$\varepsilon_T = \alpha \cdot \Delta T = 25 \times 10^{-6} \times 40 \text{ m/m} = 1.0 \times 10^{-3} \text{ m/m}$$

In order to keep the beam horizontal, the change in strain ε_σ in the beam, due to the application of P, must be such that $\varepsilon_\sigma + \varepsilon_T = 0$.
  This is the *compatibility condition*.

$$\Rightarrow \varepsilon_\sigma = -\varepsilon_T = -1.0 \times 10^{-3} \text{ m/m}$$

The corresponding change in stress in the post

$$\sigma = E\varepsilon_\sigma = -100 \times 1.0 \times 10^{-3} \text{ GPa} = -100 \text{ MPa}$$

Change in compressive force in the post

$$\Delta R = -\sigma \times \frac{\pi}{4} d^2 = 100 \times 10^6 \times \frac{\pi}{4} \times 5^2 \times 10^{-4} \text{ N} = 19.6 \times 10^4 \text{ N} = 196.0 \text{ kN}$$

From Figure 6.16b it is clear that originally, the weight of the beam (Mg) is supported by equal reactions R at its two ends (with $R = (1/2)Mg$).

The increase in reaction ΔR at each end (due to the temperature rise) will be completely balanced by P. (*Note*: Another way of getting this result is through the moment balance of the beam about the hinge.)

$$\text{Hence, } \Delta R = \frac{1}{2} P \rightarrow$$

$P = 2 \times \Delta R = 2 \times 196.0 \text{ kN} = 392.0 \text{ kN}$

b. From Chapter 5, we have $E = 2(1 + \nu)G$.

$$\rightarrow \nu = \frac{E}{2G} - 1 = \frac{100}{2 \times 38} - 1 = 0.316$$

The lateral strain of the post due to the longitudinal stress

$$\varepsilon_{lat\sigma} = -\nu \varepsilon_\sigma = -0.316 \times (-1.0 \times 10^{-3}) \text{ m/m} = 0.316 \times 10^{-3} \text{ m/m}$$

The lateral thermal strain (which should be equal to the longitudinal thermal strain, in view of the homogeneous and isotropic material)

$$\varepsilon_{latT} = \varepsilon_T = 1.0 \times 10^{-3} \text{ m/m}$$

The overall lateral strain

$$\varepsilon_{lat} = \varepsilon_{lat\sigma} + \varepsilon_{latT} = (0.316 + 1.0) \times 10^{-3} \text{ m/m} = 1.316 \times 10^{-3} \text{ m/m}$$

Change in diameter of the post

$$\Delta d = d \times \varepsilon_{lat} = 5.0 \times 1.316 \times 10^{-3} \text{ cm} = 0.066 \text{ mm (an increase, due to the +ve strain)}$$

Primary Learning Objectives
1. Solution of problems involving thermal strain
2. Proper use of the compatibility condition (in thermal problems)
3. Handling of lateral strain (combined effect, due to loading and temperature change) in thermal problems

■ **End of Solution**

6.6.1 Principle of Superposition Applied to Thermal Problems

The PoS may be applied to thermal problems as well. Specifically, as long as the cause–effect relationship is linear, the PoS may be applied to determine the overall effect due to a combination of causes. In particular, if the thermal strain is linearly related to the temperature change (which is valid for common materials and moderate temperature changes), the PoS is applicable. This is illustrated in Figure 6.17.

Axial Loading

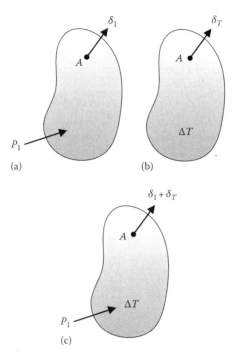

FIGURE 6.17 Application of the PoS in thermal problems: (a) deformation due to a load; (b) deformation due to a temperature change; (c) overall deformation due to the load and temperature change.

Suppose that in Figure 6.17a, the deflection δ is linearly related to the load P. Also, in Figure 6.17b, suppose that the thermal expansion δ_T is linearly related to the temperature rise ΔT. Then, when load and the temperature rise are both simultaneously applied to the system, the resulting deformation will be $\delta_1 + \delta_T$, as shown in Figure 6.17c.

Example 6.12

Consider a uniform rod of length L and area of cross section A, as shown in Figure 6.18. If an axial force P is applied to it, according to Equation 6.5, the corresponding extension is given by (Figure 6.18a)

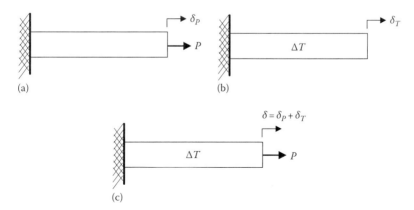

FIGURE 6.18 Principle of superposition for loading and temperature change: (a) extension due to an axial force; (b) extension due to a temperature rise; (c) extension due to combined force and temperature rise.

$$\delta_P = \frac{PL}{EA} \qquad \text{(i)}$$

Next, if the temperature of the rod is increased by ΔT, according to Equation 6.8, the corresponding extension is given by (Figure 6.18b)

$$\delta_T = \alpha \cdot \Delta T \cdot L \qquad \text{(ii)}$$

where α is the coefficient of thermal expansion.

Then, if the axial force P and the temperature rise ΔT are both applied simultaneously to the rod, the corresponding extension is given by the sum of (i) and (ii) (Figure 6.18c) as

$$\delta = \frac{P_1 L}{EA} + \alpha \cdot \Delta T \cdot L = \delta_P + \delta_T \qquad \text{(iii)}$$

Primary Learning Objectives

1. Application of the PoS to a thermal problem
2. Verification of the PoS using an axial member
3. Illustration that the PoS is applicable because the associated relationship is linear

■ **End of Solution**

Example 6.13

Now let us consider an example with thermal effects, where the PoS does not apply because the governing relationship is nonlinear. A uniform rod of length L, rectangular cross section, and side dimension b is shown in Figure 6.19. We will determine the change in area of one side of the rod due to an applied axial force and a temperature rise.

On application of an axial force P_1 (Figure 6.19a), the change in area of a side is given by (as in Example 6.7)

$$\delta A_{sP} = \frac{Lb}{AE}(1-v)P - \frac{Lbv}{(AE)^2}P^2 \qquad \text{(i)}$$

where v is the Poisson's ratio

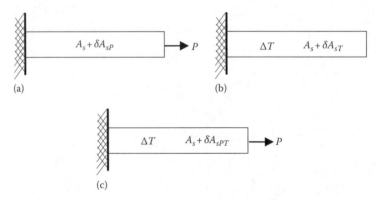

FIGURE 6.19 An example where PoS is not applicable: (a) application of a force; (b) application of a temperature rise; (c) application of the combined force and temperature rise.

Next, for a temperature change ΔT only, the new area of the side is (Figure 6.19b)

$$A_s + \delta A_{sT} = (L + \alpha \cdot \Delta T \cdot L)(b + \alpha \cdot \Delta T \cdot b) = Lb(1 + \alpha \cdot \Delta T)^2 = Lb(1 + 2\alpha \cdot \Delta T + \alpha^2 \Delta T^2)$$

Hence, the change in area is $\delta A_{sT} = Lb(2\alpha \Delta T + \alpha^2 \cdot \Delta T^2)$. (ii)

On application of the combined load P and temperature rise ΔT (Figure 6.19c), the overall area of the side is

$$A_s + \delta A_{sPT} = \left(L + \frac{LP}{AE} + \alpha \cdot \Delta T \cdot L\right)\left(b - \frac{b\nu P}{AE} + \alpha \cdot \Delta T \cdot b\right) = Lb\left(1 + \frac{LP}{AE} + \alpha \cdot \Delta T\right)\left(1 - \frac{\nu P}{AE} + \alpha \cdot \Delta T\right)$$

It is easy to show that the corresponding overall change in area is not equal to the sum of the previous two changes in area ((i) and (ii)). Specifically,

$$\delta A_{sPT} = \frac{Lb}{AE}(1-\nu)P - \frac{Lb\nu}{(AE)^2}P^2 + Lb\left(2\alpha\Delta T + \alpha^2 \cdot \Delta T^2\right) + \frac{Lb\alpha(1-\nu)}{AE}P\Delta T \neq \delta A_{sP} + \delta A_{sT}$$

Primary Learning Objectives
1. A thermal problem where PoS is not applicable
2. Consideration of nonlinear relationships of temperature change and its effect (area change)
3. Illustration that the PoS is not applicable because the associated relationship is nonlinear

■ **End of Solution**

6.7 STRESS CONCENTRATIONS

In Section 6.2, under Saint-Venant's principle, it was discussed how concentrated loads would generate sharp increases in stress in the localities of those concentrated loads. Stress concentrations will result as well at locations of sharp change in shape (i.e., nonsmooth locations) and of discontinuities (e.g., notches and ruptures) in a member. Since yielding and failure of a member are caused by a peak stress rather than an average stress, it is important to consider the degree of stress concentration when designing mechanical members and structures. In this section, we will consider stress concentration due to discontinuity or sharp changes in shape in a member.

6.7.1 NATURE OF STRESS CONCENTRATION

The relative degree of stress concentration of a member depends on the nature of its change in shape rather than the material properties. However, stress concentration is an important design consideration, particularly for *brittle material* because a brittle element can quickly fail soon after exceeding its proportional limit (see Chapter 5). In a *ductile material*, on the other hand, local yielding due to stress concentration will not cause failure in general because strain hardening (see Chapter 5) caused by yielding will in fact strengthen the material, albeit with a permanent deformation. Nevertheless, stress concentration can accelerate *fatigue failure* (Chapter 5) in ductile material as well.

Two examples of stress concentration of an axial member due to sharp change in shape are shown in Figure 6.20. Consider a uniform bar of homogeneous and isotropic material on which a square grid has been marked for the purpose of illustrating its deformation (Figure 6.20a). If two equal and opposite uniform axial loads are applied at its ends, it will extend uniformly and attain an equilibrium state, as shown in Figure 6.20b. The normal stress distribution at any cross section of the rod will be uniform in view of the uniform loading and homogeneous material. Next, suppose

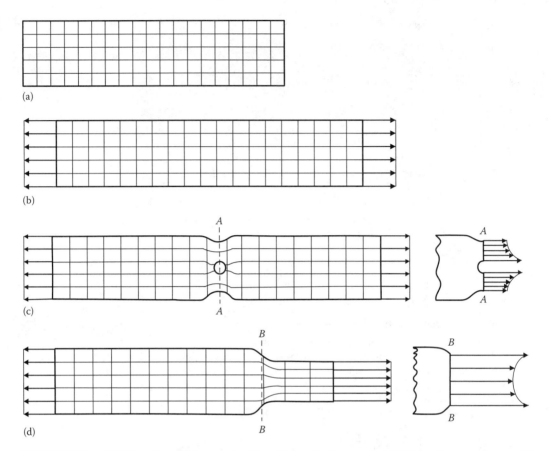

FIGURE 6.20 (a) Uniform bar with a square grid marked on it; (b) extension of the bar due to uniform axial loading; (c) deformation and stress distribution of a bar with a hole; (d) deformation and stress distribution of a two-segmented bar with shoulder fillets.

that a small hole is drilled in the bar. With uniform axial loading as discussed earlier, the stress distribution will not be uniform now, specifically in the vicinity of the hole (Figure 6.20c). For one thing, the stress has to increase to support the same load because the area of X-section has reduced due to the hole. Furthermore, due to sharp change in shape caused by the hole, the stress will peak near the edge of the hole. This is shown by the stress distribution on the cross section A–A through the hole, in Figure 6.20c. The deformation (and strain) will also be irregular in the vicinity of the hole, as shown by the shape of the deformed grid.

The second example of stress concentration is shown in Figure 6.20d. Here, the bar has two uniform segments of different areas of cross section, with shoulder fillets in the transition region of the two segments (*Note*: Shoulder fillets, in fact, will help reduce the level of stress concentration that would otherwise be caused by the sharp corners between the two segments of the bar). Since the same internal axial load exists in both segments (which is the condition of equilibrium), the average stress will be larger in the segment of smaller cross section. In addition, stress concentrations will occur in the transition region of the two segments, as shown by the stress distribution of the cross section B–B in this region (Figure 6.20d).

6.7.2 Stress Concentration Factor

For a given load, the largest average stress occurs at the location of the smallest cross-sectional area. It is in the same location that stress concentrations will likely occur (because of the change in shape

Axial Loading

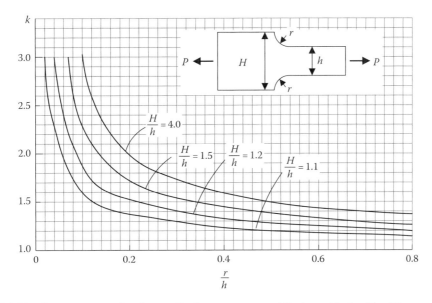

FIGURE 6.21 Stress concentration factors (k) of a two-segmented flat bar with shoulder fillets.

associated with the reduction in area). It follows that the level of stress concentration is particularly important in such locations. The stress concentration factor is the most common measure of the severity of stress concentration. Stress concentration factor k is defined as

$$k = \frac{\sigma_{max}}{\sigma_{av}} \quad (6.9)$$

where
σ_{max} is the maximum stress at the smallest X-section
σ_{av} is the average stress at the smallest X-section

The stress concentration factor depends on the shape of the member rather than the material properties. It may be determined analytically (from the principles of theory of elasticity or continuum mechanics), computationally (using the finite element method), or experimentally (by determining the local stresses, say by measuring the local strains). It is assumed that the material is elastic (i.e., proportional limit is not exceeded) and the loading is static.

Curves and tables of stress concentration factors for different shapes of members are commonly found in handbooks of machine design and structural design. Figure 6.21 presents curves of stress concentration factor k for a two-segmented uniform flat bar with shoulder fillets. Figure 6.22 presents curves of stress concentration factor k for a uniform flat bar with a circular hole, which is centrally located.

Example 6.14

A stepped flat bar with shoulder fillet is made of cast iron. A hole is drilled in the middle of the wider segment of the bar, as shown in Figure 6.23. The following parameters are given:
   Thickness of the bar: $t = 10.0$ mm
   Width of the wider segment of the bar: $H = 75.0$ mm
   Width of the narrower segment of the bar: $h = 50.0$ mm

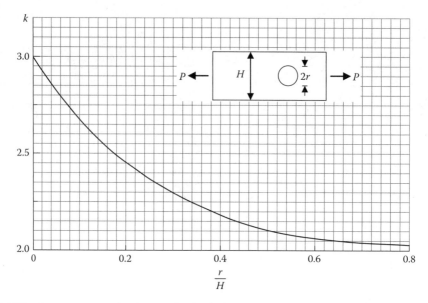

FIGURE 6.22 Stress concentration factors (k) of a flat bar with central circular hole.

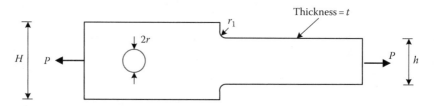

FIGURE 6.23 A cast iron bar under tensile loading with a hole and shoulder fillets.

Radius of the shoulder fillet: $r_1 = 5.0$ mm
Radius of the hole: $r = 7.5$ mm
Ultimate strength of cast iron: $\sigma_u = 170$ MPa (in tension)
Estimate the maximum tensile force P_{max} that may be supported by the bar.

Note: Assume that proportional stress of cast iron is close to the ultimate stress.

Solution

$$\text{For fillet}: \frac{H}{h} = \frac{75}{50} = 1.5 \quad \text{and} \quad \frac{r_1}{h} = \frac{5.0}{50.0} = 0.1$$

From the stress concentration data in Figure 6.21, stress concentration factor $k_{fillet} = 2.25$

For the narrower segment, average tensile stress $\sigma_{av} = \dfrac{P}{h \times t} = \dfrac{P \text{ (N)}}{50 \times 10 \text{ (mm}^2\text{)}} = \dfrac{P}{500}$ MPa

Maximum stress (at fillet) $\sigma_{max} = k_{fillet} \times \sigma_{av} = 2.25 \times \dfrac{P}{500}$ MPa ≤ 170 MPa

$$\to P \leq \frac{170 \times 500}{2.25} \text{ N} = 37.8 \times 10^3 \text{ N}$$

Axial Loading

$$\text{For hole}: \frac{r}{H} = \frac{7.5}{75} = 0.1$$

From the stress concentration data in Figure 6.22, stress concentration factor $k_{hole} = 2.67$
Average tensile stress in the segment that has the hole

$$\sigma_{av} = \frac{P}{(H-2r) \times t} = \frac{P \text{ (N)}}{(75 - 2 \times 7.5) \times 10 \text{ (mm}^2\text{)}} = \frac{P}{600} \text{ MPa}$$

$$\text{Maximum stress (at hole)}, \sigma_{max} = k_{hole} \times \sigma_{av} = 2.67 \times \frac{P}{600} \text{ MPa} \leq 170 \text{ MPa}$$

$$\rightarrow P \leq \frac{170 \times 600}{2.67} \text{ N} = 38.2 \times 10^3 \text{ N}$$

Taking the smaller of the two limiting values

$$P_{max} = 37.8 \text{ kN}$$

Primary Learning Objectives
1. Importance of stress concentration
2. Determination of the stress concentration factor of a uniform flat bar with a centrally located hole
3. Determination of the stress concentration factor of a stepped flat bar with shoulder fillets
4. Use of the stress concentration factor in component design

■ **End of Solution**

6.7.3 Residual Stresses

Consider a structure that is in the *elastic* stress–strain range. When the external loading is removed, the structure will return to its original unloaded state. At locations where the stress was originally zero (i.e., internal loading was zero), the stress will be zero again after removal of the external loading. This is true whether the structure is statically determinate or statically indeterminate.

Now consider a *statically determinate* structure. Suppose that its external loading is increased until there is *yielding* (i.e., stress exceeds the *proportional limit*, and some *plastic deformation* is produced) at some location of the structure. Then, when the external loading is removed, there will be some permanent deformation (strain) at the location that underwent yielding. However, if the stress was zero originally at that location (i.e., internal loading was zero), then the stress will return to zero when the external loading is removed, after yielding. This is because, in a statically determinate structure, internal loading will depend on statics alone (i.e., external loading alone).

This is not generally true for *statically indeterminate* structures. Even a location that originally had zero stress can have a nonzero stress once the loading is removed after yielding. This is because, in a statically indeterminate structure, the stress (and internal loading) depends not only on the external loading (i.e., statics) but also on deformation (specifically, the *compatibility condition*). So, in a statically indeterminate structure, it is possible to have a residual stress, due to the residual deformation, once the external loading is removed.

SUMMARY SHEET

Basic types of loading: Axial, shear, torsional, and bending

Statically equivalent loads: Their average (resultant) values are the same and the resultant load acts at the same point (centroid of the load distribution) in the same direction. They produce the same support reactions.

Saint-Venant's principle: In a body with the same supports (restraints), statically equivalent externals loads produce the same stress and strain conditions at sufficiently farther locations from the external load.

Minimum distance from the load for observing Saint-Venant's property = largest X-sectional dimension of the cross section in the loaded region.

Normal stress: $\sigma(x) = P(x)/A(x)$ (tensile is positive)

Normal strain: $\varepsilon(x) = \sigma(x)/E = P(x)/EA(x)$ (tensile is positive)

P is the axial force (tensile is positive); A is the area of X-section; E is the Young's modulus

Extension: $\delta = \int_0^L \varepsilon(x) dx = \int_0^L (\sigma(x)/E) dx = \int_0^L (P(x)/EA(x)) dx$

Extension: $\delta = PL/EA$ when A, E, and P are constant

Extension: $\delta = \sum_i P_i L_i / E_i A_i$ for multiple uniform segments i

Axial force diagram: Variation of the "internal" axial force of a member along the axial length

Linear elastic systems: Deflection $\propto$ Applied load; proportionality constant = stiffness

Strain $\propto$ stress; proportionality constant = modulus of elasticity

Principle of superposition (PoS): If load $P_1 \to$ deflection δ_1 and load $P_2 \to$ deflection δ_2 at the same location, then loads $\alpha_1 P_1$ and $\alpha_2 P_2$ simultaneously $\to$ deflection $\alpha_1 \delta_1 + \alpha_2 \delta_2$ (for arbitrary constants α_1 and α_2)

Note 1: When applying PoS, replace a restraint by the corresponding external load.

Note 2: If the relationship of cause and effect is not linear, then PoS does not hold.

Statically indeterminate structures: Cannot be solved using equations of equilibrium (principles of statics) alone. Conditions of deformation compatibility are needed for the solution.

Compatibility conditions: Determine the deformations (separately) of two or more parts at a common location; equate them (i.e., location did not rupture; structure remained integral).

Thermal strain $\varepsilon_T = \alpha \Delta T$: Thermal deformation per unit length (expansion is +ve; contraction is −ve); α = coefficient of thermal expansion (i.e., expansion in a line segment of unit length due to a temperature increase of 1 degree); ΔT = temperature increase (in degrees).

Thermal expansion: $\delta_T = L\varepsilon_T = L\alpha \Delta T$ (L = member length).

Stress concentration: An increase in local stress due to sharp change in shape.

Stress concentration factor: $k = \sigma_{max}/\sigma_{av}$; σ_{max} is the maximum stress, σ_{av} is the average stress at smallest X-section

Note: Depends on shape not material properties; assumes elastic behavior; more pertinent for brittle material (which fail under almost elastic conditions), or in fatigue failure of both ductile and brittle material.

Residual stresses: Occur in statically indeterminate structures, when external loading is removed after causing yielding (permanent deformation).

PROBLEMS

6.1 A block of mass M is suspended on a rigid end cap A at the top of a tube using a uniform rod (Figure P6.1). The tube rests vertically on a rigid horizontal platform. The following parameters are given.

Length of the tube $= L_t$
External diameter of the tube $= d_e$
Internal diameter of the tube $= d_i$
Young's modulus of the tube $= E_t$
Length of the rod $= L$
Diameter of the rod $= d$
Young's *modulus* of the rod $= E$

Determine the downward movement of the end cap at A, and the downward movement of the block (B) due to the weight of the suspended block.

If $M = 1000$ kg; $L = 1.0$ m; $L_t = 0.5$ m; $d = 1$ cm; $d_e = 2$ cm; $d_i = 1.5$ cm; $E = 200$ GPa; and $E_t = 100$ GPa, compute these deflections.

6.2 A vertical load $P = 1000$ N is carried at B by two uniform rods AB and BC of identical material and X-section, which are pin-jointed at B and also pin jointed to a rigid horizontal ceiling at A and C (Figure P6.2). $AC = 1$ m. Before the load is applied, $AB = 0.8$ m and $BC = 0.6$ m. Area of X-section of each rod is 4 mm² and the Young's modulus of the material is $E = 100$ GPa.

Determine the vertical movement δ_B at B due to the applied load.

6.3 A peg-shaped steel member hangs from a fixed rigid ceiling at its base (Figure P6.3). Its dimensions are: height $h = 2$ m; base $b = 1$ m; and thickness $t = 0.1$ m. Given the material properties: specific weight $\rho g = 80$ kN/m³, and Young's modulus $E = 200$ GPa, determine the deflection at the pointed end of the member due to its own weight.

6.4 A cylindrical rod of length $L = 20$ cm and diameter $d = 2$ cm, made of PVC (with Young's modulus $E = 2.5$ GPa), is pushed into a hole in a metal block (Figure P6.4) using a hydraulic

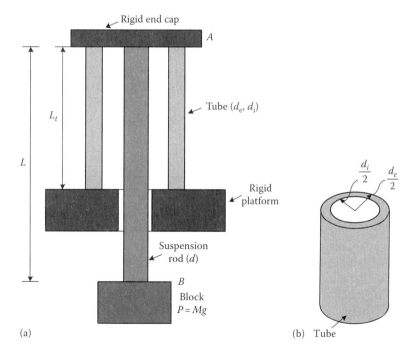

FIGURE P6.1 (a) A block suspended from a rod and supported on a tube; (b) dimensions of the tube.

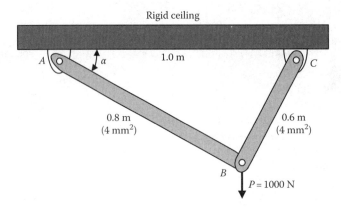

FIGURE P6.2 A load carried by a frame hung from a rigid ceiling.

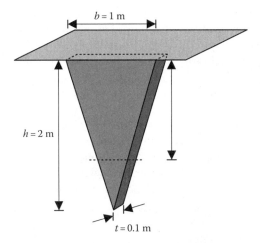

FIGURE P6.3 A heavy metal peg hanging from a rigid ceiling.

press, which applies a force of $P_0 = 1$ kN. The bottom end of the rod is exposed (i.e., there is no reaction there). The hole exerts a distributed resistance force on the rod according to the relation $p = k(L - z)$, where the top of the hole corresponds to $z = 0$ and the bottom of the hole corresponds to $z = L$. Here, p is a force per unit length.

a. Determine an expression for the normal stress $\sigma(z)$ on the X-section of the rod at a general depth z, in terms of P_0, L, and d.
b. What is the overall average contraction (i.e., reduction in length) of the rod?

Note: Neglect the effects of shear stress and shear strain.

6.5 A rigid beam ABC of length $2L$ is hinged to a fixed rigid wall at end A and horizontally supported using a vertical hanger rod hinged at the other end C. The hanger rod DC is uniform with area of X-section $A = 100$ mm², and length $l = 3$ m. The end D of the hanger rod is hinged to a fixed rigid ceiling (see Figure P6.5a). A vertical downward load $P = 70$ kN is applied at the midpoint B of the beam ($AB = L = BC$). The material of the hanger rod obeys the stress–strain curve shown in Figure P6.5b.

Determine
a. The normal (tensile) stress in the hanger rod
b. The downward movement at the point of load application (B)

Note 1: Assume that the beam is light (or its weight is included in P).

Axial Loading

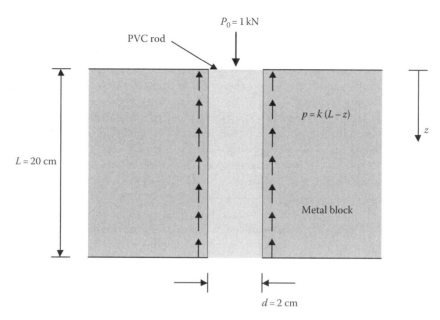

FIGURE P6.4 A PVC rod driven into a hole in a metal block using a hydraulic press.

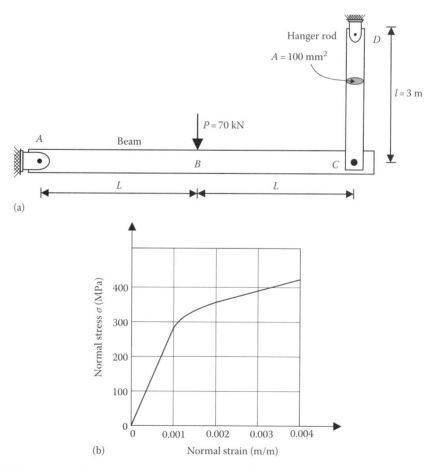

FIGURE P6.5 (a) A rigid beam supported by a hanger rod; (b) stress versus strain curve of the hanger rod.

Note 2: Assume that the movements due to P are small compared to the dimensions of the beam and the rod. (Hence, the circular arc of movement of the beam at C may be assumed to correspond to the extension of the rod DC. Similarly, the circular arc of movement of the beam at B may be assumed to be approximately vertical.)

Note 3: Under the present loading condition, the hanger rod may not be in the linear region of the stress–strain curve.

6.6 A rigid beam AB (assumed to be light) of length $L = 3$ m is maintained horizontally by using vertical hanger rods at its ends A and B, which have smooth pin joints (Figure P6.6). The hanger rods are identical in size (i.e., same length and same X-sectional area) but are made of different materials. Specifically, the hanger rod at A is made of steel (Young's modulus $E_s = 200$ GPa) and the hanger rod at B is made of bronze (Young's modulus $E_b = 100$ GPa). A vertical downward load P is applied at C, which is at a distance L_1 from A, on the beam. Determine its location (i.e., length L_1) if the beam remains horizontal after application of the load.

6.7 A brass tube of length 1 m, internal diameter 100 mm, and external diameter 150 mm is fixed horizontally to a rigid wall. A solid steel shaft of length 1 m and diameter 100 m is pressed against the brass tube using an axial force $P = 1000$ kN at the end and another axial force $P_b = 500$ kN at the rigid base of the steel shaft (see Figure P6.7).

Determine
  a. The axial force diagram for the entire system
  b. The normal stress distribution along the axis of the system
  c. The axial movement at the free end of the steel shaft due to the combined loading in the shaft and the tube

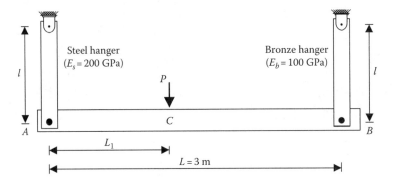

FIGURE P6.6 A beam supported horizontally by a steel hanger and a bronze hanger.

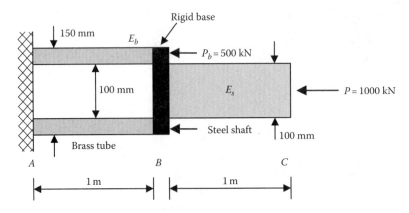

FIGURE P6.7 Steel shaft pressed against a brass tube.

Axial Loading 221

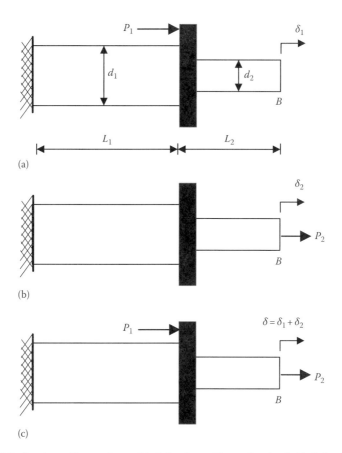

FIGURE P6.9 (a) Deflection with one force; (b) deflection with another load; (c) deflection with the combined load.

The Young's modulus of brass and steel are $E_b = 100$ GPa and $E_s = 200$ GPa, respectively. *Note*: Consider tensile force and tensile stress to be +ve.

6.8 Repeat Problem 6.7, this time using the PoS.

6.9 A device is made by connecting two shaft segments of the same material (Young' modulus E), but with different lengths L_1 and L_2, and different diameters d_1 and d_2. First, an external axial force P_1 is applied at the joint of the two segments and the end deflection δ_1 is determined (Figure P6.9a). Next, an external axial force P_2 is applied at the end of the two segments and the end deflection δ_2 is determined (Figure P6.9b).

Show that when P_1 and P_2 are applied together, the overall end deflection will be $\delta_1 + \delta_2$ as shown in Figure P6.9c. This illustrates the PoS.

6.10 Consider the two-shaft unit of Problem 6.9. Suppose that we are interested in the change in the axial cross-section area $A_1 = L_1 d_1$. Determine the change in this area as a result of application of the forces P_1 and P_2 separately. Then, determine the change in this area when P_1 and P_2 are acting simultaneously. Comment on the result.

6.11 Consider the two-shaft unit of Problem 6.9. First, an external axial force P_1 is applied at the joint of the two segments and the end deflection δ_1 is determined. Next, without P_1, the temperature of the device is increased by ΔT. Determine the corresponding extension δ_T at the end (B) of the device.

Show that when P_1 and ΔT are applied together, the overall end deflection will be $\delta_1 + \delta_T$. This illustrates the PoS.

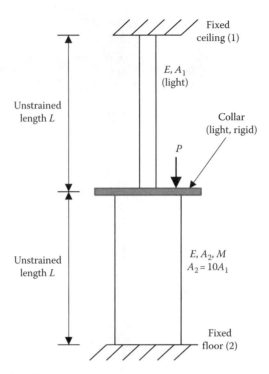

FIGURE P6.13 A vertically held two-segmented post.

6.12 Consider the two-shaft unit of Problem 6.10. Again, suppose that we are interested in the change in the axial cross-section area $A_1 = L_1 d_1$. Determine the change in this area as a result of the applied forces P_1. Next, determine the change in this area as a result of a temperature rise ΔT. Finally, determine the change in this area when P_1 and ΔT are applied simultaneously. Comment on the result.

6.13 A post is made of two uniform segments of equal length (L, under unstrained conditions) and of the same material (same E). There is a light and rigid collar at the joint of the two segments. The post is held vertically between a fixed ceiling and a fixed floor, rigidly mounted (fixed) at its two ends, and a downward force P is applied to the collar, as shown in Figure P6.13. The area of X-section of the bottom segment (A_2) is 10 times that of the top segment (A_1). Since $A_2 = 10A_1$, neglect the weight of the top segment. The weight of the bottom segment is Mg.
  a. Is this a statically determinate system? Why?
  b. Determine the axial end forces (i.e., reactions) P_1 and P_2 (magnitude and direction) at the top end and the bottom end, respectively, of the post (in terms of P and Mg only).
  c. Determine the maximum normal stress in the post (in terms of P, Mg, and A_2 only) and its location.

  Note: Assume that initially (i.e., before P is applied and before the post is released to exert the weight of the bottom segment on it) the entire post is unstrained with a total length of $2L$. The spacing between the floor and the ceiling is constant at $2L$.

6.14 A schematic diagram of the anchoring cable (steel) of a rigid/fixed sail post at Canada Place in Vancouver is shown in Figure P6.14. The cable is firmly attached to the top of the sail post at one end. The other end of the cable has a rigid end cap, the position of which may be adjusted by means of a tensioning device. The tensioning device has an identical pair of steel bolts with nuts, and is rigidly attached to a fixed support base. The bolts pass through the two holes in the end cap of the cable. Tensioning is done by tightening the two bolts by equal amounts.

Axial Loading

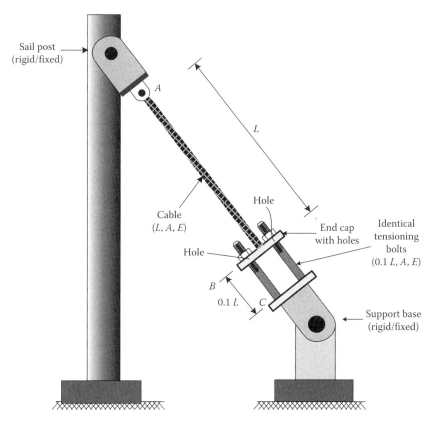

FIGURE P6.14 Anchoring cable system of a sail post.

Under unstrained conditions
  L = length of the cable
  $L/10$ = length of each bolt (measured from the position of the nut).
Also,
  δ = screw pitch of a bolt/nut
  E = Young's modulus of cable = Young's modulus of bolt.
Assume that the area of X-section of the cable = area of X-section of the bolt.
  If both nuts are tightened through n turns each, in the same direction, determine in terms of E, L, n, and δ, an expression for
  a. The tensile stress in the cable
  b. The tensile stress in each bolt

6.15 A uniform rigid beam of mass M is hinged (smooth) at one end. Two identical rods (same material, same unstrained length, and same area of X-section) are attached vertically to the beam, one at the midspan and the other at the free end, using pin joints (smooth). The other ends of the rods are attached to a rigid ceiling (Figure P6.15). Initially, the beam is held horizontally with the two rods just unstrained, and then released. Determine the resulting axial loads in the two rods and the reaction at the hinged end of the beam.

6.16 A rigid beam is placed horizontally on two identical steel end posts and a bronze midpost of equal length, which are erected on a rigid base (Figure P6.16). A vertical load P is applied downward at the midspan of the beam. There is a temperature rise of ΔT. Assuming that the beam is always in contact with the posts and is horizontal, determine the compressive loading in the posts. The following parameters are given:

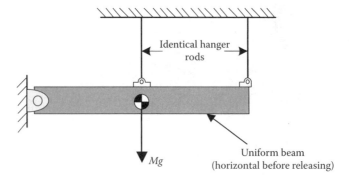

FIGURE P6.15 A hinged rigid beam supported using hanger rods.

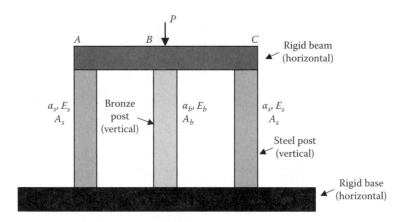

FIGURE P6.16 Rigid beam horizontally supported on three posts.

α_s = coefficient of thermal expansion of steel = $12 \times 10^{-6}/°C$
α_b = coefficient of thermal expansion of bronze = $17 \times 10^{-6}/°C$
E_s = Young's modulus of steel = 200 GPa
E_b = Young's modulus of bronze = 100 GPa
$\Delta T = 50°C$
A_s = area of X-section of a steel post = 100 mm²
A_b = area of X-section of the bronze post = 100 mm²

Note: Neglect gravity or assume that the weight of the beam is included in *P*.

6.17 A solid steel shaft of length $L_s = 0.75$ m and X-sectional area $A_s = 250$ mm² is fixed horizontally to a rigid wall and has a flange at the free end. A solid brass shaft of length $L_b = 0.5$ m and X-sectional area $A_b = 500$ mm² is fixed horizontally to an opposite rigid wall and has a flange at the free end, aligned with the flange of the steel shaft. Initially, there is a gap of $\delta = 0.5$ mm between the two flanges (Figure P6.17). Subsequently, the two flanges are welded together after stretching them using a vise (when the gap is removed).
  a. Determine the resulting internal axial force *P* and the stresses σ_s and σ_b in the steel shaft and the brass shaft.
  b. If the temperature of the shafts is increased by $\Delta T = 10°C$, determine the new axial force and the stresses in the two shafts.

Use $E_s = 200$ GPa, $E_b = 100$ GPa, $\alpha_s = 20 \times 10^{-6}/°C$, $\alpha_b = 15 \times 10^{-6}/°C$

Axial Loading   225

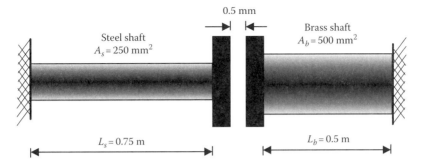

FIGURE P6.17 Two opposing horizontal shafts with a gap.

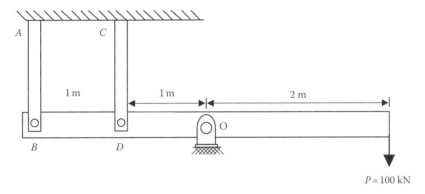

FIGURE P6.18 A hinged rigid beam held horizontally by hanger rods.

6.18 A rigid yet light beam is hinged at its midspan O and supported by two identical brass hanger rods of X-sectional area $A = 8$ cm^2, length $L = 1.5$ m each, one (CD) at a distance 1 m from O and the other (AB) at a distance 2 m from O on the same side (Figure P6.18). *Note*: B and D are pin joints. Initially, the beam is horizontal and the hanger rods are unstrained. Then a vertical load $P = 100$ kN is applied to point E of the beam, which is at 2 m from O on the other side.
  a. Determine the axial forces and the normal stresses in the hanger rods, the reaction R at O, and the vertical movement δ of point E on application of load P.
  b. What would be the values of the unknowns in part (a) if in addition, the temperature of the structure changed by $\Delta T = 40°C$?
  Young's modulus of brass $E = 100$ GPa
  Coefficient of thermal expansion of brass $\alpha = 20.0 \times 10^{-6}/°C$
  Note: Assume that the beam does not deform due to loading or temperature change.

6.19 A simplified model of a lifeguard observation deck is shown in Figure P6.19. A rigid platform (beam) is kept horizontally under extreme winter conditions (unloaded, without a lifeguard) using two pin-jointed vertical hanger rods having identical square section. The dimensions are as follows: lengths of the hanger rods: $L_1 = 2.0$ m; $L_2 = 1.0$ m; length of the beam (platform) $a = 5.0$ m
  Material properties of the hanger rods are as follows: Young's modulus $E = 70.0$ GPa; coefficient of thermal expansion $\alpha = 20.0 \times 10^{-6}/°C$
  Assume that initially (extreme winter) the hanger rods are unstrained. Under extreme summer conditions, there is a temperature rise of 50°C, and a vertical load of $P = 100$ kN due to

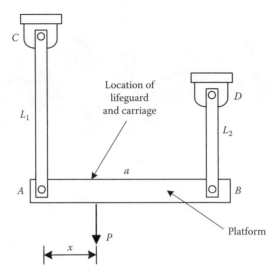

FIGURE P6.19 A liftguard deck supported using hanger rods.

the lifeguard and the carriage/chair on the horizontal beam (platform) at a distance x from the longer hanger rod.

Determine the location x such that the platform remains horizontal (under extreme summer conditions and loaded). What are the corresponding vertical movements of the two ends of the beam (from the unloaded extreme winter conditions to the loaded extreme summer conditions)?

6.20 A steel column of square X-section with side 4 cm stands snugly between a rigid ceiling and a rigid floor, both of which are assumed fixed (Figure P6.20). If the temperature rises by $\Delta T = 30°C$, determine the
   a. Resulting axial force in the column.
   b. Change in the lateral dimension of the column
   The following parameters are given: $\alpha = 12.0 \times 10^{-6}/°C$, $E = 200$ GPa, $G = 80$ GPa.

6.21 A rectangular flat bar is made of structural steel having yield strength $\sigma_Y = 250$ MPa. A hole of radius r is drilled in the midregion of the bar and a tensile force P is applied at the ends of the bar (Figure P6.21). The following dimensions are given:
   Width of the bar: $H = 100$ mm
   Thickness of the bar: $t = 5$ mm
   Radius of the hole: $r = 6$ mm
   With a factor of safety of 2, determine the maximum P that can be carried by the bar.
   Note: Use yield strength as the limiting strength, with the given factor of safety.

6.22 A stepped flat bar with shoulder fillets at the stepped corners is subjected to tensile force P (Figure P6.22). The bar is made of bronze with yield strength $\sigma_Y = 340$ MPa. The following dimensions are given:
   Width of the wider segment: $H = 160$ mm
   Width of the narrower segment: $h = 40$ mm
   Thickness of the bar: $t = 10$ mm
   Radius of a shoulder fillet: $r = 8$ mm
   With a factor of safety, FOS = 2, estimate the maximum tensile load P_{max} that can be carried by the bar.
   Note: Use yield strength as the limiting strength, with the given factor of safety.

6.23 Carry out a search on the evacuation system of the trapped miners in the Chilean Cu–Au mine.

Axial Loading

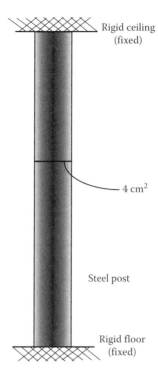

FIGURE P6.20 A steel post fitted between a ceiling and a floor.

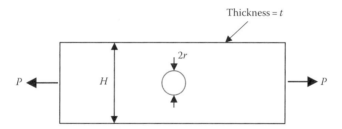

FIGURE P6.21 Flat bar of structural steel.

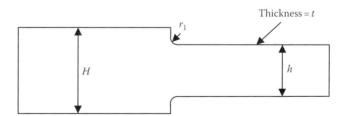

FIGURE P6.22 Stepped flat bar with shoulder fillets.

a. Sketch the mechanical parts of the evacuation system in sufficient detail and describe the operation of the system.
b. Briefly describe how each component of the system may be designed using the principles of mechanics of materials.
c. What are the most critical components and what are some of the less critical components of the system? Why?

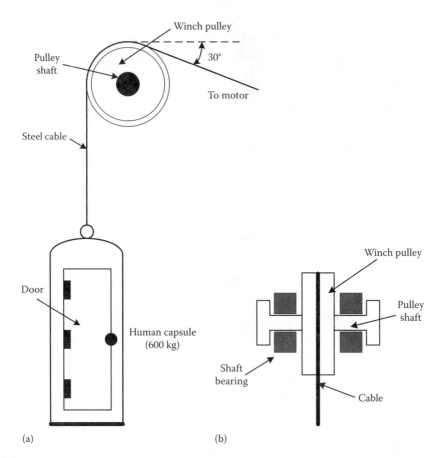

FIGURE P6.24 (a) Cable of the capsule winch; (b) pulley shaft.

6.24 Two critical components of the mine evacuation system of Problem 6.23 are the cable (Figure P6.24a) and the pulley shaft (Figure P6.24b) of the capsule winch.
The following data are given:
  Weight of the capsule with the heaviest allowable occupant = 600 kg
  Inclination of the cable segment leading to the drive motor = 30° below horizontal
  Yield strength of the cable material = 250 MPa
  Fracture stress in shear of the pulley shaft material = 350 MPa
Design the cable and the pulley shaft (i.e., determine their diameters) using the following factors of safety:
  Factor of safety for yielding of the cable = 2.0
  Factor of safety for fracture of the pulley shaft (in shear) = 4.0

Further Review Problems

6.25 A uniform shaft ABC of length $L = 0.5$ m is force fitted into a uniform hole in a fixed metal block over a segment $AB = L_1 = 0.2$ m of its length (see Figure P6.25). Area of cross section of the shaft is $A = 5$ cm^2. An axial force $P_0 = 20$ kN is applied at the free end C of the rod.
Determine
  a. The normal stress distribution along the shaft
  b. The axial movement of point C on application of the force P_0
Given: Young's modulus of the shaft, $E = 100$ GPa
Assume: Distribution of the holding force in the segment AB of the shaft is uniform.

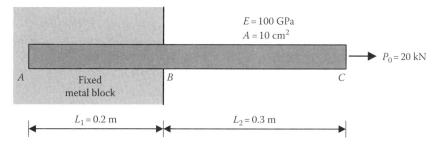

FIGURE P6.25 A force-fitted metal shaft with axial load.

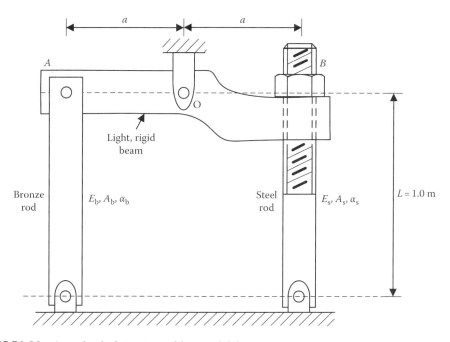

FIGURE P6.26 A mechanical structure with smooth joints.

6.26 A mechanical structure consists of an overhead beam (assumed rigid) that is hinged (smooth) at its midspan and supported on a bronze rod and a steel rod (see Figure P6.26). The top end of the bronze rod is pin-jointed (smooth) to the left end of the beam. The bottom end of the bronze rod is hinged (smooth) to a rigid base. The bottom end of the steel rod is also hinged (smooth) to the rigid base, at the same horizontal level. The top segment of the steel rod has threads and passes loosely through a hole at the right end of the beam. The two rods are vertical, and initially there are no loads on these rods (neglect the weight of the beam). The nut of the threaded segment of the steel is lightly tightened initially (with negligible force on the beam and the rod). The following parameter values are given:

Effective length of the bronze rod and the steel rod, $L = 1.0$ m
Area of X-section of the bronze rod, $A_b = 20$ mm²
Area of X-section of the steel rod, $A_s = 10$ mm²
Young's modulus of bronze, $E_b = 100$ GPa
Young's modulus of steel, $E_s = 200$ GPa

Coefficient of thermal expansion of bronze, $\alpha_b = 17 \times 10^{-6}/°C$
Coefficient of thermal expansion of steel, $\alpha_s = 12 \times 10^{-6}/°C$
Pitch of the threading on the steel rod (i.e., axial movement of the nut when rotated through a full turn), $p = 0.5$ mm

a. From the initial unstrained conditions when the nut is snugly in contact with the beam, the steel rod is tensioned by tightening the nut through two full turns ($n = 2$). Determine the resulting axial normal stresses in the bronze rod and the steel rod.

b. After the nut is tightened as in part (a), the ambient temperature of the structure is increased through $\Delta T = 20°C$. Determine the axial normal stresses in the two rods after this temperature increase.

Note: The two vertical rods are two-force members. Also, neglect the weight and any thermal deformation of the rigid beam. Angles of rotation, due to the deformation of the rods, are very small.

6.27 A rigid beam *OBC* is hinged (smooth) at *O* and horizontally supported at its midspan *B* using a vertical bronze rod pinned at *B* and firmly fixed at the base (Figure P6.27). A vertical steel cable is fixed at the same base, and passes through a vertical hole at the end *C* of the beam. The threaded end segment of the cable has a nut. Initially, the nut snugly sits on the top face of the beam, as shown in Figure P6.27, with the members unstrained. Then, the nut is tightened with $n = 2$ full turns so that the bronze rod is in compression and the steel cable is in tension. Determine the resulting normal stresses (axial) in the bronze rod and the steel cable.

Given,
Length of the bronze rod = effective length of the steel cable = $L = 0.5$ m
Area of cross section of the bronze rod = $A_b = 10$ mm²
Area of cross section of the steel cable = $A_s = 5$ mm²
Pitch of the cable thread = $P = 0.5$ mm
Young's modulus of bronze = $E_b = 100$ GPa
Young's modulus of steel = $E_s = 200$ GPa

Note: Neglect the weight of the members and the thickness of the beam (compared to the lengths).

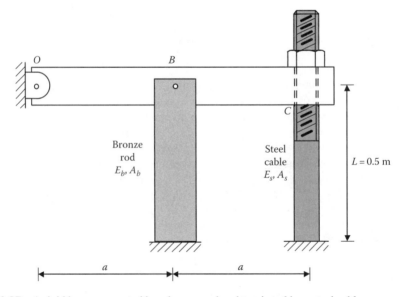

FIGURE P6.27 A rigid beam supported by a bronze rod and tensioned by a steel cable.

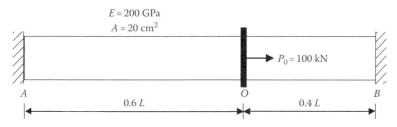

FIGURE P6.28 A rod with an axial load applied to its collar.

6.28 A uniform rod AOB of length L and area of cross section $A = 20$ cm² is firmly attached horizontally to two opposing rigid walls at its ends A and B (Figure P6.28). A force $P_0 = 100$ kN is applied to a collar of the rod at O, as shown. $AO = 0.6L$, $OB = 0.4L$, Young's modulus of the material of the rod, $E = 200$ GPa.
Determine the axial end loads P_A and P_B of the rod using
  a. The method of superposition
  b. The direct method

6.29 In Problem 6.27, suppose that the temperature of the system is increased through $\Delta T = 50°C$. Determine the corresponding normal stresses in the bronze rod and the steel cable.
The coefficients of thermal expansion of bronze and steel are $\alpha_b = 17 \times 10^{-6}/°C$ and $\alpha_s = 12 \times 10^{-6}/°C$, respectively.
Note: Neglect the deformation of the beam.

6.30 A uniform steel shaft of radius $r_1 = 2$ cm is inserted into a uniform bronze tube of equal length (L), internal radius $r_2 = 2.5$ cm and external radius $r_3 = 4$ cm, and rigidly capped at one end. The other end of the combined unit is firmly attached to a rigid ceiling (Figure P6.30) and the unit is held vertically. A vertical downward load $P_0 = 100$ kN is applied to the capped end. Determine the resulting axial normal stresses σ_s and σ_b in the steel shaft and the bronze tube, respectively.
Given
  Young's modulus of steel, $E_s = 200$ GPa
  Young's modulus of bronze, $E_b = 100$ GPa

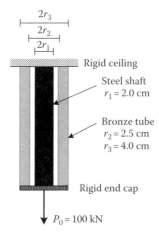

FIGURE P6.30 Steel shaft inside a bronze tube hung from ceiling and carrying a load.

6.31 In Problem 6.30, suppose that in addition to the load $P_0 = 100$ kN, there is a temperature rise, $\Delta T = 50°C$. Determine the resulting axial normal stresses σ_s^* and σ_b^* in the steel shaft and the bronze tube, respectively.

Given

Coefficient of thermal expansion of steel, $\alpha_s = 12 \times 10^{-6}/°C$
Coefficient of thermal expansion of bronze, $\alpha_b = 17 \times 10^{-6}/°C$

7 Torsion in Shafts

CHAPTER OBJECTIVES

- Analysis of uniform shafts of circular X-section
- Assumptions in the analysis
- Sign convention for the analysis
- Linear elastic case
- Sketching the internal torque distribution
- Determination of the angle of twist
- Definition and use of the polar moment of area
- Analysis of statically indeterminate torsional members
- Torsion of members with noncircular X-section
- Torsion of composite shafts

7.1 INTRODUCTION

Shafts that transmit torsional loads (torques) are common in engineering applications. A robot driven by a motor, automobile wheels driven by the engine, a milling machine in a factory, a jet engine, all have shafts of various types, shapes, and material. Uniform shafts of circular cross section are common in these applications, and they often carry purely torsional loads (torques). Even when they are subjected to other types of loads (e.g., axial forces; lateral/transverse support forces; bending moments), simultaneously, one can analyze the torsional loading separately and then add to the effects of other types of loading, by using the principle of superposition (with the assumption of linear behavior). It follows that circular shafts subjected to purely torsional loads is a topic of much importance in mechanics of materials. Their analysis is the subject of the present chapter.

As mentioned in Chapter 1, the solution of problems in mechanics of materials involves three basic considerations in general:

1. Statics (equations of equilibrium) that determine reactions at supports; and internal loads, which determine stresses
2. Nature of deformation (nature of strains)
3. Stress–strain relations (i.e., constitutive relations or physical relations). These relations are needed, for example, to determine the strains (deformations) once the stresses are known (from the knowledge of internal loads)

In Chapter 2, we reviewed the subject of statics. In Chapter 3, we studied stress and its determination from the knowledge of internal loading. In Chapter 4, we studied strain and its representation using the geometry of deformation. In Chapter 5, we studied the physical relation between stress and strain (i.e., constitutive relation), as applicable to any type of object under any type of loading. In Chapter 6, we integrated all three considerations of mechanics of materials, as listed previously, in the solution of members subjected to axial loading. In the present chapter, we integrate the same three considerations in the solution of members subjected to torsional loading. Here, the primary stresses are shear stresses, primary strains are shear strains, and the primary deformations are angles of twist. In particular, the analogies given in Table 7.1 exist between the problems of axial members (see Chapter 6) and torsional members.

TABLE 7.1
Analogies between Axial Problem and Torsional Problem

Axial Member	Torsional Member
Axial force (P)	Torque (T)
Normal stress (σ)	Shear stress (τ)
Normal strain (ε)	Shear strain (γ)
Elongation (δ)	Angle of twist (ϕ)
Young's modulus (E)	Shear modulus (G)

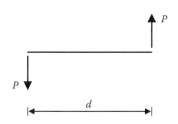

FIGURE 7.1 Illustration of torque as a couple.

7.1.1 CHAPTER OBJECTIVES

In this chapter, we specifically study the loading and the resulting deformation of torsionally loaded members. The effective loads in this case are torques. A *torque* is a load that tends to *twist* a member. A torque is equivalent to a *couple*, which is a pair of equal and opposite parallel forces. Specifically, as shown in Figure 7.1, two equal and opposite parallel forces P separated by distance d form the couple $d \times P$. In summary, a couple is equivalent to the *twisting moment* or torque T, as given by

$$T = d \times P \tag{7.1}$$

The main objective of the present chapter is to analyze internal loads, stresses, strains, and deformations of members having circular cross sections and subjected to torques. We will consider both uniform members and nonuniform members, particularly those made of several segments of different uniform cross section that are axially joined together. We will analyze to determine the distribution of internal torque along the axis, and associated stresses (shear stresses), strains (shear strains), and deformations (angles of twist). The assumptions made and the sign conventions used in the analysis will be explained. The analysis of statically indeterminate torsional members, whose internal loading cannot be determined using the equilibrium equations (statics) alone, will be presented. The analysis of torsional members having noncircular cross sections including thin-walled closed sections will be outlined. In conclusion, the analysis of composite shafts (made of two or more different material) having circular cross section will be indicated.

7.2 ANALYSIS OF CIRCULAR SHAFTS

A very large number of engineering applications employ members having circular cross sections, for transmitting torque (and speed and power). An example is shown in Figure 7.2, where a motor drives a liquid pump using a circular shaft (i.e., a shaft having circular X-section). The motor and the pump are connected to the shaft using a pair of couplings.

Suppose that the motor applies a torque T to the shaft. Since there are no other external torques acting on the shaft until it meets the pump, the internal torque at any cross section of the shaft between the motor and the pump is constant and equal to T.

Torsion in Shafts 235

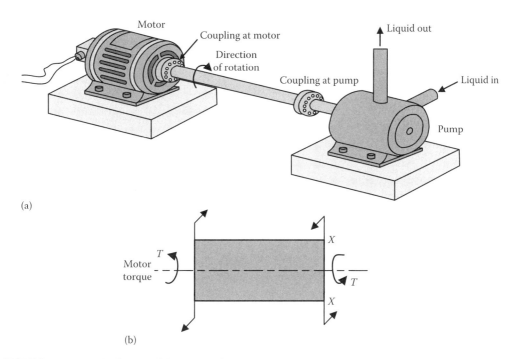

FIGURE 7.2 (a) A shaft transmitting torque from a motor to a pump; (b) a general virtual sectioning and the resulting shaft segment.

Note 1: This assumes that the speed is constant or the shaft inertia is negligible.

Note 2: This may be verified simply by making a virtual X-section at the location of interest of the shaft and considering the torque balance (under static equilibrium conditions or under steadily rotating conditions) of the shaft segment from the virtual section up to the motor, as shown in Figure 7.2b.

The pump is the "load," and it exerts a reaction torque equal to T on the shaft. The two equal and opposite torques at the two ends of the shaft will deform (twist) the shaft. In designing a shaft for this application, we should consider its stresses (shear stress), strains (shear strain), and the angle of twist. The stress will depend on the geometry and the internal torque of the shaft, while strain and the angle of twist will depend on the material properties of the shaft, in addition. These are all considerations of mechanics of materials. The present problem does not concern, for example, how the torque is generated in the motor (for instance, an ac motor depends on the rotating magnetic field produced by its stator in generating the torque, while a dc motor uses the magnetic attractions and repulsions of the poles of its stator and the rotor in generating the torque). Also, the details of the loading and stresses at the two couplings need not be addressed at this stage. Each of the couplings transmits a net torque T, as there are no external loads applied on them. If these couplings use bolt joints at their flanges (see Chapter 3), the shear forces (shear stresses) of the bolts will enable the torque transmission. Again, these details are not relevant for the analysis of the shaft segment between the two couplings.

7.2.1 Approach of the Analysis

The approach used in the analysis of a circular shaft in torsion is as follows:

1. Formulate the shear strain at a general location of an X-section of the shaft, using the geometry of deformation
2. Relate strain to stress, using the constitutive relation (which involves shear modulus, in the linear case)

3. Relate stress to internal torque of the shaft, using equilibrium conditions and geometry of the shaft
4. Relate the angle of twist to torque, using the results from steps 1 through 3

7.3 FORMULATION OF STRAIN

As the first step of torsion analysis, we will now formulate the shear strain of a shaft in torsion. For this, we need to consider the geometry of deformation of the shaft.

Assumptions

The following assumptions are made in the present formulation of strain of a uniform circular member made of homogeneous and isotropic material:

1. All cross sections along the shaft remain plane (flat) and perpendicular to the shaft axis.
2. All radii of a cross section remain straight and the radius remains unchanged (i.e., circular X-sections will remain circular with the radius unchanged).
3. There is no longitudinal displacement.

In fact, it can be shown that due to the axis-symmetry of a circular shaft made of homogeneous and isotropic material subjected to a uniform internal torque T, these assumptions are actually facts, which must hold (at least, in the ideal case that is analyzed).

Proof: See Figure 7.3.

We will prove these assumptions (properties) by contradiction.

a. Cross sections must remain plane (Figure 7.3a)
   The RHS of Figure 7.3a is obtained by simply turning the LHS figure upside down. Both figures have the same magnitude and direction of torque, and the same shaft. However, if the X-section did not remain plane on application of the torque, the deformation of the right figure would be different from that of the left figure, as shown, under identical conditions of loading. This is contradictory.
b. Radii must remain straight (Figure 7.3b)
   The RHS of Figure 7.3b is obtained by simply turning the left figure upside down. Both figures have the same magnitude and direction of torque, and the same shaft. However, if the radii did not remain straight on application of the torque, the deformation of the right figure would be different from that of the left figure, as shown, under identical conditions of loading. This is contradictory.
c. Radii must not change in length (Figure 7.3c)
   The RHS of Figure 7.3c is obtained by reversing the direction of the torque (loading) in the left figure. Both figures have the same magnitude of torque, and the same shaft. Hence, if the radius decreases in the left case, the radius should increase in the right case, as shown (since the loading is reversed). In view of the axis-symmetry of the problem, however, there is no reason why the shaft would contract in one case and would expand in the other, on application of the torque. Hence, the indicated situation is contradictory.
d. Longitudinal deformations must not exist (Figure 7.3d)
   The RHS of Figure 7.3d is obtained by reversing the direction of the torque (loading) in the left figure. Both figures have the same magnitude of torque, and the same shaft. Hence, if the shaft length increases in the left case, the shaft length should decrease in the right case (since the loading is reversed). In view of the axis-symmetry of the problem, however, there is no reason why the shaft would extend in one case and would shorten in the other, on application of the torque. Hence, the indicated situation is contradictory.

Torsion in Shafts

By implication of the stated assumptions (or facts), the following properties will hold as well:

1. Normal strain in the axial direction is zero at any X-section.
2. Normal strain in the radial direction is zero at any X-section.
3. There are no shear strains (or shear stresses) on any cylindrical section of arbitrary radius of the shaft (about the shaft axis), in the axial direction. Hence, there cannot be any shear strains or shear stresses in the radial direction on any X-section of the shaft (due to the complementarity property of shear stresses, see Chapter 3).

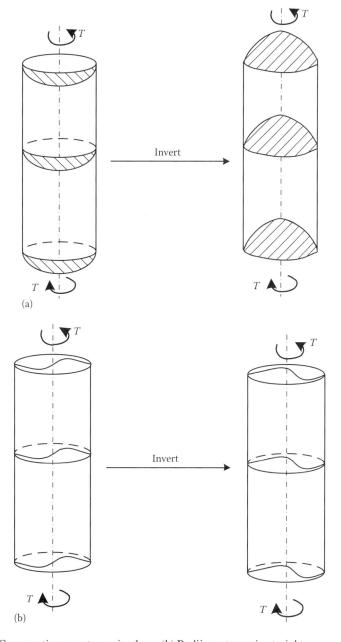

FIGURE 7.3 (a) Cross sections must remain plane. (b) Radii must remain straight.

(*continued*)

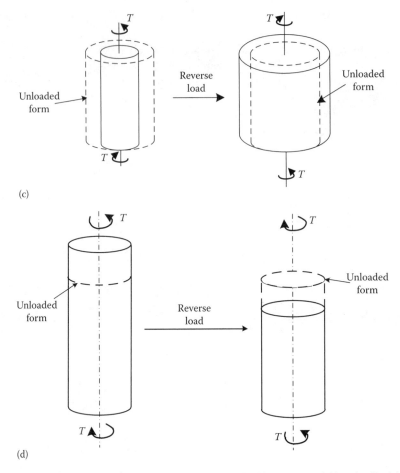

FIGURE 7.3 (continued) (c) Radii must not change in length. (d) There must not be longitudinal deformations.

4. There are no shear strains (or shear stresses) on any cylindrical section of arbitrary radius of the shaft (about the shaft axis) in the tangential direction. Hence, there cannot be any shear strains or shear stresses in the radial direction on any radial section of the shaft (due to the complementarity property of shear stresses).

It follows that the only stress (strain) that is present on a cross section of the shaft is a tangential shear stress (and the corresponding shear strain). This state of stress is illustrated in Figure 7.4, by showing small elements that are 90° apart on the shaft cross section. *Note*: The indicated shear stress in the axial direction of an element arises from the complementarity property of shear stress.

7.3.1 Sign Convention

Consider a homogeneous and isotropic shaft with uniform circular cross section of radius a, which remains in equilibrium by means of two equal and opposite torques T at its ends (see Figure 7.5). A sign convention is necessary in formulating an engineering problem, in order to indicate the positive direction of each quantity that can assume a +ve or −ve sign. Once a sign convention is established for a particular quantity, if the actual value of that quantity happens to be +ve, then the quantity is in the +ve direction of the convention. If the sign of the actual value of the quantity happens to be −ve, then the quantity is in the direction opposite

Torsion in Shafts

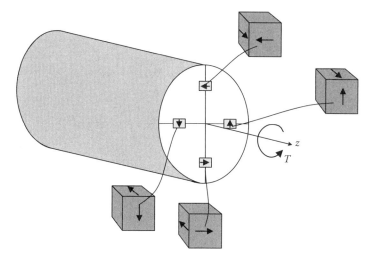

FIGURE 7.4 The state of stress on an X-section of a shaft under torsion.

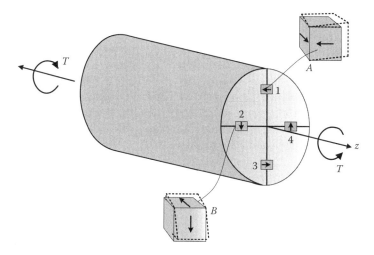

FIGURE 7.5 Sign convention for torque and shear stress in a torsional member.

to the +ve direction (i.e., in the −ve direction) of the convention. The following sign convention is used in the analysis of torsional members:

1. The positive direction of a torque is represented by an axis (z in Figure 7.5) according to the right-handed corkscrew rule (i.e., screwing out is +ve and screwing in is −ve).
2. Positive shear stress is defined so as to generate a +ve internal torque at the considered X-section. In Figure 7.5, +ve directions of shear stress are shown on four elements (1, 2, 3, and 4) that are 90° apart on the X-section.
3. Positive shear strain at a corner is one that would shrink the angle of the corner. The two adjacent shear stresses on the two planes (that meet to form the corner) should point toward the corner in this case. In Figure 7.6, the opposite corners A and C have +ve shear strain +γ, and the opposite corners B and D have −ve shear strain −γ. Similarly in Figure 7.5, corner A of element 1 has a +ve shear strain and corner B of element 2 has a −ve shear strain. Also see Chapter 4.

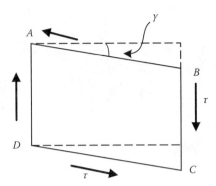

FIGURE 7.6 Sign convention for shear strain.

7.3.2 Geometry of Torsional Deformation

In formulating the strain (shear strain) of a circular member in torsion, it is useful to first understand the geometry of deformation of the member. According to the torque and the state of stress shown in Figure 7.5, it is clear that the shaft will tend to twist due to the torsional loading. Suppose that a circular grid is drawn on the shaft surface before applying the end torques. On the surface of the shaft, a line parallel to the axis (generator) before applying T will become a spiral after application of T. This is illustrated in Figure 7.7.

To further investigate the nature of deformation of a torsional member, consider a thin slice of the shaft and make two radial sections on it, as shown in Figure 7.8. Due to the nature of the twisting deformation caused by the +ve torque that is applied, the corner angle at edge A must shrink (resulting in a +ve shear strain) and the corner angle at edge A' must expand (resulting in a −ve shear strain). This is indeed consistent with the shear stresses that are present on a small element at the edge, as shown (they point toward edge A). Also, due to the nature of the twisting deformation caused by the +ve torque, the corner angle at edge B must expand (resulting in a −ve shear strain) and the corner angle at edge B' must shrink (resulting in a +ve shear strain). Otherwise, the integrity

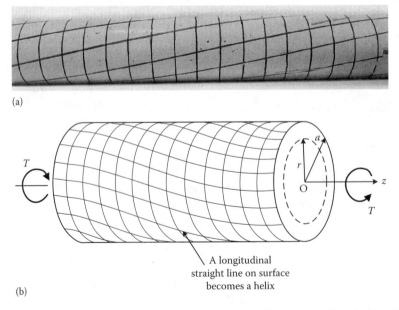

FIGURE 7.7 (a) Deformed shape of a flexible rod in pure torsion; (b) geometry of torsional deformation.

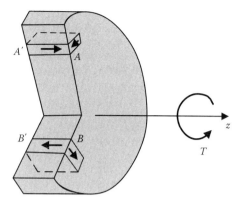

FIGURE 7.8 Nature of shear stress that is consistent with +ve twisting.

of the member would be lost, and ruptures must be formed in the material. This is also consistent with the shear stresses that are present on a small element at the edge, as shown (they point away from edge B).

7.3.3 Formulation of Strain

After understanding the geometry of torsional deformation of a circular shaft, it is now possible to formulate the shear strain in the shaft. Consider a core portion (a cylindrical element) of radius r in the shaft of radius a (*Note*: $r < a$), and take a small element of length δz from it, as shown in Figure 7.9.

As noted earlier, the straight-line segment AB of length δz along the element surface will deform into a spiral segment AB', which is almost a straight line since its length is very small.

Angular movement of the axial line segment = Shrinking of left corner = γ = Shear strain.

If the radial line OB, which meets this axial line segment AB, rotates to OB' through angle $\delta\phi$, we have (see Figure 7.9)

$$BB' = \gamma \cdot \delta z = r \cdot \delta \phi$$

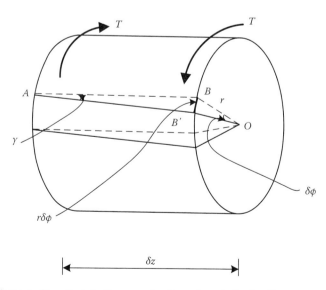

FIGURE 7.9 An element of length δz in the core of radius r in a shaft of radius a.

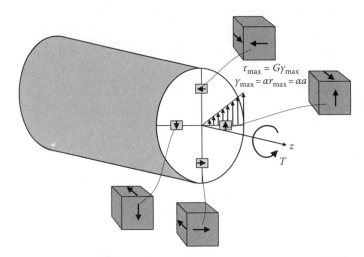

FIGURE 7.10 Linear variation of shear strain along a radius of X-section of a shaft under torsion.

Hence, in the limit (as $\delta z \to 0$)

$$\text{Shear strain at radius } r: \gamma = r\frac{d\phi}{dz} = \alpha r \tag{7.2}$$

We have defined: Angle of twist per unit length $\frac{d\phi}{dz} = \alpha$ (7.3)

According to (7.2), the strain distribution (regardless of whether the material is elastic or not) must be linear along a radius, varying from 0 at the center of the cross section to the maximum value $\gamma_{max} = \alpha r_{max} = \alpha a$ (at the outer end of the radius a). This linear variation of shear strain along a radius of a shaft cross section is shown in Figure 7.10.

7.4 FORMULATION OF STRESS

Shear stress is expressed by simply substituting the derived expression for shear strain into the stress–strain relation. We have found that the variation of the shear strain is linear along a radius of cross section. However, the variation of the shear stress will not be linear along a radius of the cross section, unless the stress–strain relationship itself is linear.

7.4.1 Linear Elastic Case

For the linear case, shear strain is related to shear stress through the shear modulus G (see Chapter 5). Hence, from Equation 7.2, we have

$$\text{Shear stress at radius } r: \tau = G\gamma = G\alpha r \tag{7.4}$$

where
  G is the shear modulus
  α is the angle of twist per unit length

It follows that in the linear elastic case, the stress distribution is also linear, varying from 0 at the center of the cross section to the maximum value $\tau_{max} = G a \alpha$ at the outer edge of radius a, as shown in Figure 7.10.

7.4.2 Polar Moment of Area

The rest of the formulation needs the geometric parameter known as the polar moment of area of a cross section, which is the second moment of area of the X-section about the central axis (i.e., axial second moment of area). Consider a small annular cross section of width δr at radius r of the shaft, as shown in Figure 7.11. The elemental area $\delta A = 2\pi r \delta r$. The polar moment of area of this area is the second moment of this area about the central axis of the shaft. It is $r^2 \times 2\pi r \delta r$. Hence, the overall polar moment of area of the X-section is

$$J = \int_A r^2 dA = \int_0^a r^2 2\pi r \cdot dr = 2\pi \int_0^a r^3 \cdot dr = \frac{\pi a^4}{2} \tag{7.5a}$$

For a hollow circular shaft (tube) of inner radius a_i and the outer radius a_o, the integration limits in (7.5a) should be from a_i to a_o. Then, we have

$$J = \frac{\pi}{2}\left(a_o^4 - a_i^4\right) \tag{7.5b}$$

Note: The formulas derived later will hold for both solid shafts and hollow shafts of circular X-section. For solid shafts, formula (7.5a) should be used for J; while for hollow shafts, formula (7.5b) should be used for J.

7.4.3 Internal Torque at a Cross Section

Consider a general cross section of a circular shaft and an elemental area δA at radius r on that X-section (Figure 7.12). Since the shear stress there is τ, the corresponding tangential force is $\tau \delta A$ and the associated moment about the central axis of the X-section is $r\tau \delta A$. Hence, at any X-section of the shaft (of radius a), the "internal" torque may be expressed as

$$\text{Torque } T = \int_A r\tau dA \tag{7.6}$$

Note: The integration is carried out over the entire X-sectional area A.

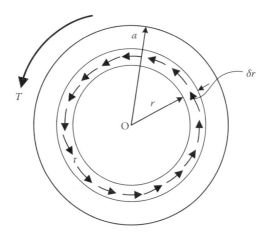

FIGURE 7.11 Annular X-sectional element and its shear stress and torque.

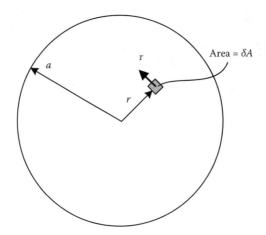

FIGURE 7.12 Formulation of internal torque using shear stress.

Substitute (7.4) in (7.6). We get

$$T = \int_A G\alpha r^2 dA = G\alpha J \tag{7.7}$$

where J is the *polar moment of area* (the axial second moment of area) of the shaft, which is defined as

$$J = \int_A r^2 dA \tag{7.5c}$$

Note: Shear modulus G is constant over the entire area of X-section, for homogeneous and isotropic material.

Substitute (7.7) in (7.4). We get

$$\tau = \frac{Tr}{J} \tag{7.8}$$

The maximum shear stress occurs when r is maximum (at $r = a$). Hence, $\tau_{max} = Ta/J$.

For a solid shaft, using the formula (7.5a) for J, we have

$$\tau_{max} = \frac{2T}{\pi a^3} \tag{7.8b}$$

7.5 ANGLE OF TWIST

The deformation of a torsional member is expressed by its angle of twist ϕ. An expression for this is obtained now.

From Equations 7.7 and 7.8

$$\frac{T}{J} = \frac{\tau}{r} = G\alpha = G\frac{d\phi}{dz} \tag{7.9}$$

By direct integration of (7.9) over the length L of the shaft, we get the total angle of twist of the shaft as

$$\text{Angle of twist } \phi = \int_0^L \frac{T}{GJ} dz \tag{7.10}$$

Torsion in Shafts

Note: We have considered here the general case where T, J, and G all may vary along the shaft (i.e., they are functions of z). For a uniform shaft of circular X-section, made of homogeneous and isotropic material, and with constant internal torque T at every X-section along the shaft (i.e., when no external torque is applied along the shaft, except at the ends), it is clear from Equation 7.9 that the angle of twist per unit length ($d\phi/dz = \alpha$) is constant throughout the shaft length. Then by performing the integration in (7.10) after taking the constant terms outside the integral, the total angle of twist of the shaft is obtained as

$$\phi = \frac{TL}{GJ} \tag{7.11}$$

Compare: Deformation of an axial member is $\delta = PL/EA$, as determined in Chapter 6. Here, P (internal force) is analogous to T (internal torque); E (Young's modulus—a material property) is analogous to G (shear modulus—a material property); and A (a geometric parameter) is analogous to J (a geometric parameter).

Note: In Equation 7.10, we allow the quantities T, J, and G to be variable along the shaft. However, the derivation of the shear strain formula in Section 7.3 assumes that the shaft is made of homogeneous and isotropic material, its circular X-section is uniform along the shaft length, and the internal torque of the shaft is constant along the shaft length. In view of this discrepancy, one may tend to argue that (7.10) is not valid whereas Equation 7.11 is valid. However, in the derivation of (7.11), we considered an elemental segment of length δz in the shaft (and the result was integrated along the shaft length). Since, by definition, the elemental length is very small, it is uniform (area of X-section is constant), its J is constant, and its G is constant. Hence, it may be concluded that the assumptions made in Section 7.3 are valid for the elemental segment as well. Consequently, Equation 7.10 should be valid as well.

Example 7.1

A lug wrench made of an aluminum alloy has a tubular stem of length L, internal radius a_i, and external radius a_e. A turning torque is manually applied to the handle of the wrench through hand forces P_1 and P_2 at turning radii l_1 and l_2 as shown in Figure 7.13a. The modulus of rigidity (shear modulus) of the wrench is G. Obtain expressions for

  a. The maximum shear stress
  b. The angle of twist in the stem of the wrench

We are given the numerical values $L = 0.5$ m, $a_i = 1$ cm, $a_e = 2$ cm, $G = 30$ GPa, yield shear stress $\tau_Y = 170$ MPa, and the maximum operating value of the applied torque, which corresponds to: $l_1 = l_2 = 0.4$ m and $P_1 = P_2 = 150$ N. What is the corresponding factor of safety (FoS) with respect to yielding under shear? What is the angle of twist of the wrench for this maximum torque?

Solution

See the free-body diagram of the stem of the wrench, as shown in Figure 7.13b.

Use Equation 7.9: $\dfrac{\tau}{r} = \dfrac{T}{J}$ with $J = \dfrac{\pi}{2}\left(a_e^4 - a_i^4\right)$ and $T = P_1 l_1 + P_2 l_2$

Maximum shear stress occurs at $r = a_e$

Hence, maximum shear stress $\tau_{max} = 2\dfrac{(P_1 l_1 + P_2 l_2) a_e}{\pi\left(a_e^4 - a_i^4\right)}$

Now use Equation 7.11: $\phi = \dfrac{TL}{GJ}$

Hence, angle of twist in the stem of the wrench $\phi = \dfrac{2(P_1 l_1 + P_2 l_2)L}{\pi G\left(a_e^4 - a_i^4\right)}$

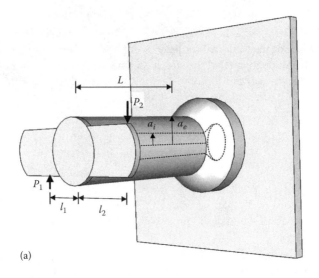

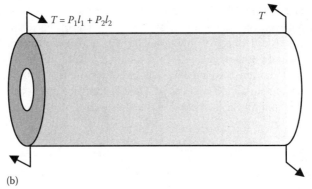

FIGURE 7.13 (a) A lug wrench; (b) free-body diagram of the wrench stem.

Substitute the given numerical values.

$$J = \frac{\pi}{2}[2^4 - 1^4] \times 10^{-8} \text{ m}^4 = 23.562 \times 10^{-8} \text{ m}^4$$

$$T = 150 \times 0.4 + 150 \times 0.4 \text{ N} \cdot \text{m} = 120 \text{ N} \cdot \text{m}$$

$$\rightarrow \tau_{max} = \frac{120 \times 2 \times 10^{-2}}{23.562 \times 10^{-8}} \text{ Pa} = 10.2 \text{ MPa}$$

With a yield shear stress $\tau_Y = 170$ MPa

$$\text{Factor of safety} = \frac{170}{10.2} = 16.7$$

Note: This FoS is rather high.

$$\text{Angle of twist } \phi = \frac{120 \times 0.5}{30 \times 10^9 \times 23.562 \times 10^{-8}} \text{ rad} = 0.0085 \text{ rad} = 0.5°$$

Torsion in Shafts

Primary Learning Objectives

1. Determination of the critical shear stress of a torsional member
2. Determination of the angle of twist of a torsional member
3. Use of the appropriate formulas in the solution of torsion problems
4. Design considerations of a torsional member
5. Interpretation of the factor of safety

■ **End of Solution**

Example 7.2

A load is driven by an induction motor from one end of a shaft. Driving of the shaft is supplemented by a belt-drive unit, which is placed on the other side of the load (Figure 7.14a). The lengths of the resulting three segments of the shaft are equal at l. The shaft is supported on a

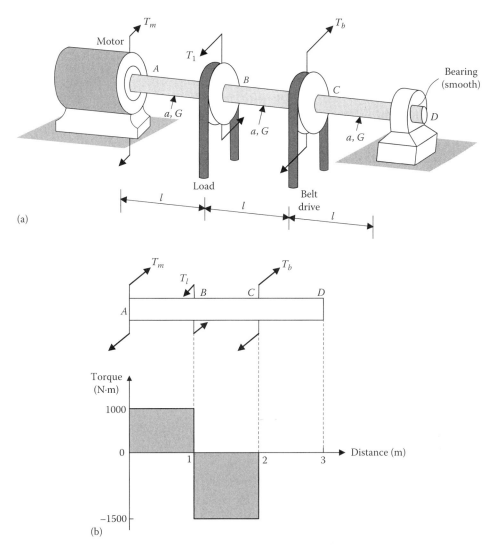

FIGURE 7.14 (a) A motor and a belt drive cooperatively driving a load; (b) FBD and the internal torque diagram of the shaft.

smooth ball bearing at the far end. The shaft has a uniform circular cross section of radius a and is made of homogeneous and isotropic material of shear modulus G. The torque applied by the motor is $T_m = 1000$ N·m in the counter-clockwise (ccw) direction looking in from the left. The drive torque of the belt-drive is $T_b = 1500$ N·m in the clockwise (cw) direction looking from the right, as shown in the figure.

a. Sketch the internal torque diagram of the shaft. What is the torque T_l transmitted to the load?
b. What is the maximum shear stress in the shaft? At what location does it exist?
c. Determine the angle twist due to torsion of the shaft
   i. At the load location (B) with respect to the motor location (A)
   ii. At the bearing location (D) with respect to the load location (B)
   iii. At the bearing location (D) with respect to the motor location (A)

Solution

a. The free-body diagram of the shaft is shown in the top part of Figure 7.14b.
Equilibrium of the shaft: $T_m - T_l + T_b = 0$

Hence, $T_l = T_m + T_b = 1000 + 1500$ N·m $= 2500$ N·m

(ccw looking in from right or cw looking in from left)
Using this information, the internal torque diagram of the shaft is sketched in the bottom part of Figure 7.14b.

b. Polar moment of area of the shaft X-section: $J = \dfrac{\pi}{2} a^4 = \dfrac{\pi}{2} \times 5^4 \times 10^{-8}$ m^4

Maximum shear stress occurs at the rim of the shaft (at radius $a = 5$ cm) in the segment BC (where the internal torque is the largest).

Using Equation 7.8: $\tau = \dfrac{Tr}{J}$, we have $\tau_{max} = \dfrac{1500 \times 5 \times 10^{-2}}{(\pi/2) \times 5^4 \times 10^{-8}}$ Pa $= 7.64$ MPa

Note: Since the shaft is solid, we could have directly used the formula (7.8b), $\tau_{max} = 2T/\pi a^3$ in this computation

c.

$$\text{Use Equation 7.11:} \quad \phi = \dfrac{TL}{GJ}$$

$$\phi_{B/A} = \dfrac{1000 \times 1}{75 \times 10^9 \times (\pi/2) \times 5^4 \times 10^{-8}} \text{ rad} = 0.0014 \text{ rad} = 0.08°$$

(ccw looking in from right)

$$\phi_{D/B} = \phi_{C/B} = \dfrac{1500 \times 1}{75 \times 10^9 \times (\pi/2) \times 5^4 \times 10^{-8}} \text{ rad} = \dfrac{3}{2} \phi_{B/A} = 0.0020 \text{ rad} = 0.06°$$

(cw looking in from right)
Note: There is no internal torque in the shaft segment CD (because of the smooth bearing)

$$\rightarrow \phi_{D/A} = \phi_{D/B} + \phi_{B/A} = 0.0020 - 0.0014 \text{ rad} = 0.0006 \text{ rad}$$

(cw looking in from right)

Torsion in Shafts

Primary Learning Objectives

1. Determination of the internal torque diagram of a torsional member
2. Determination of the critical shear stress of a torsional member
3. Determination of the relative angle of twist between two locations of a torsional member
4. Use of the appropriate formulas and sign conventions in the solution of torsional problems

■ **End of Solution**

Example 7.3

A solid shaft with circular cross section is made of a steel segment of diameter 100 mm and length 3 m, and a brass segment of diameter 50 mm and length 1 m, rigidly welded together. The steel end is rigidly attached to a wall (Figure 7.15).

a. If the allowable shear stress for the steel segment is $\tau_{allws} = 50$ MPa and the allowable shear stress for the brass segment is $\tau_{allwb} = 25$ MPa, determine the maximum torque $T_{max\,e}$ that may be applied at the free end of the shaft.
b. Determine the maximum shear stress in the steel segment $\tau_{max\,s}$ and in the brass segment $\tau_{max\,b}$ when this allowable torque is applied.
c. Determine the corresponding angular movement ϕ at the free end of the shaft.

The shear moduli of steel and brass are $G_s = 80$ GPa and $G_b = 40$ GPa, respectively.

Solution

Make a virtual cut at any general cross section of the shaft, and consider the equilibrium of the segment to the right. This will give
Internal torque T = Applied end torque T_e
→ The internal torque is uniform (i.e., constant).

a. The maximum shear stress occurs at the largest value of the radius r (i.e., at the outer end of the radius, where $r = a$).

$$\text{Apply Equation 7.8b: } \tau_{max} = \frac{2T}{\pi a^3}$$

→ The larger τ_{max} occurs in the segment with the smaller radius, which is the brass segment. Also, since the allowable τ for the brass segment is smaller, along with the fact that its maximum τ is larger, the brass segment will reach the allowable shear stress first.

$$\tau_{allwb} = \frac{2T_{max}}{\pi a_b^3}$$

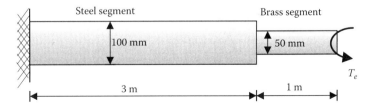

FIGURE 7.15 A two-segmented shaft made of steel and brass.

Substitute numerical values:

$$T_{max} = \frac{\pi}{2} \times (25 \times 10^{-3})^3 (m^3) \times 25 \times 10^6 (Pa) = \frac{\pi}{2} \times 25^4 \times 10^{-3} \, N \cdot m = 613.6 \, N \cdot m$$

b. In the presence of the maximum torque T_{max} as computed in part (a), the corresponding maximum torques in the brass segment and the steel segment are

$$\tau_{max\,b} = \tau_{allw\,b} = 25.0 \, MPa$$

$$\tau_{max\,s} = \frac{2T_{max}}{\pi a_s^3} = \frac{2 \times 613.6}{\pi \times (500 \times 10^{-3})^3} \, Pa = 3.125 \, MPa$$

Alternatively, since T is the same

$$\tau_{max\,s} = \left(\frac{a_b}{a_s}\right)^3 \times \tau_{max\,b} = \left(\frac{25}{50}\right) \times 25.0 \, MPa = 3.125 \, MPa$$

c. Apply Equation 7.11: $\phi = \frac{TL}{GJ}$ with $J_s = \frac{\pi}{2} \times 50^4 \, mm^4 = 98.2 \times 10^{-7} \, m^4$ and $J_b = \frac{\pi}{2} \times 25^4 \, mm^4 = 6.132 \times 10^{-7} \, m^4$.

Angle of twist in the steel shaft segment:

$$\phi_s = \frac{613.6 \times 3.0}{80 \times 10^9 \times 98.2 \times 10^{-7}} \, rad = 2.343 \times 10^{-3} \, rad$$

Angle of twist in the brass shaft segment:

$$\phi_b = \frac{613.6 \times 1.0}{40 \times 10^9 \times 6.136 \times 10^{-7}} \, rad = 25.0 \times 10^{-3} \, rad$$

Primary Learning Objectives
1. Torsional analysis of a multisegmented shaft
2. Determination of the internal torque
3. Determination of the critical shear stresses of a multisegmented torsional member
4. Design consideration of a torsional member with respect to the allowable shear stress
5. Determination of the angles of twist of a multisegmented member

■ **End of Solution**

Example 7.4

A solid cone of length $3L = 0.3$ m and cone angle $2\theta = 40°$ is cut at length $2L = 0.2$ m to form a shaft segment (frustum) as shown in Figure 7.16a. It is fixed to a rigid wall at the base and a torque of T_e is applied at the free end, as shown.

a. What is the maximum value of T_e such that the allowable shear stress $\tau_{allw} = 50$ MPa is not exceeded?
b. What is the corresponding angle of twist ϕ of the shaft?

Torsion in Shafts

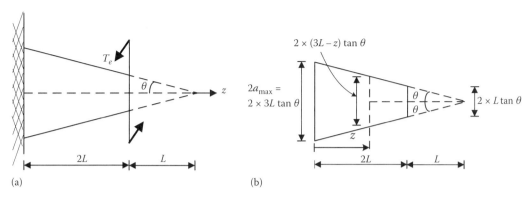

FIGURE 7.16 (a) A conical torsion member (frustum) with an end torque. (b) Geometry of the conical torsion member.

Given: $G = 80$ GPa.

Hint: In this example, even though a "uniformity" assumption (which is used in deriving the torsion formulas for a circular shaft) is not valid for the overall member (because the X-section varies along the member), the derived formulas are applicable for an elemental segment (whose length $\delta z \to 0$) where the conditions are uniform.

Solution

a. Apply Equation 7.8b: $\tau_{max} = 2T/\pi a^3$ for the smallest value of a (i.e., the free end), which corresponds to the maximum value of τ_{max}.
From Figure 7.16b

$$a_{min} = L\tan\theta = 0.1\tan 20° = 0.0364 \text{ m}$$

Since the maximum possible shear stress is the allowable shear stress τ_{allw}, we have

$$50 \times 10^6 = \frac{2 \times T_{eallw}}{\pi \times (0.0364)^3}$$

$$\to T_{eallw} = \frac{50 \times 10^6 \times \pi \times (0.0364)^3}{2} \text{N·m} = 3.788 \text{ kN·m}$$

b. Consider a cross section at distance z from the fixed end (base), as shown in Figure 7.16b. Corresponding radius of the X-section, $a(z) = (3L - z)\tan\theta$

$$\to J(z) = \frac{\pi}{2}a^4 = \frac{\pi}{2}(3L - z)^4 \tan^4\theta$$

Apply Equation 7.10: angle of twist

$$\phi = \int_0^{2L} \frac{T}{GJ(z)}dz = \frac{T_e}{G}\int_0^{2L} \frac{1}{\frac{\pi}{2}(3L-z)^4 \tan^4\theta}dz = \frac{2T_e}{\pi G \tan^4\theta} \times \frac{1}{3} \times \frac{1}{(3L-z)^3}\bigg|_0^{2L}$$

$$= \frac{2T_e}{3\pi G \tan^4\theta}\left[\frac{1}{L^3} - \frac{1}{(3L)^3}\right] = \frac{2T_e}{3\pi G L^3 \tan^4\theta}\left[1 - \frac{1}{3^3}\right] = \frac{2 \times 3.788 \times 10^3}{3\pi \times 80 \times 10^9 \times (0.1)^3 \times \tan^4 20°}\left[1 - \frac{1}{27}\right] \text{rad}$$

$$\phi = 0.551 \times 10^{-3} \text{ rad}$$

Primary Learning Objectives
1. Torsional analysis of a nonuniform shaft (whose X-section is variable)
2. Determination of the critical shear stress of a nonuniform torsional member
3. Design consideration of a nonuniform torsional member with respect to allowable shear stress
4. Determination of the angle of twist of a nonuniform torsional member

■ **End of Solution**

Example 7.5

A uniform drill-bit AC, while drilling halfway into an object suddenly encounters a hard region, which results in an end torque T_e (at C, see Figure 7.17a). Also, the drilled hole exerts a linearly distributed external torque on the contact segment BC of the drill-bit. Let

T_m = torque provided by the motor
T_e = end reaction torque from the hard material
kz = resisting torque per unit length (from the drilled hole) at location z measured from the surface of the drilled object; i.e., midpoint B of the drill-bit (*Note*: k is a constant.)
$2L$ = length of the drill-bit
J = polar moment of area of the X-section of the drill-bit
G = modulus of rigidity (shear modulus) of the drill-bit

a. Neglecting the inertia of the drill-bit, determine an expression for T_e in terms of T_m, k, and L.
b. Sketch the internal torque diagram of the drill-bit.
c. Determine the total angle of twist (over the length 2L) of the drill-bit in terms of T_m, k, L, G, and J.

Solution

a. Resistance torque in an elemental segment of length δz of the drill-bit is $kz \cdot \delta z$. Hence, the total torque from the distributed resistance of the hole

$$T_R = \int_0^L kz\, dz = \frac{1}{2}kL^2 \qquad (i)$$

Note: This distributed torque may be represented by an equivalent torque of magnitude (i) acting at the centroid of the torque distribution (which is at two-thirds of the length BC, from B).

A free-body diagram of the drill-bit is shown in Figure 7.17b, where the equivalent resisting torque T_R is used.

Equation of equilibrium: $T_m - \frac{1}{2}kL^2 - T_e = 0$

$$\to T_e = T_m - \frac{1}{2}kL^2$$

b. For the segment AB: internal torque $T = T_m$ (because there is no external torque in that segment)
For the segment BC: Make a virtual cut of the X-section at location z, and consider the drill-bit segment to the left of the cut (Figure 7.17c).
T = torque at the cut X-section = internal torque

Torsion in Shafts

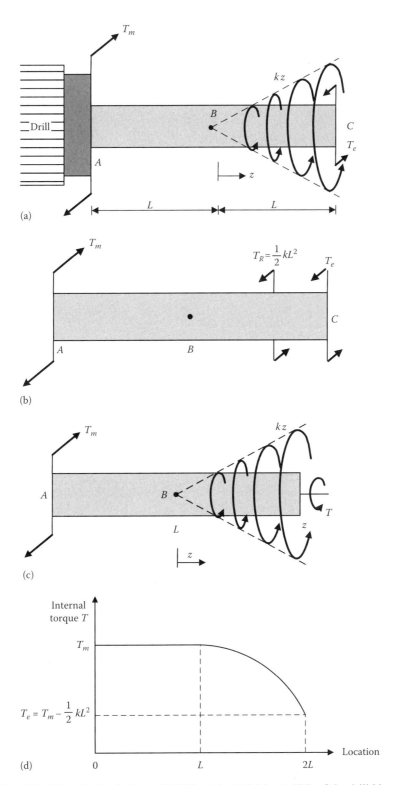

FIGURE 7.17 (a) Drilling of a hard object; (b) FBD of the drill-bit; (c) FBD of the drill-bit segment up to location z; (d) internal torque diagram of the drill-bit.

Equilibrium of the cut segment (from A up to z): $T_m - \int_0^z kz'dz' - T = 0$

$$\rightarrow T_m - \frac{kz^2}{2} - T = 0 \rightarrow T = T_m - \frac{kz^2}{2}$$

The corresponding internal torque diagram is sketched in Figure 7.17d.

c. Apply Equation 7.11 for segment AB: $\phi_{AB} = \frac{T_m L}{GJ}$

Apply Equation 7.10 for segment BC:

$$\phi_{BC} = \frac{1}{GJ} \int_0^L \left(T_m - \frac{kz^2}{2} \right) dz = \frac{1}{GJ} \left(T_m L - \frac{kL^3}{6} \right)$$

Total angle of twist:

$$\phi = \phi_{AB} + \phi_{BC} = \frac{L}{GJ} \left(2T_m - \frac{kL^2}{6} \right)$$

Primary Learning Objectives
1. Determination of the internal torque diagram of a member under distributed external torque
2. Determination of the angle of twist of a member under distributed external torque
3. Proper use of free-body diagrams in torsional analysis

■ **End of Solution**

7.6 STATICALLY INDETERMINATE TORSIONAL MEMBERS

In statically indeterminate torsional members, the torques that are needed to analyze the problem cannot be determined using the equations of equilibrium (i.e., statics) alone. Additional "compatibility" conditions must be introduced. A compatibility condition is a condition that satisfies structural integrity (or geometric continuity) of the structure. It is written in terms of the displacements at a point in a given direction. A compatibility condition in a torsional problem is as follows.

The angle of twist at a given X-section is the same (i.e., unique), regardless of which of the two segments joined at the X-section is used to determine that angle of twist.

Note: The application of the compatibility condition is rather analogous to that for axially loaded members (see Chapter 6).

Example 7.6

A uniform circular shaft AC_1 of length L and polar moment of area J is attached to a rigid wall at end A and connected to a step-down gear at the other end, C_1. A belt-drive at midspan B applies a torque T_M to the shaft. A second identical shaft C_2D is torqued by the gear at end C_2 and connected to the rigid wall at the other end, D. The shafts are supported by smooth ball bearings, which are located near the gear unit. The step down gear ratio is p.

a. Sketch the internal torque diagrams for the two shafts.
b. Determine the torques T_A and T_D of the shafts at their fixed ends A and D in terms of T_M and p.

Torsion in Shafts

Solution

Suppose that the drive gear (top gear in the figure) rotates through angle θ.
Then, the driven gear (bottom gear) rotates through angle θ/p (by the definition of p).
Circumferential movement of the drive gear $= R_1\theta$
Circumferential movement of the driven gear $= R_2\theta/p$
where
R_2 is the radius of the driven gear
R_1 is the radius of the drive gear

Since the two gear wheels are perfectly meshed (without backlash), compatibility requires

$$R_1\theta = R_2\frac{\theta}{p} \rightarrow \frac{R_2}{R_1} = p \tag{i}$$

a. The free-body diagrams of the two shaft units are shown in Figure 7.18b.
   Let F = tangential common force at the gear mesh point X.

   Note: In the indicated direction of motor torque, the top gear will push the bottom gear out of the plane with force F. Hence, there will be an equal reaction F on the top gear, into the plane.

   Then,
   Torque on gear $1 = T_1 = R_1F$ (cw direction, looking in from left)
   Torque on gear $2 = T_2 = R_2F$ (cw direction, looking in from left)
   Divide the second equation by the first:

$$\frac{T_2}{T_1} = \frac{R_2}{R_1} = p \tag{ii}$$

Equilibrium of shaft ABC_1:

$$T_A - T_M + T_1 = 0 \rightarrow T_1 = T_M - T_A \tag{iii}$$

(+ve direction is cw looking in from left)
Equilibrium of shaft C_2D:

$$T_2 - T_D = 0 \rightarrow T_2 = T_D \tag{iv}$$

(+ve when cw looking in from left)

Note: This is a statically indeterminate system because there are four unknowns T_1, T_2, T_A, and T_D with only three equations of equilibrium (ii), (iii), and (iv).

Even before determining the four unknown torques, it is possible to sketch the internal torque diagrams of the two shafts, using the information that is available thus far. These diagrams are sketched in Figure 7.18c and d.

b. Let ϕ_1 = rotation at C_1 (ccw looking in from left)
   ϕ_2 = rotation at C_2 (cw looking in from left)
   Then, by definition of the step-down gear ratio p, we have

$$\frac{\phi_1}{\phi_2} = p \tag{v}$$

Equation v is the compatibility condition at the gear transmission. We will use it to obtain the fourth equation for solving the problem.

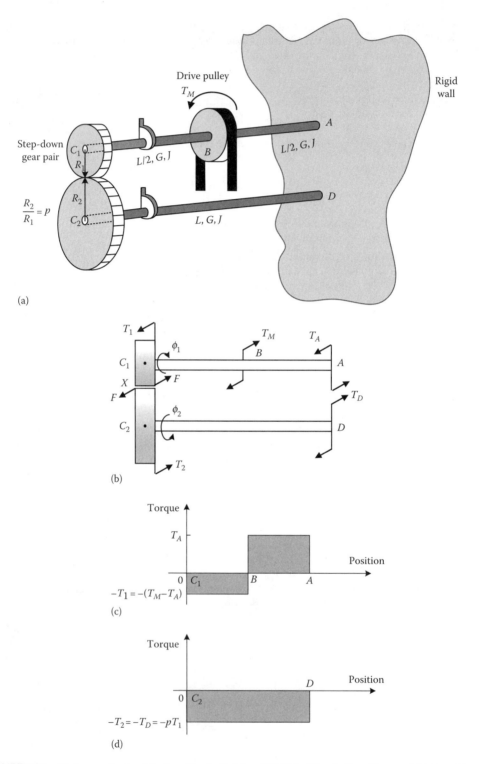

FIGURE 7.18 (a) A geared pair of shafts with a belt-drive; (b) FBD of the shafts with gear; (c) internal torque diagram of C_1A; (d) internal torque diagram of C_2D.

Torsion in Shafts

First apply (7.11): $\phi = TL/GJ$, for the two shafts. We have

$$\phi_1 = \frac{T_A L/2}{GJ} - \frac{(T_M - T_A)L/2}{GJ}$$

$$\phi_2 = \frac{T_D L}{GJ} = \frac{p(T_M - T_A)L}{GJ}$$

Substitute these two equations into (v): $\dfrac{T_A L/2}{GJ} - \dfrac{(T_M - T_A)L/2}{GJ} = \dfrac{p^2(T_M - T_A)L}{GJ}$

$$\rightarrow \frac{T_A}{2} - \frac{1}{2}(T_M - T_A) = p^2(T_M - T_A) \rightarrow (p^2 + 1)T_A = \left(p^2 + \frac{1}{2}\right)T_M$$

Hence,

$$T_A = \frac{(p^2 + 1/2)}{(p^2 + 1)} T_M$$

From (ii), (iii), and (iv): $T_D = p(T_M - T_A)$
Hence,

$$T_D = p\left[1 - \frac{(p^2 + 1/2)}{(p^2 + 1)}\right] T_M = \frac{p\left(p^2 + 1 - p^2 - (1/2)\right) T_M}{p^2 + 1}$$

$$\rightarrow T_D = \frac{p}{2(p^2 + 1)} T_M$$

Primary Learning Objectives
1. Torsional analysis of geared shafts
2. Torsional analysis of statically indeterminate systems
3. Derivation of the compatibility condition in geared systems
4. Determination of the internal torque diagram of geared shafts

■ **End of Solution**

Example 7.7

A shaft ABC of circular cross section is made by welding a steel segment AB of length $4L = 4$ m and radius $a_s = 50$ mm to a brass segment BC of length $L = 1$ m and radius $a_b = 25$ mm. The two ends A and C are rigidly fixed to walls (Figure 7.19a). A torque T_B is applied to the shaft at B.
Find:

a. The maximum torque T_{Bmax} that can be applied at B, without exceeding the allowable levels of shear stress
b. Corresponding reaction torques T_A and T_C at the two ends of the shaft
c. Corresponding maximum shear stresses in the two segments of the shaft
d. Corresponding angular movement ϕ_B at joint B

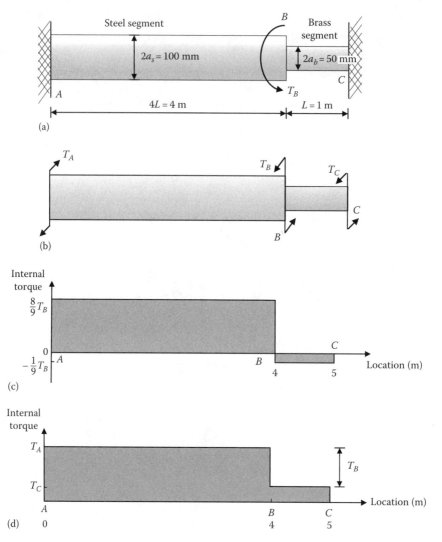

FIGURE 7.19 (a) A circular shaft with a steel segment and a brass segment; (b) free-body diagram of the shaft; (c) preliminary internal torque curve of the shaft; (d) complete internal torque curve of the shaft.

Given: The allowable shear stresses for the steel segment and the brass segment are, respectively, $\tau_{sallw} = 50$ MPa, and $\tau_{ballw} = 25$ MPa. The shear moduli of the two segments are

$$G_s = 80 \text{ GPa}, \quad G_b = 40 \text{ GPa}.$$

Solution

a. The FBD of the shaft is shown in Figure 7.19b.
   Equilibrium condition: $T_A - T_B - T_C = 0$ (i)
   → Two unknowns T_A and T_C, and one equation
   → Statically indeterminate system

   The preliminary internal torque curve is sketched in Figure 7.19c. This will give the nature of variation of the internal torque of the shaft, even though we do not know yet some key torques in order to fully determine the internal torque values.

Apply (7.11)

$$\phi = \frac{TL}{GJ}$$

Angle of twist in AB: $\phi_{AB} = \dfrac{T_A \times 4L}{G_s J_s}$ with $J_s = \dfrac{\pi}{2} a_s^4$

Angle of twist in BC: $\phi_{BC} = \dfrac{T_C \times L}{G_b J_b}$ with $J_b = \dfrac{\pi}{2} a_b^4$

Compatibility condition (in view of the fixed ends A and C):

$$\phi_{AB} + \phi_{BC} = 0$$

$$\rightarrow \frac{4 T_A L}{G_s (\pi/2) a_s^4} + \frac{T_C L}{G_b (\pi/2) a_b^4} = 0 \rightarrow \frac{4 T_A}{G_s a_s^4} + \frac{T_C L}{G_b a_b^4} = 0 \rightarrow \frac{4 T_A}{80 \times 50^4} + \frac{T_C}{40 \times 25^4} = 0$$

$$\rightarrow T_A + 8 T_C = 0 \qquad \text{(ii)}$$

(ii)−(i): $8T_C + T_B + T_C = 0$

$$\rightarrow T_C = -\frac{T_B}{9} \qquad \text{(iii)}$$

(ii): $T_A = \dfrac{8}{9} T_B \qquad \text{(iv)}$

The maximum shear stress occurs at $r = a$, the maximum radius. Apply (7.8b): $\tau_{max} = 2T/\pi a^3$ for a solid shaft.

For steel shaft:

$$\tau_{max\,s} = \frac{2 T_A}{\pi a_s^3} = \frac{2 \times (8/9) T_B}{\pi \times (0.05)^3} = \frac{2 T_B}{9\pi (0.025)^3} \qquad \text{(v)}$$

For brass shaft:

$$\tau_{max\,b} = \frac{2 T_C}{\pi a_b^3} = -\frac{2 \times (1/9) T_B}{\pi \times (0.025)^3} = -\frac{2 T_B}{9\pi (0.025)^3} \qquad \text{(vi)}$$

It is seen that the maximum shear stresses (which occur at the maximum radius of the X-section) are equal in magnitude in the two segments of shaft.

In designing a shaft, its maximum shear stress must be limited by the allowable shear stress. This requirement must be satisfied by both steel shaft and brass shaft, separately.

Since $\tau_{sallw} > \tau_{ballw}$, the brass segment will reach the allowable value of shear stress before the steel segment. Hence, the maximum allowable torque is governed by the brass shaft ($T_{max\,B}$). We have $|\tau_b| = \tau_{ballw}$

(vi):
$$\frac{2}{9\pi} \frac{T_{max\,B}}{(0.025)^3} = 25 \times 10^6$$

$$\rightarrow T_{max\,B} = \frac{9\pi \times (0.025)^3 \times 25 \times 10^6}{2} \, \text{N} \cdot \text{m} = 5.522 \, \text{kN} \cdot \text{m}$$

b. From (iv):
$$T_{max\,A} = \frac{8}{9} \times 5.522 \text{ kN} \cdot \text{m} = 4.908 \text{ kN} \cdot \text{m}$$

From (iii):
$$T_{max\,C} = -\frac{1}{9} \times 5.522 \text{ kN} \cdot \text{m} = -0.6136 \text{ kN} \cdot \text{m}$$

Since we know all the unknown torques, we can now sketch the complete internal torque curve as shown in Figure 7.19d.

c. From (v) and (vi):
$$\tau_{s\,max} = \tau_{ballw} = 25.0 \text{ MPa}$$

$$\tau_{b\,max} = -\tau_{ballw} = -25.0 \text{ MPa}$$

d. In order to determine the rotation at B, we apply $\phi = TL/GJ$ to either the steel shaft (using $T_{max\,A}$) or the brass shaft (using $T_{max\,C}$). We will get the same answer from both; one of them can be used as a check). Let us consider the steel shaft.

$$\phi_B = \frac{T_{max\,A} \times 4L}{G_s J_s} = \frac{4.908 \times 10^3 (\text{N} \cdot \text{m}) \times 4.0 \text{ (m)}}{8.0 \times 10^9 \times (\pi/2)(0.05)^4 \text{ (m}^4)} = 0.025 \text{ rad}$$

Primary Learning Objectives

1. Torsional analysis of a multisegmented shaft
2. Torsional analysis of a statically indeterminate shaft
3. Determination of the critical shear stress of a multisegmented torsional member
4. Design consideration of a multisegmented shaft with respect to allowable shear stress
5. Determination of the angles of twist of a multisegmented shaft

■ **End of Solution**

7.7 SOLID NONCIRCULAR SHAFTS

In torsion, noncircular shafts behave differently from circular shafts. For example, we know from Equation 7.8 that the maximum shear stress in a circular shaft in pure torsion occurs at the farthest point of the cross section from the center (i.e., at $r = a$). For a shaft with rectangular cross section, in contrast, it can be easily shown that the shear stress is zero at the farthest point of its cross section (i.e., at a corner). To show this, consider a small element at a corner of the section, as shown in Figure 7.20. There cannot be any shear stresses on the outer surfaces of the shaft because these surfaces are free of any external loading. Hence, the shear stresses τ_1, τ_3, τ_3', and τ_2 must be zero, because they are all on the two outer surfaces of the element. The two shear stress components on the cross-sectional plane of the element are τ_1' and τ_2'. But, τ_1' is the complementary shear stress of τ_1, which must be equal (see Chapter 3). Similarly, τ_2' is the complementary shear stress of τ_2, which also must be equal. It follows that the shear stress at a corner location of a shaft cross section must be zero.

A reason for the different behavior of a shaft of noncircular cross section is that, even under pure torsion, such a cross section does not remain plane (i.e., there will be some "warping" in the cross section). In contrast, as we showed previously, a circular cross section remains plane under torsion.

Torsion in Shafts

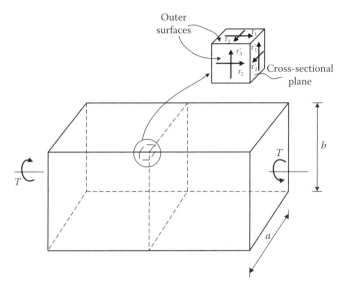

FIGURE 7.20 Zero shear stress at a corner of a rectangular X-section in pure torsion.

TABLE 7.2
Torsional Properties of Some Noncircular Sections

X-Sectional Shape	Max Shear Stress τ_{max}	J in Angle of Twist $\phi = \dfrac{TL}{GJ}$
Rectangle; sides a and b; $a > b$	$\dfrac{T}{\alpha a b^2}$	$\beta a b^3$
Triangle; equal sides a	$\dfrac{20T}{a^3}$	$\dfrac{\sqrt{3} a^4}{80}$
Ellipse; principal radii a and b; $a > b$	$\dfrac{2T}{\pi a b^2}$	$\dfrac{\pi a^3 b^3}{a^2 + b^2}$

TABLE 7.3
Parameter Values for Torsional Properties of a Rectangular Section

a/b	1.0	1.5	2.0	2.5	3.0	4.0	6.0	10	∞
α	0.208	0.231	0.246	0.258	0.267	0.282	0.298	0.312	0.333
β	0.141	0.196	0.229	0.249	0.263	0.281	0.298	0.312	0.333

For a shaft with rectangular cross section, the maximum shear stress occurs at the closest point of the edge of the section, that is, at the midpoint of the longer side of the cross section (see Figure 7.20).

The derivation of the formulas for shear stress and angle of twist, in terms of the internal torque, for solid shafts with uniform noncircular cross section is beyond the scope of the current treatment. Some results are presented in Table 7.2, without derivation.

For the rectangular section, selected values of the two parameters are listed in Table 7.3, as commonly provided in fine books and handbooks on the subject.

Example 7.8

Two shafts, one of square cross section and the other of rectangular section are made of the same material (shear modulus G), and the shaft lengths are the same (L). They have the same area of cross section (i.e., the same material content) as shown in Figure 7.21. If the two shafts are in pure torsion, subjected to the same torque T, compare their maximum shear stresses and the angles of twist.

Solution

In the formulas given in Table 7.2, for the square shaft, $a = b$. The corresponding parameter values in Table 7.3 are $\alpha = 0.208$ and $\beta = 0.141$.

$$\text{Its maximum shear stress is } \tau_{max} = \frac{T}{0.208a^3}$$

$$\text{Its angle of twist is } \phi = \frac{TL}{0.141Ga^4}$$

For the rectangular shaft, $a \to 2a$ and $b \to a/2$. This corresponds to $a/b = 4$, and the associated parameter values in Table 7.3 are $\alpha = 0.282$ and $\beta = 0.281$.

$$\text{Its maximum shear stress is } \tau_{max} = \frac{T}{0.282a^3/2} = \frac{T}{0.181a^3}$$

$$\text{Its angle of twist is } \phi = \frac{TL}{0.281Ga^4/4} = \frac{TL}{0.070Ga^4}$$

It is seen that, in the rectangular shaft, the maximum shear stress is about 15% greater and the angle of twist is twice as large. So, with respect to torsional performance, the square section is far superior to the rectangular section.

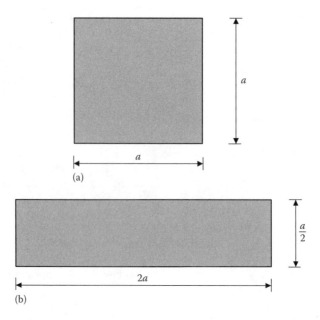

FIGURE 7.21 Two noncircular shafts with the same material volume; (a) square section; (b) rectangular section.

Torsion in Shafts

Primary Learning Objectives
1. Torsional analysis of shafts with noncircular solid cross section
2. Design consideration of shafts of noncircular X-section, with respect to maximum shear stress and angle of twist (flexibility)

■ **End of Solution**

7.8 THIN-WALLED TUBES

For shafts with hollow (closed) cross section, if the wall thickness is small compared to the cross-sectional dimensions, the torsional analysis becomes far simpler. In this section, a straightforward derivation of the shear stress relation for a thin-walled closed section (tube) subjected to a specific internal torque is given.

7.8.1 SHEAR STRESS RELATION

Consider a general cross section of a tube at an axial distance x. A general location P along the central perimeter of this hollow section (Figure 7.22a) is defined using a curvilinear coordinate s measured from a fixed point O on the perimeter. The tangential shear stress at P is τ_{xs}, and the wall thickness there is t. The perpendicular distance from the shear center C to the tangential line at P is r.

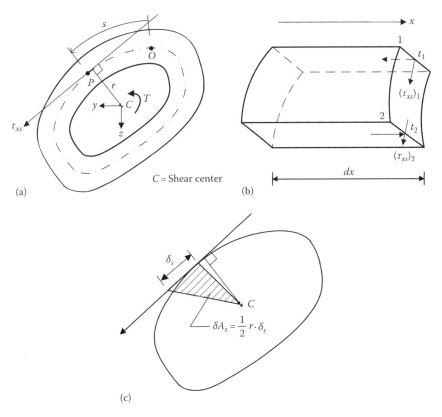

FIGURE 7.22 Nomenclature for a tube in torsion: (a) cross section; (b) axial element; (c) enclosed area element.

Shear center: The shear center of a cross section is the location where a transverse force may be applied without causing any twist (torsion) in the section. For a symmetric cross section, shear center coincides with the centroid (e.g., center of a circular section). It follows that shear center is the point about which moments should be taken of the tangential (transverse) forces in order to determine the net torque at the cross section.

Since the inside and the outside surfaces of the tube are free from stress and the wall thickness is small when compared with the perimeter, the radial shear-stress component τ_{xr} is negligible in comparison to τ_{xs}.

Isolate a small element of axial length dx and perimetric length $s_2 - s_1$ from the tube (Figure 7.22b) by cutting (virtually) the cross sections at the axial locations x and $x + dx$ and also cutting at the locations 1 and 2 of the central perimeter using axial planes (i.e., planes parallel to the x-axis) that are perpendicular to perimetric tangents at s_1 and s_2.

Equation of equilibrium: At corner 1 of the element, there is a tangential shear stress $(\tau_{xs})_1$ on the cross section, and a complementary and equal shear stress on the axial plane (in the direction opposite to x), as shown in Figure 7.22b. At corner 2 of the element, there is a tangential shear stress $(\tau_{xs})_2$ on the cross section, and a complementary and equal shear stress on the axial plane (in the direction of x), as shown.

Equilibrium in the x direction of the element: $t_2 dx (\tau_{xs})_2 - t_1 dx (\tau_{xs})_1 = 0$

$$\text{Hence, } t\tau_{xs} = \text{constant} = q \tag{7.12}$$

Shear flow: The constant q is called shear flow. It is clear from the equation of equilibrium given earlier that this is in fact the shear force on an axial section, per unit length (in the x direction). Furthermore, dividing the shear flow by the wall thickness, we get the shear stress at the location.

As shown in Figure 7.22c, consider a small tangential element of length δs and thickness t. The tangential force on the cross-sectional surface of this element is $\tau_{xs} t \delta s$. The moment of this force about the shear center is $r\tau_{xs} t \delta s$. This is the contribution of the elemental force of the considered element to the overall internal torque T at the tube cross section. The overall torque T is the summation of these elemental torques, which may be expressed in the limit using the cyclic integral over the central perimeter as $T = \oint r \tau_{xs} t \, ds$.

In view of Equation 7.12, the constant term, shear flow, may be taken outside the integral. Hence,

$$T = t \tau_{xs} \oint r \, ds \quad \text{(i)}.$$

From Figure 7.22c, it is clear that the shaded triangular area is $\delta A_s = (1/2) r \delta s$ (*Note*: Area of a triangle = Half the product of base and height). It follows that

$$A_s = \oint dA_s = \frac{1}{2} \oint r \, ds \tag{7.13}$$

where A_s is the enclosed area of the cross section (contained within the central perimeter).

On substituting Equation 7.13 into (i), we obtain

$$\tau = \frac{T}{2 A_s t} \tag{7.14}$$

This gives the shear stress (tangential) at any general point of wall thickness t of the tube section.

Torsion in Shafts

Example 7.9

Consider two thin-walled tubes, one of circular cross section of average radius a and the other with square cross section with average side length b, as shown in Figure 7.23. Both tubes have the same length and the same wall thickness t. The tubes are in pure torsion with the same internal torque T. If the material volume of the two tubes is the same, compare their shear stresses at the cross section.

Solution

Area of X-section of the circular section: $A_{sc} = \pi a^2$
Material area of X-section of the circular shaft: $A_{mc} = 2\pi a t$
Area of X-section of the square section: $A_{ss} = b^2$
Material area of X-section of the square shaft: $A_{ms} = 4bt$
In order to have the same material volume (for the same shaft length), we must have $A_{ms} = A_{mc}$

$$\rightarrow 4bt = 2\pi a t \rightarrow b = \frac{\pi}{2} a$$

The corresponding shear stresses at the wall sections are as follows (from Equation 7.14):

$$\text{Circular section: } \tau_c = \frac{T}{2\pi a^2 t}$$

$$\text{Square section: } \tau_s = \frac{T}{2b^2 t} = \frac{T}{2((\pi/2)a)^2 t} = \frac{2T}{\pi^2 a^2 t} \approx \frac{T}{1.571\pi a^2 t}$$

It is seen that, in the square tube, the shear stress at the wall section is more than 27% larger.

Primary Learning Objectives

1. Torsional analysis of thin-walled tubes
2. Design consideration of tubes of different sectional shapes

■ End of Solution

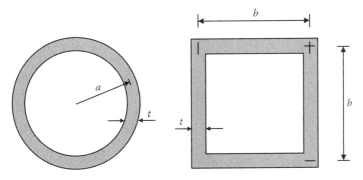

FIGURE 7.23 Two thin-walled tubes of circular X-section and square X-section.

7.9 COMPOSITE SHAFTS

Composite shafts are made of two or more materials in its different regions. A common example is a bimetallic shaft where the cylindrical core is made of one metal and the outer cylindrical segment is made of another metal. The two segments are assumed to be firmly attached together at their common cylindrical interface. Let us analyze in more detail such a composite circular shaft made of two homogeneous and isotropic materials (Figure 7.24). The core of the shaft has radius a_1 and is made of a material of shear modulus G_1. The outer segment of the composite shaft has an inner radius a_1 and outer radius a_2, and is made of a material of shear modulus G_2.

Since the two segments of the shaft are integral (firmly interconnected at the common interface) the previously derived shear strain Equations 7.2 and 7.3 are still valid. As mentioned earlier, the shear strain varies linearly from zero at the center of the cross section to the maximum value at $r = a_2$. Also, the angle of twist per unit length $\alpha = d\phi/dx$ is constant along the shaft (when the external torques exist only at the two ends of the shaft). The shear stress Equation 7.4 is valid as well, provided that the proper G is used depending on the material region that is considered.

In using the torque Equation 7.7, the two material segments have to be treated separately, as they have different shear moduli G and different polar moments of area J.

$$\text{Equation 7.7}: T = G_1 \alpha \int_0^{a_1} r^2 dA + G_2 \alpha \int_{a_1}^{a_2} r^2 dA = \alpha(G_1 J_1 + G_2 J_2)$$

$$\rightarrow \alpha = \frac{d\phi}{dx} = \frac{T}{(G_1 J_1 + G_2 J_2)} \qquad (7.15)$$

By integrating (7.15) over the shaft length L, the angle of twist may be obtained as

$$\phi = \frac{TL}{(G_1 J_1 + G_2 J_2)} \qquad (7.16)$$

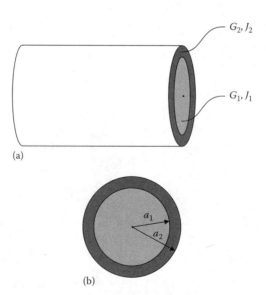

FIGURE 7.24 (a) A composite solid shaft; (b) cross section.

Torsion in Shafts

It is seen that in the torque deflection analysis, a composite shaft may be treated as a shaft of a single material with equivalent GJ as given by $(GJ)_{eq} = (G_1 J_1 + G_2 J_2)$.

Example 7.10

A bimetallic circular shaft of length L is made of steel (shear modulus $G_s = 75$ GPa) and bronze (shear modulus $G_b = 38$ GPa). Suppose that the material volumes of steel and bronze are equal. Consider two cases. Case A: steel core and bronze outer shell; Case B: bronze core and steel outer shell.

   a. Compare the angle of twist of the shaft in the two cases.
   b. Sketch the shear stress along the radius of a shaft cross section, for the two cases. Compare the shear stresses at the common interface of the two materials and at the outermost location of the shaft cross section, in the two cases.

Note: Express the answers in terms of the core radius a_1 and the internal torque T, which are the same in the two cases.

Solution

Let a_2 = outer radius of the shaft

For equal volumes of the two materials, we must have the same area of cross section for them, for a given length

$$\rightarrow \frac{\pi}{2} a_2^2 = 2 \times \frac{\pi}{2} a_1^2 \rightarrow a_2 = \sqrt{2} a_1 \tag{i}$$

a. Polar moments of area of the core (radius a_1) and the outer shell (inner radius a_1, outer radius a_2) are

$$J_1 = \frac{\pi}{2} a_1^4 \tag{ii}$$

$$J_2 = \frac{\pi}{2}\left(a_2^4 - a_1^4\right) = \frac{\pi}{2}\left(4 a_1^4 - a_1^4\right) = \frac{3\pi}{2} a_1^4 \tag{iii}$$

From Equation 7.16:

$$\text{Angle of twist } \phi = \frac{TL}{(G_1 J_1 + G_2 J_2)} = \frac{TL}{\left(G_1 \times (\pi/2) a_1^4 + G_2 \times (3\pi/2) a_1^4\right)} = \frac{2TL}{\pi a_1^4 (G_1 + 3G_2)} \tag{iv}$$

Case A: $1 \equiv$ steel; $2 \equiv$ bronze

$$\text{Equation iv: } \phi_A = \frac{2TL}{\pi a_1^4 (75 + 3 \times 38)}$$

Case B: $1 \equiv$ bronze; $2 \equiv$ steel

$$\text{Equation iv: } \phi_B = \frac{2TL}{\pi a_1^4 (38 + 3 \times 75)}$$

$$\Rightarrow \frac{\phi_B}{\phi_A} = \frac{75 + 3 \times 38}{38 + 3 \times 75} \simeq 0.72$$

It is seen that the angle of twist has reduced by about 28% in Case B. Hence, the shaft becomes more rigid when the outer shell is made of steel.

b. First we use Equation 7.15 to determine α, the angle of twist per unit length.

$$\text{Case A: } \alpha_A = \frac{T}{G_s J_1 + G_b J_2} = \frac{2T}{\pi a_1^4 (75 + 3 \times 38)}$$

$$\text{Case B: } \alpha_B = \frac{T}{G_b J_1 + G_s J_2} = \frac{2T}{\pi a_1^4 (38 + 3 \times 75)}$$

$$\Rightarrow \frac{\alpha_B}{\alpha_A} = \frac{75 + 3 \times 38}{38 + 3 \times 75} \simeq 0.72 \text{ (this is consistent with the earlier result) (v)}$$

Now use Equation 7.4 to determine the shear stresses.
Case A:
Interface stresses

$$\tau_{1s} = \alpha_A G_s a_1 = 75 \alpha_A a_1$$

$$\tau_{1b} = \alpha_A G_b a_1 = 38 \alpha_A a_1$$

Stress at outer radius, $\tau_{2b} = \alpha_A G_b a_2 = 38\sqrt{2} \alpha_A a_1 = 53.7 \alpha_A a_1$

Case B:
Interface stresses

$$\tau_{1b} = 38 \alpha_B a_1 = 38 \times 0.72 \times \alpha_A a_1 = 27.4 \alpha_A a_1$$

$$\tau_{1s} = 75 \alpha_B a_1 = 75 \times 0.72 \alpha_A a_1 = 54 \alpha_A a_1$$

Stress at outer radius, $\tau_{2s} = \alpha_B G_s a_2 = 75\sqrt{2} \times 0.72 \alpha_A a_1 = 76.4 \alpha_A a_1$

Using these results, the shear stress variation along the shaft radius is sketched in Figure 7.25.

It is seen that the overall maximum shear stress is only slightly greater in Case B while providing far greater rigidity. It follows that the use of a steel outer shell is better than using a steel core.

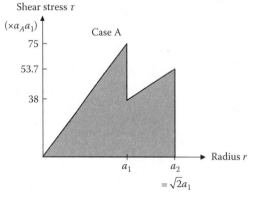 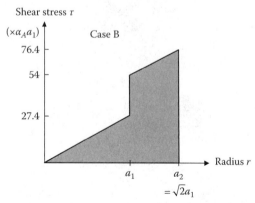

FIGURE 7.25 Shear stress variation (Case A: steel core; Case B: steel outer shell).

Torsion in Shafts

Primary Learning Objectives

1. Torsional analysis of composite shafts
2. Considerations of stiffness and maximum shear stress of a composite shaft
3. Design considerations of composite shafts

■ **End of Solution**

SUMMARY SHEET

Torque: May be represented by a couple (i.e., $T = d \times P$; equal and opposite parallel forces P separated by distance d).

Sign convention: Torque that moves a right-handed corkscrew out of the plane is +ve.

Shear strain (γ): At radius r of circular X-section; $\gamma = r\dfrac{d\phi}{dz} = \alpha r$ (*Note*: Linear variation along radius; valid for nonelastic material as well); ϕ is the angle of twist; $\dfrac{d\phi}{dz} = \alpha$ is the angle of twist per unit length).

Polar moment of area: $J = \dfrac{\pi a^4}{2}$ for solid shaft of radius a; $J = \dfrac{\pi}{2}\left(a_o^4 - a_i^4\right)$ for hollow shaft of inner radius a_i and outer radius a_o.

Two basic equations: $\dfrac{T}{J} = \dfrac{\tau}{r} = G\dfrac{d\phi}{dz}$; circular X-section; linear elastic material; T is the internal torque at X-section; τ is the tangential shear stress at r; J is the polar moment of area (second moment of area about shaft axis); G is the shear modulus (modulus of rigidity).

Maximum shear stress: Occurs at maximum radius ($r = a$). For a solid shaft, $\tau_{max} = \dfrac{2T}{\pi a^3}$; For a hollow shaft, $\tau_{max} = \dfrac{2Ta}{\pi\left(a_o^4 - a_i^4\right)}$.

Angle of twist: $\phi = \int_0^L \dfrac{T}{GJ}\,dz$; length $= L$; circular X-section)

$\phi = \dfrac{TL}{GJ}$ for uniform circular X-section (i.e., constant J), uniform internal torque (i.e., constant T), same material (i.e., constant G).

Torque diagram: Variation of the "internal" torque of a member along the axial length.

Statically indeterminate torsional members: Use compatibility condition (i.e., determine angle of twist of segments at a common point, and equate them).

Gear ratio: Step down gear ratio $p = $ [Rotation of input gear]/[Rotation of output gear] $= \dfrac{\theta_1}{\theta_2} =$ [Radius of output gear]/[Radius of input gear] $= \dfrac{R_2}{R_1} =$ [Torque of output gear]/[Torque of input gear] $= \dfrac{T_2}{T_1}$.

Shear flow in thin-walled torsional member: Product of tangential shear stress (τ) and thickness (t); $q = t\tau = $ constant.

Shear stress in thin-walled torsional member: $\tau = \dfrac{T}{2A_s t}$; A_s is the enclosed area of cross-section (contained within the central perimeter).

Composite circular shaft: Angle of twist per unit length, $\alpha = \dfrac{d\phi}{dx} = \dfrac{T}{(G_1 J_1 + G_2 J_2)}$

For uniform case and length L, $\phi = \dfrac{TL}{(G_1 J_1 + G_2 J_2)}$.

(Cylindrical core is made of one material and the tubular shell outside it is made of a different material; they have the same length).

PROBLEMS

7.1 A solid circular shaft of length $L=0.5$ m is made of a material with shear modulus $G=80$ GPa, and carries a torque $T=100$ N·m (see Figure P7.1). If the allowable maximum shear stress for the torque is $\tau_{al}=70$ MPa, determine a suitable radius a for the shaft. What is the corresponding angle of twist?

7.2 A circular hollow shaft of length $L=0.5$ m is expected to carry a torque $T=100$ N·m (see Figure P7.2). If its angle of twist should be limited to 0.02 rad and its maximum shear stress should be limited to $\tau_{al}=50$ MPa, select a suitable inner radius a_1 and an outer radius a_2 for the shaft.

7.3 A shaft ABC of length $2L$ and circular cross section is fixed to a wall at end A. From A to B, the shaft is solid, with cross-sectional radius $2a$. From B to C, the shaft is hollow, with internal radius a and outer radius $2a$ (see Figure P7.3).

G is the shear modulus of the shaft material
τ_Y is the yield shear stress of the shaft material

Suppose that a torque is applied at the free end C.
a. Determine the location at which the yielding will begin.
b. What is the maximum shear stress in the other segment of the shaft when yielding is about to begin in one segment?
c. Determine the maximum torque that may be applied at C before the shaft starts to yield.
d. When the yielding is about to begin, what is the angle of twist of the shaft (at C with respect to A)?

Note: Express your answer in terms of the given parameters only (a, L, G, τ_Y).

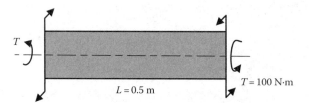

FIGURE P7.1 A uniform circular shaft with end torques.

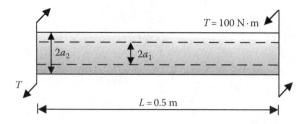

FIGURE P7.2 A uniform circular hollow shaft with end torques.

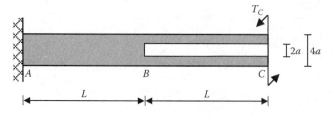

FIGURE P7.3 A shaft with a solid segment and a hollow segment.

7.4 A machine tool has a drive motor, which applies torque T to the chuck. The tool bit is firmly held by the chuck. A rigid block of material is machined (Figure P7.4). Assume the chuck to be a uniform circular tube with internal radius a_i, external radius a_e, and length L. The tool bit is assumed to be a uniform rod of circular cross section (radius a_b) and length L_b. The modulus of rigidity of the chuck is G and that of the tool bit is G_b.

The following numerical values are given:

$$a_e = 1 \text{ cm}, \quad a_i = 0.5 \text{ cm}, \quad L = 10 \text{ cm}, \quad G = 80 \text{ GPa}, \quad a_b = 0.5 \text{ cm},$$

$$L_b = 15 \text{ cm}, \quad G_b = 85 \text{ GPa}, \quad T = 10 \text{ N} \cdot \text{m}.$$

Determine:
  a. The maximum shear stress in the chuck-tool bit unit and its location
  b. The angle of twist of the chuck
  c. The angle of twist of the tool bit

7.5 The drive joint of a robot arm has a dc motor directly connected (without gearing) to the robot arm (i.e., the direct-drive configuration) using a tubular shaft, which also serves as a torque sensor element (Figure P7.5). The shaft has an internal radius a_i, external radius a_e (= 2 cm) and length L (= 2 cm). The modulus of rigidity of its material is $G = 30$ GPa. A strain

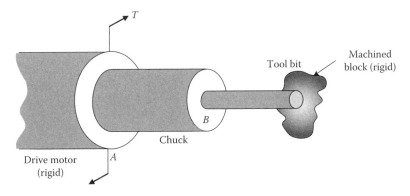

FIGURE P7.4 Machining of a hard material.

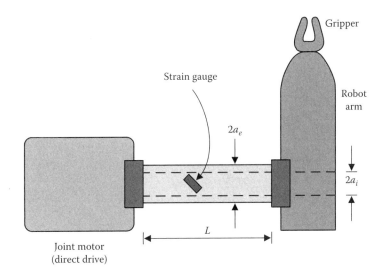

FIGURE P7.5 A direct-drive robot arm with a torque sensor.

gauge is mounted on the surface of the torque sensor tube at an angle of 45° to the axial direction. The normal strain ε measured by the strain gauge is related to the shear strain γ in the tangential direction near the shaft surface by $\gamma = 2\varepsilon$.

a. Obtain a relation to express the torque T transmitted by the tubular shaft in terms of the measured strain ε and the parameters G, a_e, and a_i.

b. From operation bandwidth considerations, the stiffness of the joint shaft K should not be below a specified value. *Note*: Stiffness $K = \dfrac{\text{Torque}}{\text{Angle of twist}} = \dfrac{T}{\phi}$. If the required $K = 150 \times 10^3$ N·m/rad, determine a suitable a_i for the joint shaft.

c. If the strain gauge measures $\varepsilon = 3000\mu\varepsilon$, what is the corresponding torque T?

7.6 A stepped steel shaft of circular cross section is shown in Figure P7.6. Its end O is fixed and end C is free. $OA = OB = 1.0$ m and $BC = 0.5$ m. The diameters of the shaft segments OA, OB, and BC are 80, 60, and 30 mm, respectively. Torques of −2.0, −2.2, and +0.2 kN·m are applied to the shaft at A, B, and C, respectively (*Note*: Use the standard sign convention for the torques). Shear modulus of steel, $G = 80$ GPa.

a. Sketch the internal torque diagram for the overall shaft.
b. Determine the angles of rotation at A, B, and C.
c. Determine the maximum shear stress in each of the three shaft segments. Which shaft segment is critical in determining the torsional strength of the overall shaft?

7.7 A load is driven jointly by a belt-drive and a chain-drive (Figure P7.7) through a common circular shaft OAB. The flat belt extends over the narrow segment OA of the shaft with radius a_1. The chain of the chain-drive is engaged with a sprocket wheel at edge A of the wide segment of the shaft having radius a_2. The driven load is attached to end B of the shaft.

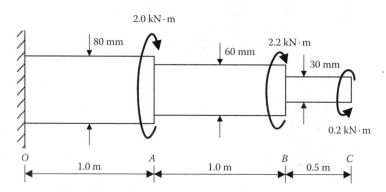

FIGURE P7.6 A stepped shaft with external torques.

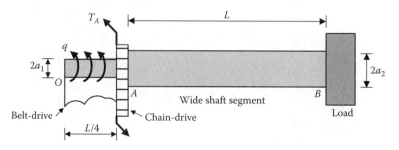

FIGURE P7.7 Load driven jointly by a belt-drive and a chain-drive.

Suppose that the belt applies a uniformly distribute torque of rate $q = 1000$ N·m/m over OA, and the chain applies a concentrated torque $T_A = 500$ N·m at A in the same direction (ccw direction, looking into the shaft from the left end). Determine
a. The internal torque diagram of the shaft
b. The maximum shear stress τ_{max} in the shaft and its location
c. The angle of twist of the shaft at B with respect to the end O

Given: $AB = L = 1$ m; $OA = L/4$; $a_1 = 1$ cm; $a_2 = 2$ cm; shear modulus of the shaft material, $G = 40$ GPa (bronze).

7.8 While drilling a block of hard material, a drill-bit experiences a distributed resistance along its entire length (near the completion of the task), which is given by $q(z) = kz$ N·m/m, where z is measured along the axis of the drill-bit, with $z = 0$ at the root (see Figure P7.8). The drill-bit is considered uniform with circular cross section of radius a and length L. The drill motor applies a point torque T_A on the drill-bit. *Given*: $T_A = 50$ N·m, $k = 3000$ N·m/m², $a = 1$ cm, $L = 20$ cm, and the modulus of rigidity of the drill-bit $G = 70$ GPa.
Determine
a. The maximum shear stress in the drill bit and its location
b. The total angle of twist in the drill bit

7.9 A concrete drill has a uniform drill-bit AE with approximately circular cross section of radius $a = 10$ mm and length $2L = 0.4$ m. When a hole BE of depth $L = 0.2$ m is drilled in a block of concrete, the drill-bit experiences a uniform resistance (due to firm contact with the hole) of $q = 100$ N·m/m. In addition, the hole exerts a resistance torque of T_e at the end E of the drill-bit due to the drilling action. The motor of the drill applies a torque $T_m = 50$ N·m on the drill-bit at A (see Figure P7.9).
Determine
a. The end torque T_e
b. The maximum shear stress τ_{max} in the drill-bit
c. The maximum shear stress τ_{emax} at the end of the drill-bit
d. The total angle of twist ϕ of the drill-bit

Given: Shear modulus of the drill-bit, $G = 80$ GPa

Hint: In this example, even though a "uniformity" assumption (which is used in deriving the formulas) is not valid for the overall member (because the internal torque T changes along the member), the derived formulas are applicable for an elemental segment (whose length $\delta z \to 0$) where the conditions are uniform.

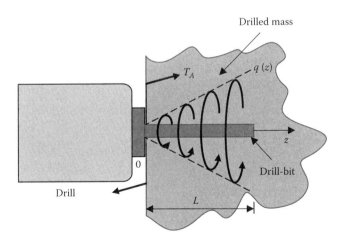

FIGURE P7.8 Drilling of a hard material.

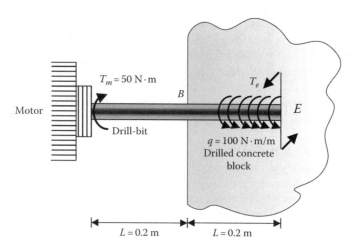

FIGURE P7.9 Drilling of a concrete block.

7.10 A frustum-shaped (i.e., conical) hollow shaft segment OA of length L has an outer radius a at the narrow end O and an outer radius $2a$ at the wide end A (Figure P7.10). The central bore of the shaft is uniform with radius a_i. The shaft segment is subjected to equal and opposite end torques T. The geometry of the frustum is such that the apex of the completing cone is at a distance L from the narrow end.
Determine
   a. The maximum shear stress $\tau_{\max}$ in the shaft segment and its location
   b. The angle of twist ϕ of the shaft segment

Given: $L=1$ m; $a=2$ cm; $a_i=1$ cm; $T=1000$ N·m; shear modulus $G=25$ GPa (aluminum alloy)

$$\text{Hint:} \int \frac{dz}{z^2 - b^2} = \frac{1}{2b} \ln \frac{z-b}{z+b} \quad \text{for } z^2 > b^2$$

$$\int \frac{dz}{z^2 + b^2} = \frac{1}{b} \tan^{-1} \frac{z}{b}$$

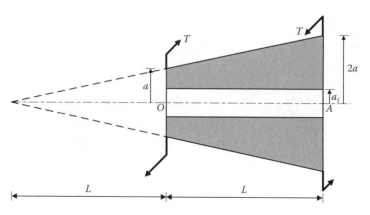

FIGURE P7.10 Frustum-shaped hollow shaft with end torques.

7.11 A motor drives a load using a step-down gear of speed ratio, motor speed: load speed $=p{:}1$, with $p=2$. The motor shaft and the load shaft both are made of steel (shear modulus $G=75$ GPa) and have a cross-sectional radius, $a=2$ cm. The length of the motor shaft is $L_m=0.3$ m and the length of the load shaft is $L_l=0.5$ m. The torque applied by the motor to its shaft is $T_m=100$ N·m (see Figure P7.11).

Determine
   a. The maximum shear stress and its location
   b. The angle of rotation of the load with respect to the motor rotor

7.12 A circular steel shaft CDE is of length $2L=2$ m and radius $a_s=50$ mm is fixed at C and free at E. A circular brass shaft AB of length $L=1$ m and radius $a_b=25$ mm is fixed at A and is kept in parallel with the steel shaft. The end B of the brass shaft has a gear of radius 10 cm. The mid-span D of the steel shaft carries a gear of radius 20 cm, which is perfectly meshed (without any backlash; i.e., no looseness) with the gear of the brass shaft (Figure P7.12).
   a. If the allowable shear stresses for the steel shaft and the brass shaft are $\tau_{sallw}=50$ MPa and $\tau_{ballw}=25$ MPa, respectively, determine the maximum torque T_{Emax} that can be applied at the free end (E) of the steel shaft such that the allowable shear stresses are not exceeded in the two-shaft unit.
   b. Determine the angle of rotation of the steel shaft at E (the point of application of the external torque) corresponding to this maximum torque.

Note: The shear moduli of the steel shaft and the brass shaft are $G_s=80$ GPa and $G_b=40$ GPa, respectively.

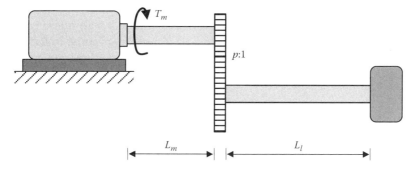

FIGURE P7.11 Load driven by a motor through a step-down gear.

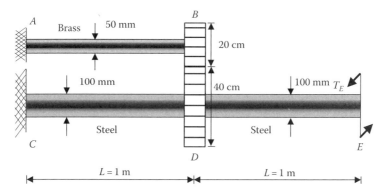

FIGURE P7.12 Two shafts coupled through a gear unit.

7.13 A brass tube AC of length $3L = 3$ m, inside diameter $d_i = 50$ mm, and outside diameter $d_o = 100$ mm is rigidly fixed at its two ends and a torque $T_B = 5$ kN·m is applied at location B where $AB = 2L = 2$ m (see Figure P7.13).

Determine
  a. The reaction torques T_A and T_C at the ends of the tube
  b. The maximum shear stress in the tube
  c. The rotation at section B

Given: Shear modulus of the tube material, $G = 40$ GPa

7.14 A uniform solid shaft $OABC$ of circular cross section is firmly fixed at its two ends O and C (Figure P7.14). A uniformly distributed torque of rate $q = 2000$ N·m/m is applied over the segment AB of the shaft.

Determine
  a. The internal torque diagram of the shaft
  b. The maximum shear stress and its location
  c. The angle of twist at A (with respect to the fixed ends)

Given: Radius of the shaft X-section, $a = 2$ cm
Shear modulus of the shaft, $G = 80$ GPa (steel)

$$OA = \frac{L}{2}; \quad AB = \frac{L}{4} \quad OC = L = 2 \text{ m}$$

7.15 A shaft of length $3L = 3$ m and circular cross section of radius $a = 10$ mm is driven with a torque $T_A = 12$ N·m at end A and is meshed with another shaft using a gear wheel pair at the other end B (see Figure P7.15). The second shaft DE of length $4L = 4$ m is meshed with the first shaft at C where $DC = L = 1$ m. The cross section of this shaft is also circular, and has a radius $1.5a = 15$ mm. The two shafts are made of the same material with $G = 40$ GPa. The two ends of the second shaft are rigidly fixed and the gear does not have any backlash. The step-down gear ratio from the first shaft to the second shaft is 3:1.

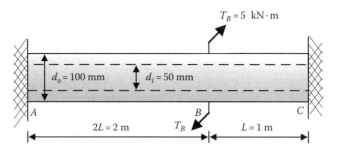

FIGURE P7.13 A tube subjected to an external torque.

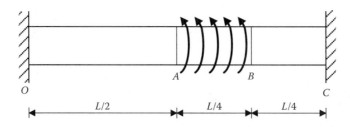

FIGURE P7.14 A shaft with fixed ends and subjected to a distributed torque.

Torsion in Shafts

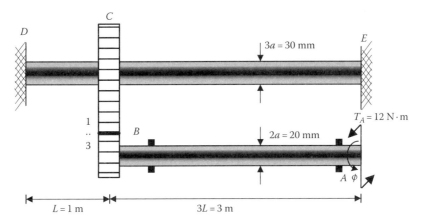

FIGURE P7.15 Two parallel shafts coupled through a gear pair.

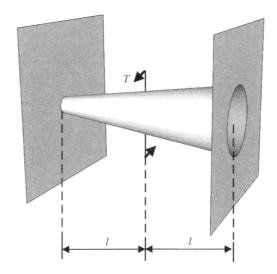

FIGURE P7.16 Torsion of a solid conical member (frustum).

Determine
  a. The reaction torques T_D and T_E of the second shaft at the fixed ends D and E
  b. The rotation ϕ at A of the first shaft due to the applied torque

7.16 A solid frustum is rigidly fixed to two vertical walls at its ends and maintained horizontally. An external torque T is applied at the midspan of the frustum, as shown in Figure P7.16. The apex angle of the conical section of the frustum is 2θ and the modulus of rigidity (shear modulus) of the material is G. Determine the holding torques at the two ends.

8 Bending in Beams

CHAPTER OBJECTIVES

- Importance of the subject
- Sign convention
- Shear diagram and moment diagram
- Graphical method for shear and moment diagrams
- Flexure formula (for bending stress)
- Composite beams and equivalent sections
- Transverse shear and shear flow at joints
- Deflection in beams
- Statically indeterminate beams

8.1 INTRODUCTION

In this chapter, we will study flexible beams subject to transverse loading. Related issues of mechanics of material will be considered, which mainly deal with stresses (strength) and strains (deflection and stiffness). The main focus of the present chapter will be the analysis of straight beams with symmetric cross-sections. The topic of bending in beams is important in view of the wide applicability of bending members (flexural members) in engineering applications. Building structures, bridges, overhead vehicle guideways, robots, airplanes, ships, ground vehicles, and construction machinery such as cranes are examples (see Figure 8.1). Even when beam elements are subjected to other types of loads (e.g., axial forces, torques), in the linear case, one can analyze the bending loading and deflections separately and then add to the effects of other types of loading, by using the principle of superposition.

Bending stresses and bending strains generated in beams are an important criterion in beam design. In engineering design, stresses govern strength, load-carrying capacity, structural integrity, and safety. Deflections may be associated with operational requirements of moving members and may also need to be limited in view of such considerations as safety, comfort, stability, vibration, noise, esthetics, and operational quality. It follows that the analysis of stresses, strains, and deflections in beams is of great importance.

In the ideal case of a beam subjected to pure bending moments, the primary stresses and strains of importance are normal stresses and normal strains. However, typically bending loads are caused by lateral forces, which also generate internal shear forces, and associated shear stresses and shear strains (or deflections due to shear). Hence, we will also study shear stresses and their effects in bending members. In statically indeterminate bending members, "compatibility conditions" in terms of deflections have to be used in addition to the equations of equilibrium (statics) to determine support reactions and internal loads. Such problems will be discussed as well in this chapter.

8.2 SHEAR AND MOMENT DIAGRAMS

The knowledge of internal bending moments and internal shear forces is essential in the analysis (determination of stresses and deflections) of bending members (beams). Hence, the starting step of the beam bending problem is the development of the shear diagram and the moment diagram of the analyzed member.

FIGURE 8.1 Examples of the engineering application of bending members; (a) a building structure; (b) boom of a crane; (c) a water wheel; (d) a cantilevered balcony. (Courtesy of C.W. de Silva.)

Shear diagram: This shows how the "internal" shear force (in the transverse direction, i.e., perpendicular to the axial direction) varies along the beam.

Moment diagram: This shows how the "internal" bending moment (at beam cross section) varies along the beam.

8.2.1 Steps of Derivation of Shear and Moment Diagrams

1. Draw a free-body diagram for the bending member and mark all the loads on it, including those at the supports.
2. Write equations of equilibrium and determine the unknown loads (typically support loads).
3. Divide the member into segments (based on the support locations and external load locations/discontinuities).
4. Starting from a convenient end of the member, make a virtual cross section (cut) within the first segment of the member. By writing equilibrium equations for the isolated part of the segment, determine the shear force and the bending moment at the cut section.
5. Repeat Step 4 for the remaining segments of the member.
6. Sketch the shear and moment diagrams as a function of the location along the beam. Mark the significant values of shear force and bending moment on the two diagrams.

Bending in Beams

8.2.2 Sign Convention

A suitable sign convention has to be established for the loading, shear force, and bending moment before equations are derived for the shear and moment diagrams. Specifically, the following sign convention is used.

1. Distributed load $w(x)$ per unit length acting upward is +ve.
2. A shear force that tends to rotate the considered element in the clockwise (cw) direction is +ve.
3. A bending moment that tends to bend the ends of the considered element upward (i.e., forming "bucket" shape or "smiling mouth" shape) is +ve.

Note: A point load is a special case of a distributed load, and its sign convention is consistent with what is given above for a distributed load.

8.2.3 Governing Relations

Consider a beam subjected to a general distributed load $w(x)$. Take a small element of length δx at location x along the beam as shown in Figure 8.2a. Suppose that the shear force and the bending moment (both "internal") are denoted by V and M, respectively, on the left cross section (at x), as shown in Figure 8.2b. Then, they are $V + \delta V$ and $M + \delta M$ on the right cross section, where δV and δM are the increments of the values over the small length δx. The distributed force, shear force, and bending moment all three are indicated in the +ve direction in Figure 8.2b, according to the sign convention.

Note: For convenience, we may use the term "moment" to mean "bending moment" and the term "shear" to denote "shear force."

We write the equations of equilibrium.
Force balance (upward is taken +ve):

$$\uparrow \sum F_y = 0 \Rightarrow V - (V + \delta V) + w(x) \cdot \delta x = 0 \Rightarrow \delta V = w(x) \cdot \delta x$$

Note: $w(x)$ may be assumed uniform over the elemental length δx, which is very small by definition and approaches zero.

In the limit:

$$\frac{dV}{dx} = w(x) \tag{8.1}$$

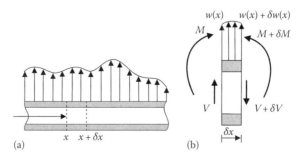

FIGURE 8.2 (a) A beam subjected to a distributed external force; (b) a general element of the beam and associated loading.

Moment about the right section (counter-clockwise is taken +ve):

$$\circlearrowleft \sum M = 0 \Rightarrow -M + (M + \delta M) - V \cdot \delta x = 0 \Rightarrow \delta M = V \cdot \delta x$$

Note: In this equation too, we have neglected the terms smaller than $\delta(\cdot)$; i.e., "order 2 terms" or $O(2)$ terms, which are insignificant in the limit (as $\delta x \to 0$). In particular, we have neglected the moment of the lateral force $w(x) \cdot \delta x$. Here, both the force and the distance (moment arm) are "δ" terms, and hence the moment is a product of two "δ" terms, which is $O(2)$.

In the limit:

$$\frac{dM}{dx} = V \tag{8.2}$$

8.2.4 Effect of Point Load on Shear and Moment

When crossing an external concentrated force on the beam, it is important to know whether this point force contributes positively or negatively to the internal shear force. Similarly, when crossing an external point moment on the beam, it is important to know whether this point moment contributes positively or negatively to the internal bending moment. The answer can be easily established by considering the equilibrium of a small beam element that includes the point load.

From Figure 8.3a, with an upward external point force F_0, the force balance of the element gives

$$V_{x+\delta x} = V_x + F_0 \tag{8.3}$$

Note: If F_0 is downward, the corresponding result is obtained by simply changing F_0 to $-F_0$ in Equation 8.3.

Hence, we can state the following for a point force:

1. Crossing from left to right, for internal shear, add upward external force or subtract downward external force.
2. Crossing from right to left, for internal shear, subtract upward external force or add downward external force.

From Figure 8.3b, with a clockwise (cw) external point moment M_0, the moment balance of the element gives

$$M_{x+\delta x} = M_x + M_0 \tag{8.4}$$

Note: If M_0 is in the counter-clockwise (ccw) direction, the corresponding result is obtained by simply changing M_0 to $-M_0$ in Equation 8.4.

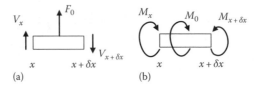

FIGURE 8.3 (a) Effect of external point force on internal shear; (b) effect of external point moment on internal moment.

Bending in Beams

Hence, we can state the following for a point moment:

1. Crossing from left to right, for internal moment, add cw external moment or subtract ccw external moment.
2. Crossing from right to left, for internal moment, subtract cw external moment or add ccw external moment.

Example 8.1

A beam AD of length $2L = 8$ m is hinged (frictionless) at end A and supported on a smooth roller at its midpoint B (see Figure 8.4a). An L-shaped rigid attachment is firmly welded to the over-hung end D of the beam, as shown. A vertical downward force $P_C = 100$ kN m is applied to end C of the rigid attachment where C falls exactly midway between B and D. A uniformly distributed downward force $w_0 = 50$ kN/m is applied over the segment AB of the beam.

Determine and sketch the shear diagram and the moment diagram of the beam.

Note: You must include the true shapes of the curves and also give the key numerical values on them.

Solution

Free-body diagram of the systems is shown in Figure 8.4b.
First, we determine the unknown support reactions by using the equations of equilibrium

$$\uparrow \sum F = 0: R_A + R_B - 200 - 100 = 0 \rightarrow R_A + R_B = 300 \tag{i}$$

$$\circlearrowleft \sum M_A = 0: -200 \times 2 + R_B \times 4 - 100 \times 6 = 0$$

$$\rightarrow R_B = 250 \text{ kN}$$

(i): $R_A = 50$ kN

Now we will use the *method of sections* to obtain the shear diagram and the moment diagram.

Segment AB (Figure 8.4c)

Make a virtual section at X, between A and B at distance x, and consider the segment to the left, as shown in Figure 8.4c.

Equilibrium:

$$\uparrow \sum F = 0: R_A - w_0 x - V = 0 \rightarrow V = R_A - w_0 x = 50 - 50x \text{ kN}$$

$$\circlearrowleft \sum M_X = 0: M + w_0 x \times \frac{x}{2} - R_A x = 0 \rightarrow M = R_A x - \frac{w_0 x^2}{2}$$

$$\rightarrow M = 50x - 25x^2 \text{ kN·m}$$

Point B: Due to the upward concentrated force R_B at B, the internal shear force V will jump by $R_B = 250$ kN, at B.

Note: Internal bending moment M will not change at B, because there is no external concentrated moment there.

Segment BC (Figure 8.4d)

Make a virtual section at X, between B and D at distance x from A, and consider the segment to the left, as shown in Figure 8.4d.

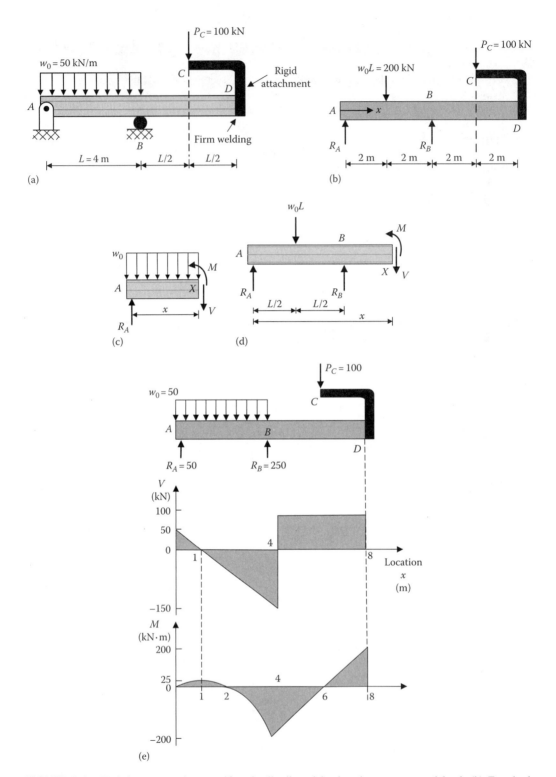

FIGURE 8.4 (a) A beam carrying a uniformly distributed load and a concentrated load. (b) Free-body diagram of the system. (c) Free-body diagram of the sectioned segment for the analysis of *AB*. (d) Free-body diagram of the sectioned segment for the analysis of *BD*. (e) Shear diagram and moment diagram.

Bending in Beams

Equilibrium:

$$\uparrow \sum F = 0: R_A + R_B - w_0 L - V = 0 \to V = R_A + R_B - w_0 L$$

$$\to V = 300 - 200 = 100 \text{ kN}$$

$$\circlearrowleft \sum M_X = 0: M + w_0 L \times \left(x - \frac{L}{2}\right) - R_A x - R_B \times (x - L) = 0$$

$$\to M = R_A x + R_B \times (x - L) - w_0 L \times \left(x - \frac{L}{2}\right) = 50x + 250(x - 4) - 200(x - 2)$$

$$\to M = 100x - 600 \text{ kN·m}$$

The obtained expressions for V and M are sketched in Figure 8.4e.

Primary Learning Objectives

1. Determination of internal loading when an external load is applied indirectly
2. Determination of expressions for internal shear force and internal bending moment at a general location along a beam, using the *method of sections*
3. Change of the internal shear force as a result of an external point force
4. Sketching of shear diagram and moment diagram

■ **End of Solution**

Example 8.2a

Two equal beam segments AB and BD, each of length $L = 5$ m, are pin-jointed (smooth) at B. End A is supported on smooth rollers and end D is clamped. A uniformly distributed downward force of rate $w_0 = 40$ kN/m is applied along AB. A downward concentrated force $P_C = 100$ kN is applied at the midspan C of the segment BD (Figure 8.5a).

Determine the shear diagram and the moment diagram of the beam.

Solution

Support load at A is a vertical reaction force R_A. Support load at D consists of a vertical reaction force R_D and a bending moment M_D.

Since there are three unknowns, to determine them, the two beam segments have to be considered separately. (*Note*: If the entire beam is considered, we get only two equations of equilibrium from which the three unknowns cannot be determined.) The free-body diagrams of the two beam segments are shown in Figure 8.5b.

Equilibrium of AB:

$$\uparrow \sum F = 0: R_A - w_0 L + R_B = 0 \to R_A + R_B = w_0 L = 40 \times 5 = 200 \text{ kN} \qquad \text{(i)}$$

$$\circlearrowleft \sum M_A = 0: R_B \times L - w_0 L \times \frac{L}{2} = 0 \quad \to \quad R_B = \frac{w_0 L}{2} = \frac{40 \times 5}{2} = 100 \text{ kN} \qquad \text{(ii)}$$

Substitute (ii) in (i): $R_A = 200 - 100 = 100$ kN

Equilibrium of BD:

$$\uparrow \sum F = 0: -R_B - P_C + R_D = 0 \rightarrow R_D = R_B + P_C = 100 + 100 = 200 \text{ kN} \quad \text{(i)}$$

$$\circlearrowleft \sum M_B = 0: -P_C \times \frac{L}{2} + R_D \times L - M_D = 0 \rightarrow M_D = R_D L - \frac{P_C L}{2} \quad \text{(ii)}$$

$$\rightarrow M_D = 200 \times 5 - \frac{100 \times 5}{2} = 750 \text{ kN} \cdot \text{m}$$

Next, we determine the shear and moment diagrams for different segments of the beam.

Segment AB (Figure 8.5c)
Make a virtual section at X, which is located at a distance x from A.
Equilibrium:

$$\uparrow \sum F = 0: R_A - w_0 x - V = 0 \rightarrow V = R_A - w_0 x \rightarrow V = 100 - 40x \text{ kN}$$

$$\circlearrowleft \sum M_X = 0: M - R_A \times x + w_0 x \times \frac{x}{2} = 0 \rightarrow M = R_A x - \frac{w_0 x^2}{2}$$

$$\rightarrow M = 100x - 20x^2 \text{ kN} \cdot \text{m}$$

Segment BC (Figure 8.5d)
Make a virtual section at X, which is located between B and C, at a distance x from A.

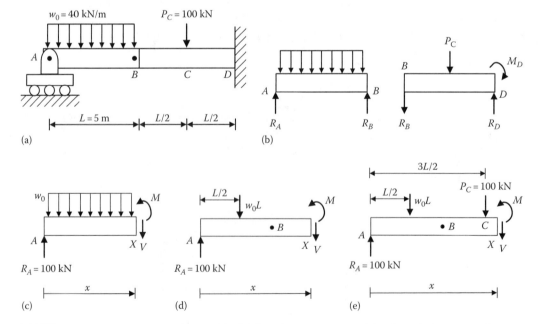

FIGURE 8.5 (a) Two-segmented beam with distributed and concentrated loading. (b) Free-body diagrams of the beam segments. (c) Beam segment with sectioning between A and B. (d) Beam segment with sectioning between B and C. (e) Beam segment with sectioning between C and D.

Bending in Beams

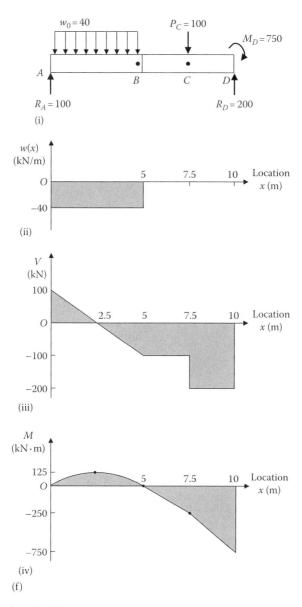

FIGURE 8.5 (continued) (f) (i) Free-body diagram of the beam; (ii) distributed load; (iii) shear diagram; (iv) moment diagram.

Equilibrium:

$$\uparrow \sum F = 0: R_A - w_0 L - V = 0 \to V = R_A - w_0 L \to V = 100 - 40 \times 5 = -100 \text{ kN}$$

$$\circlearrowleft \sum M_X = 0: M + w_0 L \times \left(x - \frac{L}{2}\right) - R_A \times x = 0$$

$$\to M = R_A x - w_0 L \left(x - \frac{L}{2}\right) = 100x - 200\left(x - \frac{5}{2}\right)$$

$$= -100x + 500 \text{ kN} \cdot \text{m}$$

Segment CD (Figure 8.5e)

Make a virtual section at X, which is located between C and D, at a distance x from A.
Equilibrium:

$$\uparrow \sum F = 0: R_A - w_0 L - P_C - V = 0 \rightarrow V = R_A - w_0 L - P_C$$

$$\rightarrow V = 100 - 200 - 100 = -200 \text{ kN}$$

$$\circlearrowleft \sum M_X = 0: M + P_C \times \left(x - \frac{3L}{2}\right) + w_0 L \times \left(x - \frac{L}{2}\right) - R_A \times x = 0$$

$$\rightarrow M = 100x - 100\left(x - \frac{3L}{2}\right) - 200\left(x - \frac{L}{2}\right) = -200x + 1250 \text{ kN}$$

The shear and moment diagrams are sketched in Figure 8.5f.

Primary Learning Objectives
1. Determination of the support reactions and internal loading when there is an internal pin joining two beam segments
2. Determination of expressions for internal shear force and internal bending moment at a general location along a multisegmented beam, using the *method of sections*
3. Sketching of shear diagram and moment diagram

■ **End of Solution**

8.2.5 Area Method (Graphical Method) for Shear Diagram and Moment Diagram

It is clear from the integrated versions of Equations 8.1 and 8.2 that

1. Change in internal shear force in an interval = Area of the distributed force per unit length curve in that interval
2. Change in internal bending moment in an interval = Area of the shear diagram in that interval

So, once a starting value is established, the *area method* (*graphical method*) can be used to construct the shear diagram from the distributed load curve, and the moment diagram from the shear diagram using the integrated forms of (8.1) and (8.2):

$$V(x) = V_0 + \int_0^x w(x)\,dx \qquad (8.5)$$

$$M(x) = M_0 + \int_0^x V(x)\,dx \qquad (8.6)$$

It follows that the shear diagram can be graphically determined by adding the area under the $w(x)$ curve to the starting value of shear.

Similarly, the moment diagram can be graphically determined by adding the area under the shear curve $V(x)$ to the starting value of moment.

These are illustrated in Figure 8.6.

Bending in Beams

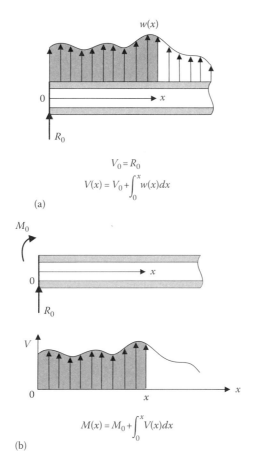

FIGURE 8.6 Area method (a) to generate the shear diagram; (b) to generate the moment diagram.

Example 8.2b

Example 8.2a may be solved using the "area method" (or, the graphical method) as well. Here, we start with the "distributed force" curve $w(x)$ as shown in part (ii) of Figure 8.5f. Then, starting from the initial values of V and M (at A), we add the area under the $w(x)$ curve to obtain V, and add the area under the V curve to obtain M.

Starting Values

Since $R_A = 100$ kN (upward), according to the sign convention, $V_0 = 100$ kN. (This may be confirmed by making a section just right of A and considering the equilibrium of the resulting left segment.) Also, since end A cannot support a bending moment (because of the pin support), we have $M_0 = 0$.

Segment AB

$$w(x) \text{ is constant} \rightarrow V \text{ varies linearly}$$

End value of V (at B) = V_0 + Area of the $w(x)$ curve = $100 + (-40 \times 5) = -100$ kN
Accordingly, the V curve (joining the points 100 and -100) will pass through the midpoint of AB (i.e., at 2.5 m). This point corresponds to maximum M because it is where the +ve V ends. After that, M will decrease (because of $-$ve V).

Since the variation of V is linear, the variation of M is quadratic, and symmetric about $x = 2.5$ m in view of the nature of the V curve.

Segment BC

$w(x) = 0 \rightarrow V$ will be constant, with the value -100 kN
$\rightarrow M$ will decrease linearly (because of $-$ve V)
End value of M (at C) = M_B + Area of V curve from B to C

$$= 0 + (-100 \times 2.5) = -250 \text{ kN·m}$$

Point C: Internal shear V will jump down by $P_C = 100$, to value $-100 - 100 = -200$ kN because P_C is acting downward.

Note: M will not jump at C (because there is no concentrated external moment there).

Segment CD

$w(x) = 0 \rightarrow V$ will remain constant, at -200 kN
$\rightarrow M$ will decrease linearly (because of $-$ve V)
$M_D = M_C$ + Area of V curve from C to D

$$= -250 + (-200 \times 2.5) = -750 \text{ kN·m}$$

In this manner, we get the overall shear curve shown in Figure 8.5f, part (iii); and the overall moment curve shown in Figure 8.5f, part (iv). The curves obtained in this manner are identical to those obtained by the method of sections (and the resulting analytical expressions) in Example 8.2a.

Primary Learning Objectives

1. Application of the area method to determine the shear diagram and the moment diagram
2. Determination of the support reactions and internal loading when there is an internal pin joining two beam segments
3. Determination of expressions for internal shear force and internal bending moment of a multisegmented beam

Example 8.3

Two identical beam segments AC and CD, each of length L are joined by a smooth pin at C. The overall beam is fixed horizontally at end A and supported horizontally on a smooth roller at the other end D (see Figure 8.7a). A concentrated downward force $2P$ is applied at the midspan B of the segment AC (*Note:* $AB = L/2$). A distributed downward force, which varies linearly from zero at C to w_o/length at D, is applied to the segment CD.

Taking $w_o L = 6P$, and using the graphical method (area method), determine the shear diagram and the moment diagram for the overall beam. Mark the key values of shear and moment on the curves.

Solution

The FBD of the beam segment CD is shown in Figure 8.7b.

Note: There is no moment at C because it is pin-jointed.

Equations of equilibrium:

Note: Represent the distributed force by the equivalent point force of magnitude $w_o L/2 = 3P$ acting at the loading centroid, which is at $2L/3$ from C.

$$\uparrow \sum F_y = 0: R_C - 3P + R_D = 0 \Rightarrow R_C + R_D = 3P \qquad \text{(i)}$$

$$\circlearrowleft \sum M_C = 0: -3P \times \frac{2L}{3} + R_D \times L = 0 \Rightarrow R_D = 2P \qquad \text{(ii)}$$

$$\text{(i)} \Rightarrow R_C = 3P - R_D = P \qquad \text{(iii)}$$

Bending in Beams

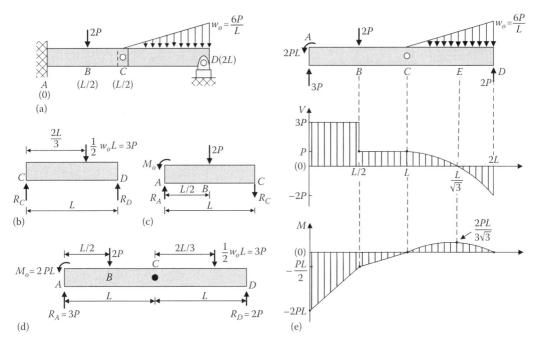

FIGURE 8.7 (a) Two-segmented beam with internal pin; (b) free-body diagram of segment CD; (c) free-body diagram of segment AC; (d) free-body diagram of the entire beam; (e) shear and moment diagrams.

The FBD of the beam segment ABC is shown in Figure 8.7c.

Note: The fixed end A has both a reaction R_A and a bending moment M_o.

Equations of equilibrium:

$$\uparrow \sum F_y = 0 : R_A - 2P - R_C = 0 \Rightarrow R_A = 2P + R_C$$

$$\text{(iii)} \Rightarrow R_A = 2P + P = 3P \tag{iv}$$

$$\curvearrowleft \sum M_A = 0 : M_o - 2P \times \frac{L}{2} - R_C \times L = 0$$

$$\text{(iii)} \Rightarrow M_o = PL + R_C L = PL + PL = 2PL \tag{v}$$

A check for the obtained results may be made by considering the FBD of the entire beam (Figure 8.7d). Note that the forces balance, and the moments about any point also balance.

The graphical method (area method) of obtaining the shear curve and the moment curve may proceed as follows:

Shear Curve

Starting from the known shear force value at A ($= R_A = 3P$), add the area of the distribute force (or subtract if the force is downward).

Note: A downward point force causes a drop in the shear force by its value (see Figure 8.7e).

Moment Curve

Starting from the known bending moment value at A ($= -M_o = -2PL$), add the area of the shear force curve (see Figure 8.7e).

Check: The bending moment should be zero at the pin joint C and also at the free end D.
The maximum or minimum value of M occurs when $dM/dx = 0$ (i.e., when $V = 0$). The corresponding point (E in Figure 8.7e) is obtained as follows:

Shear force at $C = P$

Let the shear force is zero at a distance l from C.

$$\text{Corresponding area of the downward distributed force} = \frac{1}{2} \times l \times \frac{l}{2} w_o = \frac{l^2}{2L} w_o = \frac{l^2}{2L} \times \frac{6P}{L} = \frac{3l^2 P}{L^2}$$

Shear force $V = 0$ at E. We must have

$$P - \frac{3l^2 P}{L^2} = 0 \Rightarrow l^2 = \frac{L^2}{3} \Rightarrow l = \frac{L}{\sqrt{3}}$$

The corresponding value of M = area of the V curve from C to E.
This can be obtained by integration as follows:
Consider a point at a distance x from C.
Area of the distributed force curve from C to x

$$= \frac{1}{2} \times x \times \frac{x}{L} w_o = \frac{1}{2} \times x \times \frac{x}{L} \times \frac{6P}{L} = 3P \frac{x^2}{L^2}$$

Shear force at x:

$$V_x = P - \frac{3Px^2}{L^2}$$

Integrate: bending moment at x:

$$M_x = 0 + Px - \frac{Px^3}{L^2}$$

$$\text{At } x = \frac{L}{\sqrt{3}}, M_{\max} = P \cdot \frac{L}{\sqrt{3}} - \frac{P}{L^2} \times \left(\frac{L}{\sqrt{3}}\right)^3$$

$$\Rightarrow M_{\max} = PL \left(\frac{1}{\sqrt{3}} - \frac{1}{3\sqrt{3}} \right) = \frac{2PL}{3\sqrt{3}}$$

Primary Learning Objectives
1. Determination of the support reactions and internal loading when there is an internal pin joining two beam segments
2. Sketching of shear diagram and moment diagram of a beam using the *graphical method* (*area method*)

■ **End of Solution**

8.2.6 Coordinate-Reversal Method

In some problems of generating the shear curve and the moment curve, it may be convenient to analyze the beam from the right end to the left end. Then, the coordinate axis that defines the direction of traverse along the beam may be defined from right to left instead of the conventional +ve direction of left to right. This is known as *reversal of the direction of traverse*.

Equivalently, a "coordinate transformation" may be applied to the problem while keeping the +ve direction as left to right. After obtaining the results with the transformed coordinates, however, they are converted into the required form by reversing the coordinate transformation in the results. The following steps should be used in this approach of coordinate transformation:

Bending in Beams

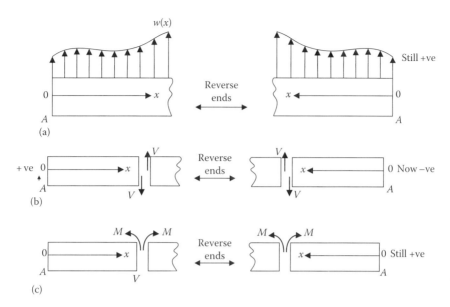

FIGURE 8.8 Coordinate reversal; (a) sign of $w(x)$ is unchanged; (b) sign of $V(x)$ is reversed; (c) sign of $M(x)$ is unchanged.

1. Reverse ends of the system (i.e., take the *mirror image*) and define the x coordinate to extend from the new left end to the new right.
2. Generate expressions for V and M for the transformed problem, in terms of the new x coordinate (any approach may be used here).
3. In the resulting expressions for V and M, change x to $L - x$, where L is the beam length.
4. Reverse the sign of the result for V only.

Rationale for Step 3: See Figure 8.8a. The location x of the transformed problem (right) is equal to the location $L - x$ of the original problem (left). *Note*: In fact, this is precisely the necessary coordinate transformation for end reversal.

Rationale for Step 4: It is clear from Figure 8.8b, according to the sign convention for V (cw rotation is +ve), the sign of V in the left figure is +ve while the sign of V in the right figure is −ve (hence, sign should be reversed in the result for V). Furthermore, according to the sign convention for M (bending of the ends upward is +ve), the sign of M in the left figure is +ve and the sign of M in the right figure is also +ve (hence, sign should not be changed in the result for M).

Example 8.4

A continuous cantilever beam ABC of length $2L$ is fixed horizontally at A. A downward concentrated force P_B is applied at the midspan B and a distributed downward force that linearly varies from w_o/length at B to zero at C is applied as well (Figure 8.9a).

Using the coordinate reversal method, determine analytical expressions for the shear force variation and the internal bending moment variation along the beam, as functions of x (the distance from A to a general location along the beam). Also, sketch the shear and moment diagrams.

Solution

The mirror image of Figure 8.9a is shown in Figure 8.9b.
Note: New $x = 2L -$ old x; Also, sign of shear force V is reversed.
Now, we analyze the transformed problem (mirror image).

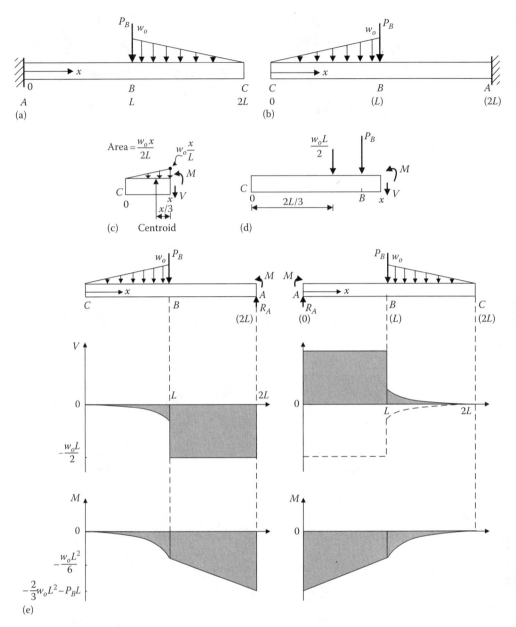

FIGURE 8.9 (a) A cantilever with a distributed force and a concentrated force; (b) mirror image (coordinate transformation) of the given problem; (c) sectioning between C and B; (d) sectioning between B and A; (e) shear and the moment diagrams for the transformed problem (left) and the original problem (right).

Segment CB

Cut at x in between C and B and consider the segment to the left (Figure 8.9c). *Note*: In this problem, we do not need to determine the support reactions at A.

Equations of equilibrium:

$$V = -\frac{1}{2} \times \frac{w_o x}{L} \times x = -\frac{w_o x^2}{2L}; \quad M = -\frac{w_o x^2}{2L} \times \frac{x}{3} = -\frac{w_o x^3}{6L}$$

Bending in Beams

Segment *BA*

Cut at x in between B and A, and consider the segment to the left (Figure 8.9d). *Note*: The distributed force is now represented by an equivalent point force.

Equations of equilibrium:

$$\uparrow \sum F_y = 0: -\frac{w_o L x}{2} - P_B - V = 0 \rightarrow V = -\frac{w_o L}{2} - P_B$$

$$\curvearrowleft \sum M_x = 0: M + P_B \times (x - L) + \frac{w_o L}{2} \times \left(x - \frac{2L}{3}\right) = 0$$

$$\rightarrow M = -\frac{w_o L}{2}\left(x - \frac{2L}{3}\right) - P_B(x - L)$$

Transform the results back to the original coordinate frame.

$$AB: V = \frac{w_o L}{2} + P_B$$

$$M = -\frac{w_o L}{2}\left(2L - x - \frac{2L}{3}\right) - P_B(2L - x - L) = -\frac{w_o L}{2}\left(\frac{4L}{3} - x\right) - P_B(L - x)$$

$$BC: V = \frac{w_o}{2L}(2L - x)^2$$

$$M = -\frac{w_o}{6L}(2L - x)^3$$

The shear and moment diagrams for the transformed problem and the original (given) problem are sketched in Figure 8.9e.

Primary Learning Objectives

1. Reversal of the direction of traverse along a beam
2. Coordinate reversal for the analysis of mirror image of beam
3. Determination of expressions for internal shear force and internal bending moment at a general location for the transformed problem and the original problem
4. Sketching of shear diagram and moment diagram

■ **End of Solution**

8.3 FLEXURE FORMULA

The flexure formula is used for determining the bending stress at a cross section of a beam. Its derivation is given now.

8.3.1 Neutral Surface and Neutral Axis

When an upward bending moment (i.e., a +ve bending moment, according to the sign convention) is applied at a cross section of a beam, the top part of the cross section will be compressed and the bottom part of the cross section will be extended. Consequently, the top part will experience "compressive" bending stresses and the bottom part will experience "tensile" bending stresses. Hence, on the cross section there will be a line separating these two regions on which the bending stresses will be zero. This line is called the *neutral axis* (z). The surface formed by all such neutral lines along the beam is called the *neutral surface*. The axis joining the centroids of all cross sections

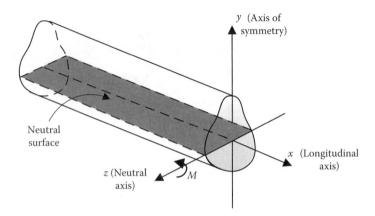

FIGURE 8.10 Illustration of the neutral axis, neutral surface, longitudinal axis, and the axis of symmetry.

along the beam is called the longitudinal axis (x). It can be shown that this axis falls on the neutral surface. These definitions are illustrated in Figure 8.10.

8.3.2 Assumptions

Initially, the following assumptions are made in the present analysis:

1. Beam cross section is symmetric. The *axis of symmetry* (y) is an axis on the cross section perpendicular to the neutral axis (NA). Hence, y passes through the centroid of the cross section (see Figure 8.10).
2. Resultant bending moment at a beam cross section is exerted about an axis perpendicular to the axis of symmetry. This axis is indeed the NA (see Figure 8.10).

Strictly, these assumptions are not necessary, and can be relaxed as follows:

1. In the analysis of beams with nonsymmetric sections, a *principal axis* with respect to the second moment of area of the cross section and passing through the centroid of the cross section may be used in place of the axis of symmetry. For proper (single-plane) bending, the resultant bending moment is exerted about the other (perpendicular) principal axis passing through the centroid.
2. Suppose that the bending moment is exerted about an arbitrary axis through the centroid of the cross section. In using the formulas derived in the present analysis, then, resolve the moment into two components about the two principal axes.

Further Assumptions:

The following further assumptions are required in the present analysis:

1. Plane sections (plane cross sections) will remain plane after bending.
2. The size and shape of the cross section will not change due to bending.
3. The original cross sections will remain perpendicular to the longitudinal axis (on the neutral surface) after bending.

8.3.3 Flexure Analysis

Consider a beam in planar bending, as shown in Figure 8.11a. An elemental slice of length δx of the beam at location x along the beam is shown in Figure 8.11b, as formed by the cross sections at x and $x + \delta x$, which are parallel. Before bending, the axial length (i.e., thickness) of this slice is the same

Bending in Beams

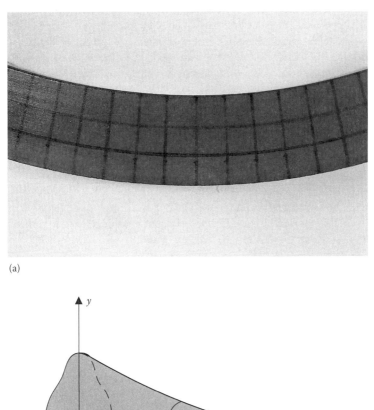

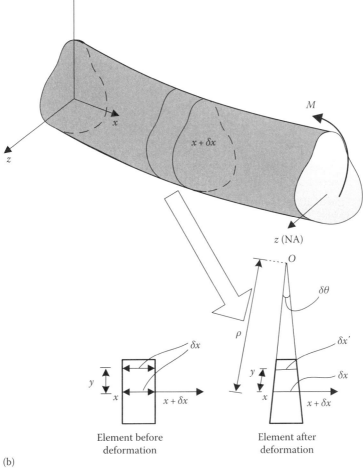

FIGURE 8.11 (a) A bent beam; (b) deformation of an elemental slice of beam.

(δx) at every location along the y axis (axis of symmetry). After bending (upward, according to the sign convention), suppose that the slice thickness at height y from the NA shrinks (due to compressive stress that is present there) to $\delta x'$ as shown in the zoomed-in segment of Figure 8.11b. The corresponding tapered lines on the two sides of the deformed slice meet at O, which is the *center of curvature* of the bent beam element. Also, ρ = radius of curvature (measured from the NA to center O)

Bending Strain and Bending Stress

By definition, normal strain (bending strain) at y is

$$\varepsilon = \frac{\delta x' - \delta x}{\delta x} \tag{i}$$

Note: Since neutral surface does not deform, the element length there will remain at δx after bending.

From geometry, we have $\delta x = \rho \delta \theta$ and $\delta x' = (\rho - y)\delta \theta$,
  where $\delta \theta$ is the angle formed by the deformed element at its center of curvature.
  By substituting these expressions into (i), we get

$$\varepsilon = \frac{(\rho - y)\delta \theta - \rho \delta \theta}{\rho \delta \theta}$$

or

$$\varepsilon = -\frac{y}{\rho} \tag{8.7}$$

This result indicates that the bending strain varies linearly along the axis perpendicular to the NA, with zero value at the NA and the maximum value ε_{max} at $y = c$, the farthest point on the cross section from the NA. Hence,

$$\varepsilon = -\frac{y}{c}\varepsilon_{max} \tag{8.8}$$

Note: This result will hold regardless of whether the material is linear elastic or not. Compare this with the analogous result obtained for a shaft in torsion (Chapter 7).

Now suppose that the material is linear elastic. Then Hooke's law applies, and we have $\sigma = E\varepsilon$. Then,

$$\sigma = -\frac{y}{c}\sigma_{max} \tag{8.9}$$

Note

σ_{max} is the magnitude (nonnegative) of the maximum stress
c is the magnitude (nonnegative) of its location

It follows that in the linear elastic case, bending stress varies linearly along the axis that is perpendicular to the NA (i.e., the axis of symmetry of the section), with zero value at the NA and maximum value σ_{max} at $y = c$, the farthest point on the cross section from the NA (see Figure 8.12).

Flexure Formula

Consider an elemental area δA at height y on the beam cross section, as shown in Figure 8.13a. The bending moment at the cross section is M (about the NA). The axial force on this area is $\delta F = \sigma \cdot \delta A$ (Figure 8.13b). Since there is no resultant axial force on the cross section, we must have

$$\int_A dF = 0 = \int_A \sigma dA = \int_A -\frac{y}{c}\sigma_{max} dA = -\frac{\sigma_{max}}{c}\int_A y dA$$

Bending in Beams

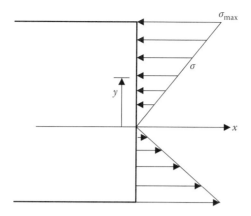

FIGURE 8.12 Linear variation of bending stress from the neutral axis.

Hence, we have

$$\int_A y \, dA = 0 \tag{8.10}$$

This result confirms that the NA (z axis) indeed passes through the centroid of the cross section.

The internal bending moment at the cross section (about the NA) is obtained by summing the moments of the elemental axial forces δF about the NA (Figure 8.13b). We have

$$M = -\int_A y \, dF = -\int_A y \sigma \, dA = -\int_A \left(-\frac{y^2}{c} \sigma_{max}\right) dA = \frac{\sigma_{max}}{c} \int_A y^2 \, dA = \frac{\sigma_{max}}{c} I$$

Note: We have substituted Equation 8.9 for bending stress. Also, as before, σ_{max} is the "magnitude" of the maximum stress value.

We have denoted the *second moment of area* of the beam cross section about the NA as I:

$$I = \int_A y^2 \, dA \tag{8.11}$$

Note: Compare this with the polar moment of area in the torsion formula (Chapter 7).

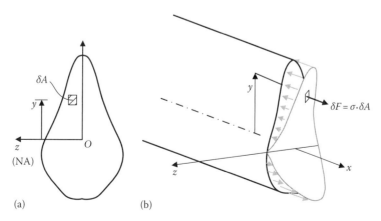

FIGURE 8.13 (a) An area element of a beam X-section; (b) distribution of axial force and bending stress on the beam X-section.

It follows

$$\sigma_{max} = \frac{Mc}{I} \qquad (8.12a)$$

or

$$\sigma = -\frac{My}{I} \qquad (8.12b)$$

Either of these two results (8.12) is known as the *flexure formula*.
Note: Compare (8.12) with the torsion formula (7.9).

8.3.4 Application of the Flexure Formula

The flexure formula can be used in several ways.

Direct application: For a particular external loading, find the max moment and its location (from the moment diagram); compute I for the beam X-section; determine the maximum magnitude of y ($=c$); substitute in the flexure formula and compute maximum normal stress σ_{max}.

Design application 1 (For a given beam, what is the allowable load?): Compute I for the beam; determine the maximum magnitude of y ($=c$); for known yield strength (or ultimate strength) and using a suitable factor of safety, determine the allowable normal stress; substitute in the flexure formula and compute the allowable moment; compute the allowable external load corresponding to this moment.

Design application 2 (For a given loading, what is a suitable beam X-section?): For the given external loading, find the maximum moment and its location (from the moment diagram); for known yield strength (or ultimate strength) and using a suitable factor of safety, determine the allowable normal stress; determine a suitable beam section (with the corresponding I and c) that satisfies the flexure formula.

8.3.5 Parallel Axis Theorem

It is seen that the second moment of area I about the NA (i.e., about the centroid of the cross section) is needed for determining the bending stress. A complex section may be subdivided into simpler areas (e.g., rectangular segments) whose second moment of area can be easily determined about their individual centroidal axes. Then, to obtain the corresponding second moment of area about the centroidal axis of the overall cross section, the parallel axis theorem may be used.

Parallel axis theorem

$$I_D = I_G + A \cdot d^2 \qquad (8.13)$$

where
I_G is the second moment of area about the centroidal axis
I_D is the second moment of area about an axis located at a distance d from the centroidal axis
A is the area of cross section

8.3.6 Area Removal Method

Integration operation is linear. Hence, in particular, we have

$$\int_A (P-Q)dA = \int_A PdA - \int_A QdA \text{ for any operands } P \text{ and } Q \qquad (8.14)$$

Bending in Beams

When a cross section O is equivalent to a simpler cross section A minus another simpler cross section B, according to the property (8.14), the following relation exists for the first moments of area about the same axis:

$$\text{Area moment of } O = \text{Area moment of } A - \text{Area moment of } B$$

Also, Area moment = Area × Distance to centroid. Hence, with regard to the location of the centroid of the three areas that are involved here, we have

$$\bar{y}_O \times (A_A - A_B) = \bar{y}_A \times A_A - \bar{y}_B \times A_B \tag{8.15}$$

Since the centroid falls on the NA, the formula (8.15) may be used to determine the location of the NA of the original section (O).

Furthermore, according to the property (8.14), the following relation exists for the second moments of area of the three sections O, A, and B, about a common axis:

$$I_O = I_A - I_B \tag{8.16}$$

This formula (along with the parallel axis theorem (8.13)) may be used to determine the second moment of area of the original section about its NA.

Often, the facts (8.13), (8.15), and (8.16) considerably simplify the computation of the second moment of area of a cross section in application of the flexure formula, as seen from the worked examples given later.

Example 8.5

A finger of a robotic hand is made of a light tube of length 100 mm, internal radius 5 mm, and external radius 10 mm. The finger joint at one end of the tube is assumed to have a smooth pin support. A tendon attached externally to the midpoint of the finger applies a tension of 5 N at 45° in a given position. The other end of the finger pushes on a smooth cylindrical object. The finger is maintained horizontal in equilibrium (see Figure 8.14a).

Sketch the shear and moment diagrams of the finger. Determine the maximum "bending" stress in the finger and its location.

Solution

Consider the free-body diagram of the finger, as shown in Figure 8.14b.
There are no moments at A and C (because of smooth supports).
Equations of equilibrium:

$$\rightarrow \sum F = 0: F_x - \frac{5}{\sqrt{2}} = 0 \Rightarrow F_x = \frac{5}{\sqrt{2}} \text{ N}$$

$$\uparrow \sum F = 0: F_y - \frac{5}{\sqrt{2}} + R = 0 \Rightarrow F_y = \frac{5}{\sqrt{2}} - R$$

$$\circlearrowleft \sum M_A = 0: -\frac{5}{\sqrt{2}} \times 50 - \frac{5}{\sqrt{2}} \times 10 + R \times 100 = 0$$

$$\Rightarrow R = \frac{5}{\sqrt{2}} \times \frac{60}{100} \text{ N} = \frac{3}{\sqrt{2}} \text{ N}$$

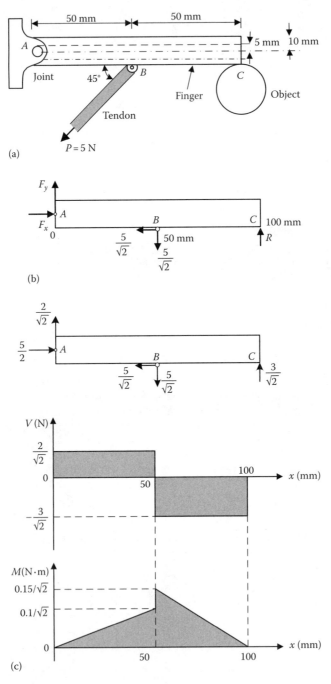

FIGURE 8.14 (a) Finger of a robotic hand touching an object. (b) Free-body diagram of the robotic finger. (c) Shear and moment diagrams of the robotic finger.

Substitute

$$F_y = \frac{5}{\sqrt{2}} \frac{3}{\sqrt{2}} = \frac{2}{\sqrt{2}} \text{N}$$

By considering the equilibrium of the segments of the finger from A to B and then from B to C, we get the shear and moment diagrams, as usual, as shown in Figure 8.14c.

Note: There is a jump in moment at B because of the added moment due to the x component of the tendon force there. Also, the moment curve segments are linear because the shear force is constant in each segment.

We observe that the maximum moment is $0.15/\sqrt{2}$ N·m and it occurs at a section just right of B.

The maximum "bending" stress is at the top edge (compressive) and bottom edge (tensile) at this section

$$\sigma_{max} = \frac{M_{max}}{I} c$$

$$M_{max} = \frac{0.15}{\sqrt{2}} \text{ N·m}, c = 10 \text{ mm} = 0.01 \text{ m}$$

From Appendix A:

$$I = \frac{\pi}{4}(10^4 - 5^4) \text{ mm}^4 = \frac{\pi}{4}(10^4 - 5^4) \times 10^{-12} \text{ m}^4$$

Substitute:

$$\sigma_{max} = \frac{0.15 \times 0.01}{\sqrt{2} \times (\pi/4) \times (10^4 - 5^4) \times 10^{-12}} \text{ Pa}$$

$$= \frac{4 \times 0.15}{\sqrt{2} \times \pi \times (1^4 - 0.5^4)} \times 10^6 \text{ Pa} = 6.67 \times 10^6 \text{ Pa}$$

$$\sigma_{max} = 6.67 \text{ MPa}$$

Note: Only the bending stress is considered in this example. The normal stress due to the axial force component (from the tendon) is not included.

Primary Learning Objectives

1. Determination of the shear and moment diagrams in the presence of external loading that generates a point moment
2. Application of the flexure formula to determine the maximum bending stress

■ **End of Solution**

Example 8.6

A cantilever beam of length 4 m (Figure 8.15a) is made of a channel section with dimensions as given in Figure 8.15b. A uniformly distributed load of 100 N/m is applied on the beam. Determine the maximum bending stress and its location in the beam.

Solution

Cut the beam at x and consider the segment from 0 to x, as shown in Figure 8.15c.

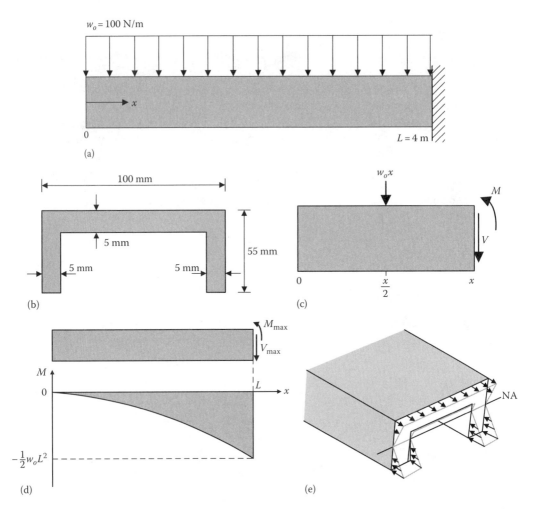

FIGURE 8.15 (a) A cantilever with uniformly distributed lateral force; (b) dimensions of the channel section. (c) Free-body diagram of a virtual left segment of beam. (d) Moment curve of the beam. (e) Bending stress distribution on a channel section.

Equation of equilibrium:

$$\sum M_x = 0: M + w_o x \times \frac{x}{2} = 0$$

$$\Rightarrow M = -\frac{1}{2}w_o x^2$$

The moment diagram is obtained from this result, as given in Figure 8.15d.
Maximum (absolute) bending moment,

$$M_{max} = -\frac{1}{2}w_o L^2 = -\frac{1}{2} \times 100 \times 4^2 \text{ N·m} = -800 \text{ N·m}$$

This occurs at the root (clamped end) of the cantilever. The location of the NA is the centroid of the section. Let

$$\bar{y} = \text{Height of the centroid (from the bottom of the section)}$$

Bending in Beams

Area Addition Method

Note that the channel section consists of three rectangular sections of dimension 5 mm × 50 mm, 5 mm × 50 mm, and 100 mm × 5 mm.

Take moments of the area segments from the bottom edge:

$$\bar{y} \times [2 \times 50 \times 5 + 100 \times 5] = 2 \times 25 \times 50 \times 5 + 52.5 \times 100 \times 5 \text{ mm}^3$$

$$\Rightarrow \bar{y} = \frac{25 \times 5 + 52.5}{5 + 5} \text{ mm} = 38.75 \text{ mm}$$

The same result may be obtained by the area removal method, as follows.

Area Removal Method

We notice that the channel section is equivalent to a rectangular section of 100 mm × 55 mm from which a rectangular section of 90 mm × 50 mm is removed. Then, taking moments of the area from the bottom edge, we have

$$\bar{y} \times [100 \times 55 - 90 \times 50] = \frac{55}{2} \times 100 \times 55 - \frac{50}{2} \times 90 \times 50 \text{ mm}^3$$

$$\Rightarrow \bar{y} = \frac{(55/2) \times 100 \times 55 - (50/2) \times 90 \times 50}{100 \times 55 - 90 \times 50} \text{ mm} = 38.75 \text{ mm}$$

This is identical to the previous result. Next we will determine the 2nd moment area of the channel section about its centroid (neutral axis) by using the two methods (area addition method and area removal method).

Area Addition Method

Apply parallel axis theorem to each of the three rectangular segments of the channel section (and use the formula for the 2nd moment area of a rectangle about its centroid, as given in Appendix A):

$$I = 2 \times \left[\frac{1}{12} \times 5 \times 50^3 + 5 \times 50 \times (38.75 - 25)^2 \right] + \frac{1}{12} \times 100 \times 5^3 + 100 \times 5 \times (52.5 - 38.75)^2 \text{ mm}^4$$

$$= [104.17 + 94.53 + 1.04 + 94.53] \times 10^3 \text{ mm}^4$$

$$= 0.294 \times 10^6 \text{ mm}^4 = 0.294 \times 10^{-6} \text{ m}^4$$

Area Removal Method

Since the channel section is equivalent to a rectangular section of 100 mm × 55 mm from which a rectangular section of 90 mm × 50 mm is removed, we have (by applying the parallel axis theorem)

$$I = \frac{1}{12} \times 11 \times 55^3 + 100 \times 55 \times \left(38.75 - \frac{55}{2}\right)^2 - \frac{1}{12} \times 90 \times 50^3 - 90 \times 50 \times \left(38.75 - \frac{50}{2}\right)^2 \text{ mm}^4$$

$$= [13.865 + 6.961 - 9.375 - 8.508] \times 10^5 \text{ mm}^4$$

$$= 0.294 \times 10^6 \text{ mm}^4 = 0.294 \times 10^{-6} \text{ m}^4$$

This is identical to the previous result.

The farthest point of the section from the NA is its bottom point. Hence, $c = 38.75$

We get

$$\sigma_{max} = \frac{M_{max} c}{I} = -\frac{800 \times 0.03875}{0.294 \times 10^{-6}} \text{ Pa} = -105.4 \text{ MPa} \quad \text{(a compressive stress)}$$

The bending stress distribution at the section is sketched in Figure 8.15e.

Primary Learning Objectives

1. Determination of the moment diagram of a cantilever subjected to a uniformly distributed force
2. Application of the flexure formula to determine the maximum bending stress on a channel section
3. Application of the "area addition method" to determine the centroid and the second moment of area about the NA of a channel section
4. Application of the "area removal method" to determine the centroid and the second moment of area about the NA of a channel section
5. Sketching of the bending stress distribution on a channel section

■ **End of Solution**

8.4 COMPOSITE BEAMS

Composite beams made of two or more layers of different materials that are firmly bonded together may be analyzed using the present theory that has been developed for a single material. This is done by transforming the composite beam section in to an equivalent section made of just one of the materials. This method is presented now.

8.4.1 Transformed Section Method

In this method, only the width dimension of the cross section of the composite beam is transformed into an equivalent cross section made of just one of the composite materials (with the corresponding value for E).

Note: The cross-sectional width rather than the height is transformed (into a common material) because the width is linear in the internal bending moment expression (i.e., bending moment increases linearly with the width when the other conditions such as strain and stress are kept the same) while the height is not.

Specifically, the beam cross-sectional width is transformed so that the transformed section

1. Will have the same deformation (i.e., strain or deflection)
2. Will carry the same load (i.e., same bending moment) as the original section

From an analytical point of view, the transformed beam section will satisfy the following properties:

a. Only the width of the X-section is transformed, and represented by a common single value of material property (Young's modulus E)
b. Bending strains (normal) of the X-section remain unchanged
c. The normal force at each level (y) of the X-section, and hence its moment, remains unchanged

Note: Since the bending stress values of the transformed material will not be the actual stresses of the original material (even though the strain values are), they have to be transformed back, by taking into account the true Young's modulus.

Consider a composite beam whose core (inner part) is made of material A with $E = E_A$ and the outer layer is made of material B with $E = E_B$. Consider an elemental strip of thickness δy at height y from the NA and parallel to the NA, as shown in Figure 8.16.

Bending in Beams

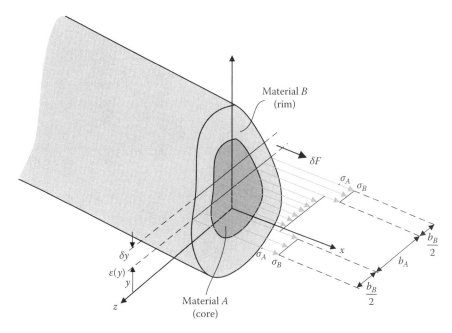

FIGURE 8.16 Transformation of a composite beam section.

The normal (axial) force on the strip

$$\delta F = \sigma_A b_A \delta y + \sigma_B b_B \delta y = \varepsilon E_A b_A \delta y + \varepsilon E_B b_B \delta y = \varepsilon E_A b_A \delta y + \varepsilon n E_A b_B \delta y = \varepsilon E_A (b_A + n b_B) \delta y$$

Internal bending moment at the X-section

$$M = \int_A y dF = \int_A y \varepsilon E_A (b_A + n b_B) dy \qquad (8.17)$$

Where, transformation factor

$$n = \frac{E_B}{E_A} \qquad (8.18)$$

Cross-sectional transformation: According to (8.17), cross-sectional transformation is as follows:
Width (in the z-dimension) of the material layer B has to be multiplied by the factor n, while the height dimension (in y-direction) is not changed.

Note 1: Now it is clear why width (not height) is the appropriate transformation dimension, because of its linear association with the normal force (and M) at the X-section (just like E), and also the fact that bending strains will be unchanged in the transformation.

Note 2: The resulting stresses in material B, in the transformed section B, have to be transformed back by multiplying them by n.

Example 8.7

A composite beam of rectangular cross section is made of a rectangular beam of wood reinforced with a plate of steel uniformly screwed on its upper surface (Figure 8.17a). The cross-sectional dimensions are shown in Figure 8.17b. If a bending moment of −2 kN·m is applied on the beam, determine the bending stresses at the top and the bottom edges of the section and at the interface of wood and steel. The Young's modulus of wood, $E_w = 10$ GPa. The Young's modulus of steel, $E_s = 200$ GPa.

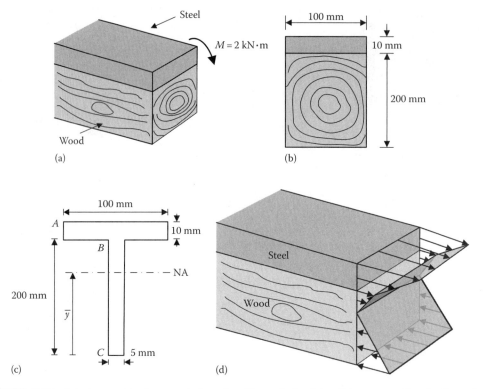

FIGURE 8.17 (a) A wooden beam reinforced with a steel plate; (b) cross-sectional dimensions. (c) Transformed section with equivalent steel. (d) Bending stress profile of the composite beam.

Solution

To transform the wood into equivalent steel having the same section height and the same load capacity, the width of wood has to be multiplied by the transformation factor:

$$n = \frac{E_w}{E_s} = \frac{10}{200} = 0.05$$

Equivalent width of wood = 100×0.05 mm = 5 mm

The corresponding transformed section (with equivalent steel) is shown in Figure 8.17c.

The NA is at the centroid of the section (say, at a height of $\bar{y}$ from the bottom edge of the section). Take moments of area about the bottom:

$$\bar{y} \times (5 \times 200 + 100 \times 10) = 100 \times (5 \times 200) + 205 \times (100 \times 10)$$

$$\Rightarrow \bar{y} = \frac{100 + 205}{1 + 1} \text{ mm} = 152.5 \text{ mm}$$

The second moment of area about the NA (from parallel axis theorem),

$$I = \frac{1}{12} \times 5 \times 200^3 + 5 \times 200 \times (152.5 - 100)^2 + \frac{1}{12} \times 100 \times 10^3 + 100 \times 10 \times (205 - 152.5)^2 \text{ mm}^4$$

$$= (3.33 + 2.76 + 0.008 + 2.76) \times 10^6 \text{ mm}^4$$

$$= 8.85 \times 10^6 \text{ mm}^4 = 8.85 \times 10^{-6} \text{ m}^4$$

Stresses in the transformed section:

$$\sigma_A = -\frac{M}{I}c_A = \frac{2\times10^3 \times (210-152.5)\times 10^{-3}}{8.85\times 10^{-6}} \text{ Pa} = 13.0 \text{ MPa}$$

$$\sigma_B = -\frac{M}{I}c_B = \frac{2\times10^3 \times (200-152.5)\times 10^{-3}}{8.85\times 10^{-6}} \text{ Pa} = 10.7 \text{ MPa}$$

$$\sigma_C = -\frac{M}{I}c_C = \frac{2\times10^3 \times (-152.5)\times 10^{-3}}{8.85\times 10^{-6}} \text{ Pa} = -34.5 \text{ MPa}$$

Actual values of the stresses:

$$\sigma_{A\,Steel} = \sigma_A = 13.0 \text{ MPa}$$

$$\sigma_{B\,Steel} = \sigma_B = 10.7 \text{ MPa}$$

$$\sigma_{B\,Wood} = \sigma_B \times n = 10.7 \times 0.05 \text{ MPa} = 0.535 \text{ MPa}$$

$$\sigma_{C\,Wood} = \sigma_C \times n = -34.5 \times 0.05 \text{ MPa} = -1.73 \text{ MPa}$$

The actual stress profile is sketched in Figure 8.17d.

Primary Learning Objectives
1. Transformation of the X-section of a composite beam into an equivalent X-section of a single material (X-section width and the material are transformed)
2. Determination of bending stresses in a composite beam
3. Sketching of the true profile of bending stress in a composite beam

■ **End of Solution**

Example 8.8

A reinforced concrete beam has a rectangular cross section of width 0.4 m. Three steel rods each of diameter 50 mm are placed at a depth 1 m from the top edge of the beam section for reinforcement (Figure 8.18a). According to the structural loading that is present under normal conditions, the maximum bending moment experienced by the beam is $M_{max} = +300$ kN·m. With respect to both yielding of the reinforcing steel and fracture of the concrete, estimate the overall factor of safety of the beam design. Given

Young's modulus of steel, $E_s = 20$ GPa
Young's modulus of concrete, $E_c = 20$ GPa
Yield stress of steel, $\sigma_{Y\,Steel} = 250$ MPa
Fracture stress of concrete, $\sigma_{F\,Concrete} = 15$ MPa

Solution

The tensile stress that unreinforced concrete can withstand is negligible. Hence, we neglect the concrete segment below the NA of the beam section.
Area of X-section of steel

$$A_s = 3 \times \frac{\pi}{4} \times (0.05)^2 \text{ m}^2$$

Equivalent area in concrete (at E_c)

$$A_{c\,eq} = n \times A_s$$

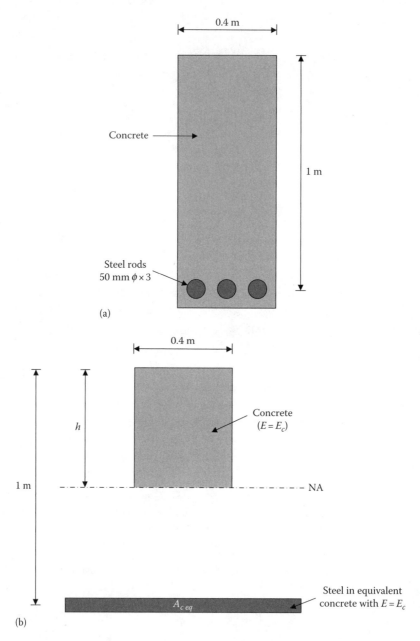

FIGURE 8.18 (a) Section of a reinforced concrete beam. (b) Beam section transformed into equivalent concrete.

where

$$n = \frac{E_s}{E_c} = \frac{200}{20} = 10$$

We have

$$A_{c\,eq} = 10 \times 3 \times \frac{\pi}{4} \times (0.05)^2 = 0.06 \text{ m}^2$$

The corresponding equivalent section in concrete (at $E = E_c$) is shown in Figure 8.18b.

Bending in Beams

Let the NA of the equivalent section be at a depth h below the top edge of the section. We determine h by taking moments of area about NA. Since NA is the centroid of the transformed section, the moments should sum to zero:

$$\frac{h}{2} \times (0.4 \times h) - (1.0 - h) \times A_{c\,eq} = 0$$

$$\rightarrow 0.2h^2 - 0.06 \times (1-h) = 0$$

$$\rightarrow h^2 + 0.3h - 0.3 = 0$$

$$\rightarrow h = \frac{-0.3 \pm \sqrt{0.3^2 + 4 \times 1 \times 0.3}}{2} \text{ m}$$

Take the +ve root: $h = 0.42$ m

We apply the parallel axis theorem to determine I_{NA}.

Note: Since the thickness of transformed steel (0.05 m) is negligible compared to the other key dimensions of the transformed section, in the parallel axis theorem, I_G of that segment (about its own centroid) can be neglected compared to the $d^2 \times A$ component.

$$I_{NA} = \frac{1}{12} \times 0.4 \times h^3 + \left(\frac{h}{2}\right)^2 \times (0.4 \times h) + (1-h)^2 \times A_{c\,eq}$$

$$= \frac{1}{12} \times 0.4 \times 0.42^3 + (0.21)^2 \times 0.4 \times 0.42 + (1-0.42)^2 \times 0.06 \text{ m}^4 = 0.03 \text{ m}^4$$

Apply $\sigma_{max} = M_{max}c/I_{NA}$ for concrete and steel separately.

Note: To obtain the actual σ_{max} for the steel, the transformed σ_{max} must be multiplied by n (because transformed σ_{max} is with respect to E_c while the true σ_{max} is with respect to E_s, at the same strain value).

$$\sigma_{max\,c} = \frac{300 \times 10^3 \text{ (N·m)} \times 0.42 \text{ (m)}}{0.03 \text{ (m}^4\text{)}} = 4.2 \times 10^6 \text{ Pa} = 4.2 \text{ MPa}$$

$$\sigma_{max\,s} = 10 \times \frac{300 \times 10^3 \text{ (N·m)} \times (1-0.42) \text{ (m)}}{0.03 \text{ (m}^4\text{)}} = 58.0 \times 10^6 \text{ Pa} = 58.0 \text{ MPa}$$

$$\rightarrow (FOS)_c = \frac{\sigma_{F\,Concrete}}{\sigma_{max\,c}} = \frac{15}{4.2} = 3.57$$

$$\rightarrow (FOS)_s = \frac{\sigma_{Y\,Steel}}{\sigma_{max\,s}} = \frac{250}{58.0} = 4.31$$

$$\rightarrow \text{Overall FOS} = 3.5$$

Also, it follows that the concrete segment is critical in the present system (since it corresponds to the lower FOS).

Primary Learning Objectives
1. Transformation of the X-section of a reinforced concrete beam into an equivalent X-section of concrete (X-section width and the material are transformed)
2. Application of the parallel axis theorem when the thickness of the area is relatively small
3. Determination of bending stresses in a reinforced concrete beam

■ **End of Solution**

8.5 TRANSVERSE SHEAR

The diagram of internal shear force confirms that a beam in bending experiences an internal shear force V in the transverse (lateral) direction (y) at a general cross section of the beam, in addition to an internal bending moment (M). So, there must be shear stresses (τ) in the same direction (y) at any point on the cross section. These are the transverse shear stresses. For a small element at a particular point of the cross section to remain in equilibrium, we know that there must be a "complementary" shear stress in an appropriate direction that is perpendicular to the transverse direction (see Chapter 3). In the case of a beam element in bending, this "complementary" direction is the axial direction. This fact is illustrated in Figure 8.19.

We have shown (see Equation 8.2) that V and M are related (through $V = dM/dx$). Hence, intuitively, the transverse shear stress (τ) at a point on the beam cross section (which must be equal to the complementary shear stress on the axial section at that point) must be related to M as well as V. This relation is called the shear formula, which is derived next.

8.5.1 Shear Formula

Consider a very small axial slice of length δx from the beam (see Figure 8.19). At a height y (along the axis of symmetry) from the NA of the section, make an axial section (i.e., a section parallel to the neutral surface).

Let

$$\text{Width of the beam } X\text{-section at height } y \text{ from the NA} = t.$$

If the transverse shear stress at y is τ, then (due to complementary shear) the shear stress along the axial section at y is also τ. This axial shear stress acts in the +ve axial (x) direction on the bottom surface of the top segment (at y), as shown in Figure 8.19.

Note: On the top surface of the bottom segment formed by the axial section at y, the axial shear stress is in the −ve x direction.

There will be normal bending stresses as well (as noted when deriving the shear formula) on the two vertical sides of this beam segment above y. They generate forces on the two vertical sides of the element segment in the axial (x) direction, which (along with the force due to the "complementary" axial shear stress) will keep the segment in equilibrium in the x direction (see Figure 8.20).

Axial shear force on the segment above y due to shear stress $= \tau \cdot t \cdot \delta x$ (acting in the +ve x direction)

For a small area element $\delta A'$ at height y' above the NA, where the normal stress is σ', the force will be $\sigma' \delta A'$.

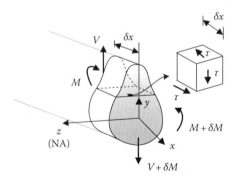

FIGURE 8.19 Shear stress and complementary shear stress on an element of a beam X-section.

Bending in Beams

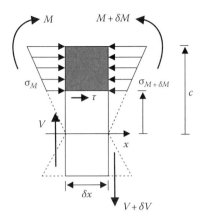

FIGURE 8.20 Axial equilibrium of the beam slice segment above y (shaded).

Apply the flexure formula (8.12): $\sigma=-(M/I)y$ for the two sides where the bending moments are M and $M+\delta M$. The equation of equilibrium of the segment above y in the x-direction:

$$\int_{A_t} \frac{M}{I} y' dA' - \int_{A_t} \frac{(M+\delta M)}{I} y' dA' + \tau \cdot t \cdot \delta x = 0$$

Here, the integration is performed over the area from height y (where the width is t) up to the farthest height of the X-section. This area is denoted by A_t.

Cancelling the common term, we get

$$\frac{\delta M}{I} \int_{A_t} y' dA' = \tau \cdot t \cdot \delta x$$

In the limit (as $\delta x \to 0$), we have

$$\tau = \frac{1}{It} \frac{dM}{dx} \int_{A_t} y' dA'$$

Denote by Q, the moment of the area segment A_t (above y) about the NA of the cross section:

$$Q = \int_{A_t} y' dA' \qquad (8.19)$$

Also use the fact that $V = dM/dx$ (Equation 8.2). We get the shear formula

$$\tau = \frac{VQ}{It} \qquad (8.20)$$

where
  V is the internal shear force on the X-section
  Q is the first moment of the area of the X-section above y, about NA
  I is the second moment of area of the entire X-section about NA (as used in flexure formula)
  t is the width of the X-section at y

Note: An easy way to recall the shear formula (8.20) is to remember "we (V) compute (Q) it (It)"

Example 8.9

A uniform beam of box section and length L (m) is clamped at one end and carries a uniform transverse load of P/L (N/m) along it (see Figure 8.21a). The outer dimensions of the box section are $2a \times 2a$ (m × m) and the inner dimensions are $a \times a$ (m × m), as shown in Figure 8.21b.

  a. Sketch the shear diagram and the moment diagram of the beam and mark the key values of the internal shear force and the internal bending moment on the two curves.
  b. Determine the maximum bending (normal) stress on the beam cross section and its location.
  c. Determine the maximum transverse shear stress on the beam cross section and its location.

Note: Express the stresses in terms of the given quantities only (P, a, and L).

Solution

  a. Define the longitudinal coordinate x of the beam from right (free end) to left. Make a virtual cut of the beam at distance x from the free end, and separate the segment to the right, as shown in Figure 8.21c.

$$\text{Equivalent external force (transverse) on the segment} = \frac{Px}{L}$$

From force balance, the internal shear force at the cross section of virtual cut,

$$V = \frac{Px}{L}$$

From the moment balance (about the point of virtual cut),

$$-M - \frac{Px}{L} \times \frac{x}{2} = 0 \rightarrow M = -\frac{Px^2}{2L}$$

From these results, the shear curve and the moment curve of the beam may be sketched, as in Figure 8.21d.

Note: The maximum internal shear force $= P$. It occurs at the clamped end.

The maximum internal bending moment (magnitude) $= PL/2$. It also occurs at the clamped end.

  b. For the box section, the NA is the horizontal axis of symmetry (i.e., the centroidal axis) of the cross section.

Second moment of area,

$$I = \frac{1}{12}[(2a)^4 - a^4] = \frac{15}{12}a^4 = \frac{5}{4}a^4$$

Apply

$$\sigma = -\frac{My}{I}$$

The maximum bending stress occurs at $y = a$ (i.e., top edge of the cross section).

The bending moment of maximum magnitude occurs at the clamped end, where $M = -(PL/2)$.

We have

$$\sigma_{max} = -\left(-\frac{PL}{2}\right)\frac{a}{(5/4)a^4} = \frac{2}{5}\frac{PL}{a^3}$$

Note: This occurs at the top edge of the beam cross section at the clamped end, as shown in Figure 8.21e.

Bending in Beams

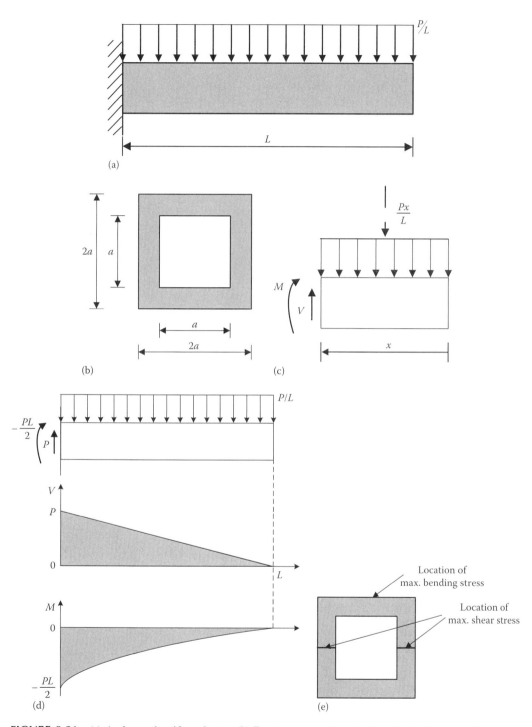

FIGURE 8.21 (a) A clamped uniform beam. (b) Beam cross section. (c) Free-body diagram of the beam segment to the right of the cut. (d) Shear diagram and the moment diagram of the beam. (e) Locations of the maximum stresses on an X-section.

c. Apply $\tau = \dfrac{VQ}{It}$

The maximum moment of area occurs about the NA.
Corresponding sectional thickness $t = 2a - a = a$
Corresponding moment of area

$$Q = (2a \times a) \times \frac{a}{2} - \left(a \times \frac{a}{2}\right) \times \frac{a}{4} = \frac{7a^3}{8}$$

The maximum internal shear force occurs at the clamped end, where $V = P$. Hence,

$$\tau_{max} = \frac{P \times (7a^3/8)}{(5/4)a^4 \times a} = \frac{7}{10}\frac{P}{a^2}$$

Note: This occurs at the NA level (i.e., horizontal axis of symmetry) of the cross section at the clamped end, as indicated in Figure 8.21e.

Primary Learning Objectives
1. Application of the flexure formula
2. Application of the shear formula
3. Determination of the maximum bending stress
4. Determination of the maximum shear stress due to bending

■ **End of Solution**

8.5.2 Shear Flow

In Chapter 7, we encountered shear flow of tubes in torsion. Now, we will discuss shear flow of beams in bending.

As noted earlier, if we make an axial section (parallel to the neutral surface) in a very small axial slice of length δx of the beam (see Figure 8.22), the shear force on the cut surface along the axial (x) direction is $\delta F = \tau \cdot t \cdot \delta x$, where t is the width of the X-section at the cut.

The shear force per unit length is called the *shear flow* and is denoted by q. We then have

$$q = \frac{dF}{dx} = \tau \cdot t \quad (8.21)$$

Substituting *shear formula* as derived previously, we have

$$q = \frac{VQ}{I} \quad (8.22)$$

Since q gives the shear force per unit length in a cut surface, it is a good measure of the shear resistance experienced by material (e.g., at a joint that is glued, nailed, riveted, etc.).

Example 8.10

A wooden beam of T-section is made by firmly nailing together two planks, as shown in Figure 8.23a. The cross section of the top plank has a width of 0.5 m and a thickness of 0.06 m. The bottom plank has a cross-sectional height of 0.4 m and a thickness of 0.06 m (Figure 8.23b). The beam is fixed at one end, as a cantilever, and a downward force $P = 5$ kN is applied at the free end. Each nail has a diameter 4 mm and an allowable shear stress 25 MPa. Estimate the number of nails per meter length of the beam that should be used so that they will not exceed their allowable shear stress.

Bending in Beams

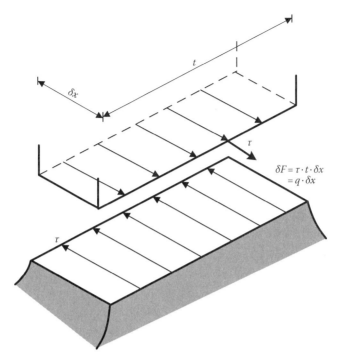

FIGURE 8.22 Definition of shear flow of a beam in bending.

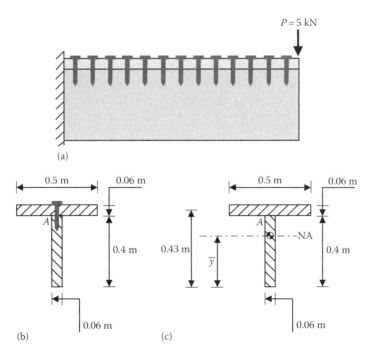

FIGURE 8.23 (a) A wooden beam made by nailing two planks together. (b) Beam X-sectional dimensions. (c) Location of the NA and the centroid of the segment above the nail joint.

Solution

Steps:

1. Find the location ($\bar{y}$) of the NA
2. Find the second moment of area about the NA (I_{NA})
3. Find the first moment of area (Q_A) of the segment above the nailed joint A (i.e., moment of the X-section of top plank) about NA
4. Apply the shear flow formula to determine the axial (shear) force per unit length at the nailed joint (i.e., q_A)
5. Determine the maximum shear force that is allowed for each nail. Using that, determine the number of nails that are needed to support the axial shear force per meter (q_A).

The beam cross section is shown in Figure 8.23c, where the height of the centroid of the cross section (i.e., NA) from the bottom edge is indicated.
Take moments of area about the bottom edge of the cross section:

$$\bar{y} \times [0.06 \times 0.4 + 0.5 \times 0.06] = \frac{0.4}{2} \times 0.06 \times 0.4 + 0.43 \times 0.5 \times 0.06$$

$$\rightarrow \bar{y} = \frac{0.2 \times 24 + 0.43 \times 30}{24 + 30} = 0.328 \text{ m}$$

Apply the parallel axis theorem to the two rectangular cross-sectional segments to find the second moment of area about NA:

$$I_{NA} = \frac{1}{12} \times 0.06 \times 0.4^3 + (0.328 - 0.2)^2 \times 0.06 \times 0.4$$

$$+ \frac{1}{12} \times 0.5 \times 0.06^3 + (0.328 - 0.43)^2 \times 0.5 \times 0.06 \text{ m}^4$$

$$= (0.032 + 0.039 + 0.001 + 0.031) \times 10^{-2} \text{ m}^4 = 1.03 \times 10^{-3} \text{ m}^4$$

First moment of area of the cross-sectional segment above the nail joint (A):

$$Q_A = (0.43 - 0.328) \times 0.5 \times 0.06 \text{ m}^3 = 3.06 \times 10^{-3} \text{ m}^3$$

Shear flow at A

$$q_A = \frac{VQ_A}{I_{NA}}$$

where shear force $V = 5$ kN

$$\rightarrow q_A = \frac{5 \times 10^3 \text{ (N)} \times 3.06 \times 10^{-3} \text{ (m}^3\text{)}}{1.03 \times 10^{-3} \text{ (m}^4\text{)}} = 14.85 \times 10^3 \text{ N/m}$$

$$= \text{Allowable shear force per nail } \frac{\pi}{4} \times (0.004)^2 \times 25 \times 10^6 \text{ N} = 314 \text{ N}$$

$$\text{Required number of nails per meter length of the beam} = \frac{14.85 \times 10^3}{314} = 47.3$$

We will use 50 nails/m.

Bending in Beams

Primary Learning Objectives
1. Understanding of the directions of the shear forces in a beam in bending
2. Application of the shear flow formula
3. Estimation of the axial shear force of a beam in bending

■ **End of Solution**

8.6 BEAM DEFLECTION

Strictly speaking, "flexure formula" should be called "stress formula" since it deals with bending stress rather than bending deformation. However, when deriving the so-called flexure formula, we first derived a formula for bending strain. It would have been more logical to analyze beam deflection at that point, as beam deflection is directly related to strain. Nevertheless, we are now better equipped to follow that track, as we have already related beam strain to beam stress, and beam stress to beam loading. By beam deflection, we mean the deflection of the elastic curve of the beam.

Elastic curve: The line joining the centroids of all cross sections of the beam is called the elastic curve. This coincides with the x axis when the beam is unstrained (i.e., before bending). Upon loading the beam, the elastic curve bends with the beam.

Sign convention: Positive deflections (v) of the elastic curve are in the positive direction of y (i.e., upward).

8.6.1 BENDING MOMENT–DEFLECTION RELATION

A common way to determine the deflection of a beam (the elastic curve) is to first obtain a relationship for internal bending moment (M) in terms of the deflection (v). Then, with the knowledge of M, as we have learned how to obtain, we can determine v.

Assumptions: Apart from the assumptions that we made in deriving the flexure formula, now we assume the following: (1) beam deflections (v) are very small compared to the beam length; (2) slope of the elastic curve (θ) is small. *Note*: In fact, assumption 2 follows from assumption 1.

Let v = upward transverse (lateral, in y-direction) deflection of the elastic curve at a general axial point x.

Slope of the elastic curve at x: $dv/dx = \theta$

Its derivative

$$\frac{d\theta}{dx} = \frac{d^2v}{dx^2}$$

Using the fact that (from calculus), in the limit, $\delta\theta = (d\theta/dx)\delta x$, we have $\delta\theta = (d^2v/dx^2)\delta x$.

Hence, the slope of the elastic curve "increases" from point x to point $x + \delta x$ by

$$\delta\theta = \frac{d^2v}{dx^2}\delta x \tag{i}$$

This angle $\delta\theta$ is also the angle between the radius of curvature of the first point (x) and the radius of curvature of the second point ($x + \delta x$) in the elastic curve (see Figure 8.24). Hence,

Length of the elemental segment of the elastic curve

$$\delta s = \rho \cdot \delta\theta \tag{ii}$$

where ρ is the radius of curvature of the elastic curve at x

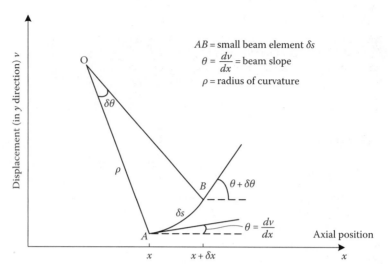

FIGURE 8.24 Deflection of an element of the beam elastic curve.

Note: δs is the length of the considered beam element along the elastic curve after deformation, and δx is the corresponding length along the x axis (i.e., before deformation). Since we neglect axial deformation of the beam, we can write $\delta x \approx \delta s \cos\theta \approx \delta s$. We will further justify this approximation as follows: since we assume that slope θ is small, we can write $\delta x \approx \delta s \cos\theta \approx \delta s$.

Consequently, (ii) can be approximated to (for small slopes, as $\delta x \to 0$)

$$\delta x = \rho \cdot \delta \theta \tag{iii}$$

(i) and (iii):
$$\frac{1}{\rho} = \frac{d^2 v}{dx^2} \tag{8.23}$$

From the previous flexure result (see (8.7) and (8.12))

$$\frac{1}{\rho} = -\frac{\varepsilon}{y} = -\frac{\sigma}{Ey} = \frac{M}{EI} \tag{8.24}$$

where EI is the flexural rigidity
(8.23) and (8.24):

$$M = EI \frac{d^2 v}{dx^2} \tag{8.25}$$

Also, from Equation 8.2, we have

$$V = \frac{d}{dx} EI \frac{d^2 v}{dx^2} \tag{8.26}$$

Note: Equation 8.26 allows for nonuniform beams where I is not constant.

8.6.2 Slope Relation

From (i), we have $\delta\theta = (d^2v/dx^2)\delta x = (M/EI)\delta x$. Then, by integration, we obtain the following result for slope of the beam (elastic curve):

TABLE 8.1
Analogies among Deflection Relationships

Quantity	Axial Member	Torsional Member	Bending Member
Deflection	Extension δ	Angle of twist ϕ	Change in slope θ
Internal load	Axial force P	Torque T	Bending moment M
Material parameter	Young's modulus E	Shear modulus G	Young's modulus E
Geometric parameter	Area of X-section A	Polar moment of area J	Second moment of area about NA I

Rotation (change in slope) of the beam segment from A to B

$$\theta_{B/A} = \int_A^B \frac{M}{EI} dx = \text{Area under the } M/EI \text{ curve from } A \text{ to } B \quad (8.27)$$

It is interesting to compare this result with those for an axial member and a torsional member, respectively, as reproduced in the following (see Chapters 6 and 7):

$$\delta = \int_A^B \frac{P}{EA} dx \text{ for a tensile member}$$

$$\phi = \int_A^B \frac{T}{GJ} dz \text{ for a torsion member}$$

There is a direct analogy among these three results, as summarized in Table 8.1.

There are several methods to determine the deflection of the elastic curve of a beam. Two common methods are *deflection by integration* and *deflection by superposition*. In the first method (deflection by direct integration), the result (8.25) is integrated twice to determine the deflection (v) of a beam. In the second method, with the knowledge of beam defection for simple and basic external loading conditions, deflection for more a complex loading condition (i.e., a linear combination of the basic loads) is determined by simply adding the basic results according to the same rule by which the basic loads are added. This superposition holds for linear systems. We will present both methods. However, the first method, which is fundamental, will be discussed in more detail. The second approach is rather "mechanical."

8.6.3 Boundary Conditions

When integrating the expression (8.25) to determine the beam deflection, we have to substitute the boundary conditions (end conditions) of the beam. Some common boundary conditions and the corresponding expressions for the deflection variable at the boundary are given in Table 8.2.

Note: Among what is presented in Table 8.2, only the boundary conditions for deflection and slope are useful in the integration process of the expression for bending moment in determining the deflection. The boundary conditions for bending moment and shear force are not useful in this exercise even though they are used in other applications.

TABLE 8.2
Boundary Conditions for Beams in Bending

1. Simply supported end: (Pinned)	Deflection $= 0 \Rightarrow v = 0$
	Bending moment $= 0 \Rightarrow \dfrac{d^2v}{dx^2} = 0$
2. Clamped end: (Fixed)	Deflection $= 0 \Rightarrow v = 0$
	Slope $= 0 \Rightarrow \dfrac{dv}{dx} = 0$
3. Free end:	Bending moment $= 0 \Rightarrow \dfrac{d^2v}{dx^2} = 0$
	Shear force $= 0 \Rightarrow \dfrac{d}{dx} EI \dfrac{d^2v}{dx^2} = 0$
4. Vertical sliding end:	Slope $= 0 \Rightarrow \dfrac{dv}{dx} = 0$
	Shear force $= 0 \Rightarrow \dfrac{d}{dx} EI \dfrac{d^2v}{dx^2} = 0$

8.6.4 Deflection by Integration

The two main steps associated with the method of direct integration in determining beam deflection are as follows:

1. Determine the moment curve (i.e., express the internal bending moment M as a function of the axial position (x) along the beam).
2. Integrate (8.25) twice, after substituting the expression for M. Use appropriate boundary conditions of the beam to determine the two constants of integration.

We will now present several examples to illustrate the method of determining deflection of the beam elastic curve by integration.

Example 8.11

A uniform beam AB of length L, second moment of area about the NA I, and Young's modulus E is horizontally fixed at end A while the other end (B) is kept free. A uniformly distributed downward force of w_0 per unit length is applied throughout the length of the beam (Figure 8.25a). Obtain an expression for the deflection of the beam (elastic curve) at a distance x along the beam axis from end A.

Solution (Method 1)

Consider the free-body diagram of the beam as shown in Figure 8.25b.
Equations of equilibrium:

$$\uparrow \sum F = 0 : R_A - w_0 L = 0 \rightarrow R_A = w_0 L$$

$$\circlearrowleft \sum_A M = 0 : M_A - w_0 L \times \frac{L}{2} = 0 \quad \rightarrow \quad M_A = \frac{w_0 L^2}{2}$$

Bending in Beams

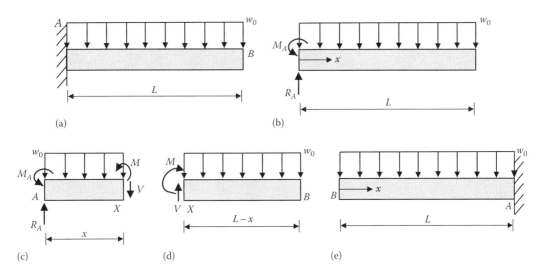

FIGURE 8.25 (a) A cantilever with a uniformly distributed force. (b) Free-body diagram of the beam. (c) Beam segment to the left from a general cut at X. (d) Beam segment to the right of a general cut at X. (e) Transformed beam.

Cut at a general cross section of the beam at distance x from A (Figure 8.25c) and consider the segment to the left.
Moment balance about the X-section (to avoid V)

$$\circlearrowleft \sum_X M = 0 : M + M_A - R_A x + w_0 x \times \frac{x}{2} = 0$$

$$\rightarrow M = -M_A + R_A x - \frac{w_0 x^2}{2}$$

We proceed with the method of direct integration of the bending moment, as follows:

$$EI\frac{d^2v}{dx^2} = -M_A + R_A x - \frac{w_0 x^2}{2}$$

$$\text{Integrate: } EI\frac{dv}{dx} = -M_A x + \frac{R_A x^2}{2} - \frac{w_0 x^3}{6} + A$$

$$\frac{dv}{dx} = 0 \text{ at } x=0 \rightarrow A=0$$

$$\text{Integrate: } EIv = -\frac{M_A}{2}x^2 + \frac{R_A x^3}{6} - \frac{w_0 x^4}{24} + B$$

$$v = 0 \text{ at } x=0 \rightarrow B=0$$

$$EIv = -\frac{M_A x^2}{2} + \frac{R_A x^3}{6} - \frac{w_0 x^4}{24}$$

$$\text{Substitute for } R_A \text{ and } M_A: v = \frac{1}{EI}\left[-\frac{w_0}{4}L^2 x^2 + \frac{w_0 L x^3}{6} - \frac{w_0 x^4}{24}\right]$$

Solution (Method 2)

Cut at cross section X and consider the segment to the right (Figure 8.25d).
 Moment balance about the X-section (to avoid V)

$$\circlearrowleft \sum_X M = 0 : -M - w_0(L-x) \times \frac{(L-x)}{2} = 0$$

$$\to M = -\frac{w_0}{2}(L-x)^2$$

We proceed with the direct integration method as follows:

$$EI\frac{d^2v}{dx^2} = -\frac{w_0(L-x)^2}{2}$$

Integrate: $EI\dfrac{dv}{dx} = +\dfrac{w_0(L-x)^3}{6} + A_1$

$\dfrac{dv}{dx} = 0$ at $x = 0 \to \quad \dfrac{w_0 L^3}{6} + A_1 = 0 \to A_1 = -\dfrac{w_0 L^3}{6}$

Integrate: $EIv = -\dfrac{w_0(L-x)^4}{24} + A_1 x + B_1$

$v = 0$ at $x = 0 \to \quad -\dfrac{w_0}{24}L^4 + B_1 = 0 \to B_1 = \dfrac{w_0 L^4}{24}$

$$v = \frac{1}{EI}\left[-\frac{w_0(L-x)^4}{24} - \frac{w_0 L^3 x}{6} + \frac{w_0}{24}L^4\right]$$

$$= \frac{1}{EI}\left[-\frac{w_0}{24}(L^4 - 4L^3 x + 6L^2 x^2 - 4Lx^3 + x^4) - \frac{w_0 L^3 x}{6} + \frac{w_0 L^4}{24}\right]$$

$$= \frac{1}{EI}\left[-\frac{w_0 L^2 x^2}{4} + \frac{w_0 L x^3}{6} - \frac{w_0 x^4}{24}\right]$$

This result for beam deflection is identical to the previous result.

Solution (Method 3)

Transform the beam into its mirror image about a vertical plane, and define a new axial coordinate from end B to the right (Figure 8.25e).

Note: New $x = L -$ old $x \to$ old $x = L -$ new x

Now we proceed as in Method 1.

$$EI\frac{d^2v}{dx^2} = -\frac{w_0 x^2}{2}$$

$$EI\frac{dv}{dx} = -\frac{w_0 x^3}{6} + A_2$$

$\dfrac{dv}{dx} = 0$ at $x = L \to \quad -\dfrac{w_0 L^3}{6} + A_2 = 0 \to A_2 = \dfrac{w_0 L^3}{6}$

Bending in Beams

$$EIv = -\frac{w_0 x^4}{24} + A_2 x + B_2$$

$v = 0$ at $x = L \rightarrow$
$$-\frac{w_0}{24} L^4 + A_2 L + B_2 = 0 \rightarrow B_2 = \frac{w_0 L^4}{24} - \frac{w_0 L^4}{6} = -\frac{w_0 L^4}{8}$$

$$\rightarrow v = \frac{1}{EI}\left[-\frac{w_0 x^4}{24} + \frac{w_0 L^3 x}{6} + \frac{w_0}{8} L^4 \right]$$

Change back coordinates:
$$v = \frac{1}{EI}\left[-\frac{w_0}{24}(L-x)^4 + \frac{w_0 L^3}{6}(L-x) - \frac{w_0 L^4}{8} \right]$$

$$\rightarrow v = \frac{1}{EI}\left[-\frac{w_0}{24}(L-x)^4 - \frac{w_0 L x^3}{6} + \frac{w_0 L^4}{24} \right]$$

This result for beam deflection is identical to what we obtained from Method 2.

Primary Learning Objectives
1. Determination of beam deflection by the method of integration
2. Formulation of the bending moment
3. Avoiding support reactions of a cantilever in applying the integration method
4. Coordinate reversal to facilitate the application of the integration method

■ **End of Solution**

Example 8.12

A beam AB of length L, uniform thickness h, and a symmetric triangular (i.e., wedge-shaped) plan view is clamped at end A, and a downward force P is applied at the other end (B), as shown in Figure 8.26a and b. The root width of the beam is a_0. Determine an expression for the lateral deflection v of the beam (elastic curve) at a general location X where $AX = x$.

Solution

Make a virtual cut at X where $AX = x$ and consider the beam segment to the right (Figure 8.26c). Equilibrium gives

$$M = -P \times (L - x) \quad \text{(i)}$$

From geometry (see Figure 8.26d), beam width a at X is given by

$$a = \frac{a_0}{L} \times (L - x)$$

The second moment of area about the NA of the cross section at X is

$$I(x) = \frac{1}{12} a h^3 = \frac{1}{12} \times \frac{a_0}{L} \times (L-x) h^3 = \frac{a_0 h^3}{12 L} \times (L-x) \quad \text{(ii)}$$

Use $EI \dfrac{d^2 v}{dx^2} = M(x)$

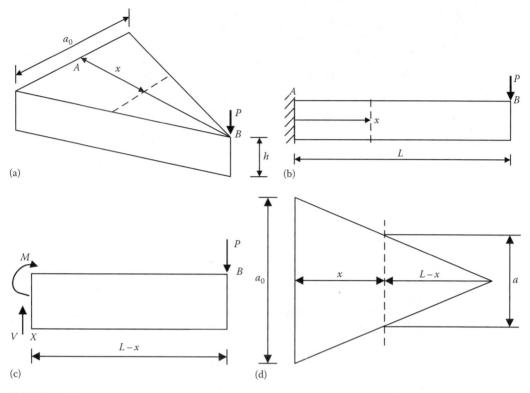

FIGURE 8.26 A cantilever with nonuniform X-section; (a) 3-D view; (b) side view; (c) beam segment to the right of the cut; (d) geometry of the beam.

Substitute (i) and (ii)

$$E \times \frac{a_0 h^3}{12L} \times (L-x) \times \frac{d^2 v}{dx^2} = -P \times (L-x)$$

$$\rightarrow \frac{d^2 v}{dx^2} = -\frac{12LP}{Ea_0 h^3} = -\alpha \text{ (say)}$$

Integrate

$$\frac{dv}{dx} = -\alpha x + A$$

$$\frac{dv}{dx} = 0 \text{ at } x = 0 \rightarrow A = 0$$

Integrate

$$v = -\frac{\alpha}{2}x^2 + B$$

$$v = 0 \text{ at } x = 0 \rightarrow B = 0$$

Bending in Beams

$$\to v = -\frac{\alpha}{2}x^2 = -\frac{12LP}{2Ea_0h^3}x^2$$

$$\to v = -\frac{6LP}{Ea_0h^3}x^2$$

Primary Learning Objectives
1. Determination of the second moment of area of a beam of nonuniform X-section
2. Formulation of the bending moment for a beam of nonuniform X-section
3. Determination of beam deflection by the method of direct integration

■ **End of Solution**

Example 8.13

A cantilever beam is made of two uniform axial segments of the same material and of equal length L, but having different second moments of area I_1 and I_2. A human load $P = Mg$ is applied at the free end of the cantilever (Figure 8.27). Determine the equation of the elastic curve of the beam.

Solution (Method 1)

Note: In this problem, there is no need to determine the reaction load (force and moment) at A, just as in Methods 2 and 3 of Example 8.11, as you will see. That is why a free-body diagram is not drawn first. However, a free-body diagram may be used to determine the support reactions, and as a check for the moment value given by the general moment expression that you will derive.

For AB
Moment at x

$$EI_1 \frac{d^2v}{dx^2} = -P(2L - x)$$

Integrate

$$EI_1 \frac{dv}{dx} = -2PLx + \frac{Px^2}{2} + C_1$$

$$\frac{dv}{dx} = 0 \text{ at } x = 0 \Rightarrow C_1 = 0 \to EI_1 \frac{dv}{dx} = -2PLx + \frac{Px^2}{2}$$

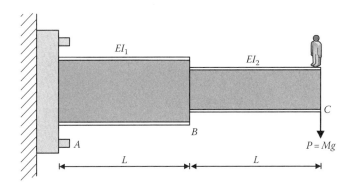

FIGURE 8.27 A two-segmented beam in bending.

Integrate again

$$EI_1 v = -PLx^2 + \frac{Px^3}{6} + C_2$$

$$v = 0 \text{ at } x = 0 \Rightarrow C_2 = 0$$

Hence,

$$EI_1 v = -PLx^2 + \frac{Px^3}{6} \text{ for } x = 0 \text{ to } L$$

For BC

Moment at x

$$EI_2 \frac{d^2 v}{dx^2} = -P(2L - x)$$

Integrate

$$EI_2 \frac{dv}{dx} = -2PLx + \frac{Px^2}{2} + D_1$$

Integrate again

$$EI_2 v = -PLx^2 + \frac{Px^3}{6} + D_1 x + D_2$$

for x = 0 to 2L
dv/dx is the same at B in the two segments (due to continuity of the two segments of the beam).

$$\Rightarrow \frac{1}{EI_1}\left[-2PL^2 + \frac{PL^2}{2}\right] = \frac{1}{EI_2}\left[-2PL^2 + \frac{PL^2}{2} + D_1\right]$$

$$\Rightarrow D_1 = \frac{3}{2} PL^2 \left(1 - \frac{I_2}{I_1}\right)$$

v is the same at B in the two segments (due to continuity of the two segments of the beam).

$$\Rightarrow \frac{1}{EI_1}\left[-PL^3 + \frac{PL^3}{6}\right] = \frac{1}{EI_2}\left[-PL^3 + \frac{PL^3}{6} + D_1 L + D_2\right]$$

$$\Rightarrow D_2 = \frac{5}{6} PL^3 \left(1 - \frac{I_2}{I_1}\right) - D_1 L = \frac{5}{6} PL^3 \left(1 - \frac{I_2}{I_1}\right) - \frac{3}{2} PL^3 \left(1 - \frac{I_2}{I_1}\right)$$

$$\Rightarrow D_2 = -\frac{2}{3} PL^3 \left(1 - \frac{I_2}{I_1}\right)$$

Solution (Method 2)

In this method, integration for both segments of the beam is done in parallel, in a partitioned manner. In this way, we can exploit the similarity of the bending moment expressions in the two segments of the beam.

Bending in Beams

Note: Method 1 is a sequential method whereas Method 2 is a parallel method, as shown in the following.

Operation	AB	BC
$\frac{1}{EI}M = \frac{d^2v}{dx^2} =$	$-\frac{P}{EI_1}(2L-x)$	$-\frac{P}{EI_2}(2L-x)$
Integrate: $\frac{dv}{dx} =$	$\frac{P}{2EI_1}\left[(2L-x)^2 + A_1\right]$	$\frac{P}{2EI_2}\left[(2L-x)^2 + A_2\right]$
$\frac{dv}{dx}$ is the same at B from both expressions:	$\frac{P}{2EI_1}[L^2 + A_1] =$	$\frac{P}{2EI_2}[L^2 + A_2]$ $\to A_2 = -L^2 + \frac{I_2}{I_1}(L^2 + A_1)$
$\frac{dv}{dx} = 0$ at $x=0$:	$\frac{P}{2EI_1}\left[(2L)^2 + A_1\right] = 0$ $\to A_1 = -4L^2$	$\to A_2 = -\left(1 + \frac{3I_2}{I_1}\right)L^2$
Integrate: $v =$	$\frac{P}{2EI_1}\left[-\frac{1}{3}(2L-x)^3 + A_1 \times (x-L) + B_1\right]$	$\frac{P}{2EI_2}\left[-\frac{1}{3}(2L-x)^3 + A_2 \times (x-L) + B_2\right]$
v is the same at B from both expressions:	$\frac{P}{2EI_1}\left[-\frac{1}{3}L^3 + B_1\right] =$	$\frac{P}{2EI_2}\left[-\frac{1}{3}L^3 + B_2\right]$
$v=0$ at $x=0$:	$\frac{P}{2EI_1}\left[-\frac{1}{3}(2L)^3 - A_1L + B_1\right] = 0$ $\to B_1 = -\frac{4}{3}L^3$	$\to B_2 = \frac{1}{3}L^3 + \frac{I_2}{I_1}\left(-\frac{1}{3}L^3 + B_1\right)$ $\to B_2 = -\frac{1}{3}L^3\left(1 - 5\frac{I_2}{I_1}\right)$
Final result: $v =$	$\frac{P}{2EI_1}\left[-\frac{1}{3}(2L-x)^3 - 4L^2 \times (x-L) - \frac{4}{3}L^3\right]$	$\frac{P}{2EI_2}\left[-\frac{1}{3}(2L-x)^3 - \left(1 + 3\frac{I_2}{I_1}\right)L^2 \times (x-L) + \frac{1}{3}L^3\left(1 - 5\frac{I_2}{I_1}\right)\right]$

Solution (Method 3)

In this method, the terms in the moment expression are expanded and the resulting individual terms are integrated. It provides some advantages in the presence of particular types of boundary conditions (specifically, clamped end), as seen from the procedure given in the following.

Operation	AB	BC
$\frac{1}{EI}M = \frac{d^2v}{dx^2} =$	$-\frac{P}{EI_1}(2L-x)$	$-\frac{P}{EI_2}(2L-x)$
Integrate: $\frac{dv}{dx} =$	$\frac{P}{EI_1}\left[-2Lx + \frac{1}{2}x^2 + A_1\right]$	$\frac{P}{EI_2}\left[-2Lx + \frac{1}{2}x^2 + A_2\right]$
$\frac{dv}{dx}$ is the same at B from both expressions:	$\frac{P}{EI_1}\left[-\frac{3}{2}L^2 + A_1\right]$	$= \frac{P}{EI_2}\left[-\frac{3}{2}L^2 + A_2\right]$ $\to A_2 = \frac{3}{2}L^2 + \frac{I_2}{I_1}\left(-\frac{3}{2}L^2 + A_1\right)$

$\dfrac{dv}{dx}=0$ at $x=0$:	$A_1=0$	$\rightarrow A_2=\dfrac{3}{2}\left(1-\dfrac{I_2}{I_1}\right)L^2$
Integrate: $v=$	$\dfrac{P}{EI_1}\left[-Lx^2+\dfrac{1}{6}x^3+B_1\right]$	$\dfrac{P}{EI_2}\left[-Lx^2+\dfrac{1}{6}x^3+A_2x+B_2\right]$
v is the same at B from both expressions:	$\dfrac{P}{EI_1}\left[-\dfrac{5}{6}L^3+B_1\right]$	$\dfrac{P}{EI_2}\left[\left(\dfrac{2}{3}-\dfrac{3}{2}\dfrac{I_2}{I_1}\right)L^3+B_2\right]$
		$\rightarrow B_2=\dfrac{2}{3}\left(-1+\dfrac{I_2}{I_1}\right)L^3+\dfrac{I_2}{I_1}B_1$
$v=0$ at $x=0$:	$B_1=0$	$\rightarrow B_2=\dfrac{2}{3}\left(-1+\dfrac{I_2}{I_1}\right)L^3$
Final result: $v=$	$\dfrac{P}{EI_1}\left[-Lx^2+\dfrac{1}{6}x^3\right]$	$\dfrac{P}{EI_2}\left[-Lx^2+\dfrac{1}{6}x^3+\dfrac{3}{2}\left(1-\dfrac{I_2}{I_1}\right)L^2x+\dfrac{2}{3}\left(-1+\dfrac{I_2}{I_1}\right)L^3\right]$

Primary Learning Objectives

1. Formulation of the bending moment for a beam having multiple segments
2. Determination of beam deflection by the method of direct integration
3. Partitioning of the beam into segments and performing the integrations for the segments in parallel
4. Use of deflection continuity in determining deflection of a beam having multiple segments
5. Organization of the integrand prior to integration in order to simplify the analytical expressions

■ **End of Solution**

8.6.5 Deflection by Superposition

In the theory of beam deflection, we have made several assumptions. In particular, we assumed that stress–strain (load–deflection) relation is linear and also the deflections and slopes of the beam (elastic curve) are small (in comparison to the beam dimensions). In view of the assumption of "linearity," the *principle of superposition* must hold. We can use this principle to determine the deflection of the beam elastic curve under somewhat complex loading conditions that are linear combinations of simpler (basic) loads. The approach is as follows.

Suppose that the expressions for beam deflection for rather simple (basic) external loads are known. (Tables are commonly available with this information.) Also, suppose that a more complex external loading condition can be represented as a linear combination of these basic loads. Then, the beam deflection under the complex loading condition is determined by simply adding the known basic results of deflection according to the same rule by which the basic loads are added to generate the complex loading condition. An illustrative example for this approach is given now.

Example 8.14

A uniform beam AB of length L, second moment of area about the NA I, and Young's modulus E is horizontally fixed at end A, while the other end (B) is kept free. A uniformly distributed downward force of w_0 per unit length is applied throughout the length of the beam. Also, a concentrated

Bending in Beams

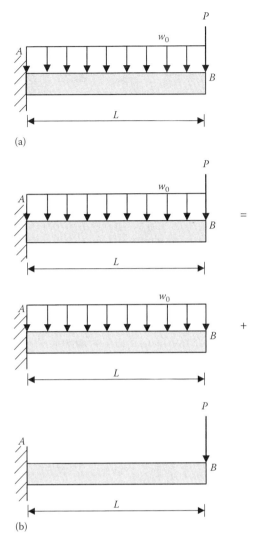

FIGURE 8.28 (a) A cantilever with a distributed force and a concentrated force. (b) Illustration of the principle of superposition.

downward force P is applied at the free end (Figure 8.28a). Obtain an expression for the deflection of the beam (elastic curve) at distance x along the beam axis from end A.

Solution

The given problem can be represented as the sum of two basic problems, as shown in Figure 8.28b. Now from Example 8.11, the deflection due to the distributed force alone may be expressed as

$$v_w = \frac{w_0 x^2}{24EI}\left[-6L^2 + 4Lx - x^2\right]$$

Next, from Example 8.13 (for the segment AB, by setting $l_1 = l$ and $2L \to L$), the deflection due to the concentrated force alone may be expressed as

$$v_P = \frac{Px^2}{6EI}[-3L + x]$$

Then, from the principle of superposition, the deflection of the beam shown in Figure 8.28a is given by

$$v = v_w + v_P = \frac{w_0 x^2}{24EI}\left[-6L^2 + 4Lx - x^2\right] + \frac{Px^2}{6EI}[-3L + x]$$

or

$$v = \frac{x^2}{24EI}\left[w_0(-6L^2 + 4Lx - x^2) + 4P(-3L + x)\right]$$

Primary Learning Objectives

1. Determination of beam deflection by the method of superposition
2. Decomposition of a given external loading condition into a combination of basic external loads

■ **End of Solution**

8.7 STATICALLY INDETERMINATE BEAMS

In a statically indeterminate bending member, the unknown loads (e.g., forces and moments at the support locations) that are needed to analyze the problem cannot be determined using static equations of equilibrium alone. In addition, "compatibility" conditions must be introduced.

A compatibility condition is a condition that satisfies structural integrity (or geometric continuity) of the structure. It is written in terms of the deflections at a point in a given direction.

A compatibility condition of displacement in a beam bending problem is as follows.

The lateral displacement v of the elastic curve of the beam at a given axial location x is the same (i.e., unique) regardless of which of the two beam segments joined at x is used to determine the displacement.

A compatibility condition of the angle of slope in a beam bending problem is as follows.

The slope dv/dx of the elastic curve of the beam at a given axial location x is the same (i.e., unique) regardless of which of the two beam segments joined at x is used to determine the slope.

Note: The application of the compatibility condition for a beam is rather analogous to that for an axially loaded member or a torsional member.

Example 8.15

A uniform beam *ABC* of length 2*L* is rigidly clamped at end *A* and supported by a smooth pin at end *C*. A uniformly distributed load of *w* per unit length is applied downward along the entire beam. Also a point moment M_o is applied in the clockwise direction at the midpoint *B* of the beam (Figure 8.29a).

  a. Write expressions for the internal bending moment in a general cross section of the beam at a distance x from *A*, for the two segments *AB* and *BC*. By successively integrating these expressions and using the boundary conditions, determine a complete expression for the lateral deflection of the elastic curve of the beam.
  b. Determine the unknown reactions at the support locations of the beam. Check your results using the static equations of equilibrium.

Solution

  a. The free-body diagram of the beam is shown in Figure 8.29b. Make a virtual section at location x between *A* and *B*, and consider the segment to the left. From the equilibrium of that segment, we can determine the internal bending moment *M* in that segment.

Bending in Beams

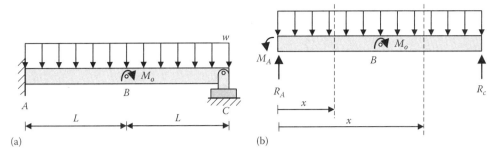

FIGURE 8.29 (a) A cantilever with a pinned end; (b) the free-body diagram.

For a location x between B and C, the expression for the bending moment should be modified by simply adding the external moment M_o at B.

These expressions are tabulated in the first row of Table 8.3.

Note: $M = EI(d^2v/dx^2)$

where v is the downward deflection of the elastic curve

The results of successive integration are given in Table 8.3.

b. Apply the remaining boundary conditions to determine the unknown reaction force R_A and the bending moment M_A.

Specially,

$$M = 0 \text{ at } x = 2L \Rightarrow -\frac{w}{2} \times (2L)^2 + R_A \times 2L - M_A + M_o = 0$$

$$\Rightarrow M_A - 2R_A L = M_o - 2wL^2 \qquad (i)$$

TABLE 8.3
Steps of Determining the Beam Deflection

Quantity	AB	BC
$M = EI\dfrac{d^2v}{dx^2}$	$-\dfrac{wx^2}{2} + R_A x - M_A$	$-\dfrac{wx^2}{2} + R_A x - M_A + M_o$
Integrate: $EI\dfrac{dv}{dx}$	$-\dfrac{wx^3}{6} + \dfrac{R_A x^2}{2} - M_A x + A_1$	$-\dfrac{wx^3}{6} + \dfrac{R_A x^2}{2} - M_A x + M_o(x-L) + B_1$
Apply:	$\dfrac{dv}{dx}$ is the same at B $(x = L)$ from both expressions $\Rightarrow A_1 = B_1$	
Apply:	$\dfrac{dv}{dx} = 0$ at $x = 0 \Rightarrow A_1 = 0$	$\Rightarrow B_1 = 0$
Integrate: EIv	$-\dfrac{wx^4}{24} + \dfrac{R_A x^3}{6} - \dfrac{M_A x^2}{2} + A_2$	$-\dfrac{wx^4}{24} + \dfrac{R_A x^3}{6} - \dfrac{M_A x^2}{2} + \dfrac{M_o}{2}(x-L)^2 + B_2$
Apply:	v is the same at B from both expressions $\Rightarrow A_2 = B_2$	
Apply:	$v = 0$ at $x = 0 \Rightarrow A_2 = 0$	$\Rightarrow B_2 = 0$
Result: EIv	$-\dfrac{wx^4}{24} + \dfrac{R_A x^3}{6} - \dfrac{M_A x^2}{2}$	$-\dfrac{wx^4}{24} + \dfrac{R_A x^3}{6} - \dfrac{M_A x^2}{2} + \dfrac{M_o}{2}(x-L)^2$

$$v = 0 \text{ at } x = 2L \Rightarrow -\frac{wL}{24} \times (2L)^4 + \frac{R_A}{6} \times (2L)^3 - \frac{M_A}{2} \times (2L)^2 + \frac{M_o}{2} \times (2L-L)^2 = 0$$

$$\Rightarrow M_A - \frac{2}{3} R_A L = \frac{M_o}{4} - \frac{wL^2}{3} \quad \text{(ii)}$$

$$\text{(ii)} - \text{(i)}: -\frac{2}{3} R_A L + 2 R_A L = \frac{M_o}{4} - M_o - \frac{wL^2}{3} + 2wL^2$$

$$\Rightarrow \frac{4}{3} R_A L = -\frac{3}{4} M_o + \frac{5}{3} wL^2$$

$$\Rightarrow R_A = -\frac{9}{16L} M_o + \frac{5}{4} wL \quad \text{(iii)}$$

Substitute back into (i):

$$M_A = 2L \times \left(-\frac{9}{16L} M_o + \frac{5}{4} wL\right) + M_o - 2wL^2$$

$$\Rightarrow M_A = -\frac{M_o}{8} + \frac{wL^2}{2} \quad \text{(iv)}$$

Check using equations of equilibrium:

$$\uparrow \sum F_x = 0 \Rightarrow R_A + R_C - 2wL = 0 \quad \text{(v)}$$

$$\curvearrowleft \sum M_A = 0 \Rightarrow M_A - M_o - 2wL^2 + 2LR_C \quad \text{(vi)}$$

(iii) and (v):

$$R_C = 2wL - R_A = 2wL - \left(-\frac{9}{16L} M_o + \frac{5}{4} wL\right) = \frac{3}{4} wL + \frac{9}{16L} M_o \quad \text{(vii)}$$

Substitute (iv) and (vii) into (vi):

$$-\frac{M_o}{8} + \frac{wL^2}{2} - M_o - 2wL^2 + 2L \times \left(\frac{3}{4} wL + \frac{9}{16L} M_o\right) = -\frac{M_o}{8} - M_o + \frac{9}{8} M_o + \frac{wL^2}{2} - 2wL^2 + \frac{3wL^2}{2} = 0$$

⇐ Checks

Primary Learning Objectives
1. Determination of the deflection of a statically indeterminate beam by the method of direct integration
2. Partitioning of the beam into segments and performing the integrations for the segments in parallel
3. Use of deflection continuity in determining deflection of a beam having multiple segments
4. Determination of the support reactions of a beam using the deflection results

■ **End of Solution**

SUMMARY SHEET

Moment: May be represented by a couple (i.e., $M = d \times P$; equal and opposite parallel forces P separated by distance d).

Internal shear force (V): Lateral force on a virtual cross section of a beam.

Internal bending moment (M): Bending moment on a virtual cross section of a beam.

Sign convention:

1. Force: Upward (+ve y axis) is +ve (concentrated force, F; or distributed force, $w(x)$ per unit length)
2. Shear Force: That tends to rotate element clockwise (cw) is +ve.
3. Bending Moment: That tends to bend ends of element upward (i.e., forming "bucket" shape or "smiling mouth" shape) is +ve.

Shear diagram: Variation of the internal shear force along the beam (i.e., V vs. x).

Moment diagram: Variation of the internal bending moment along the beam (i.e., M vs. x).

Relations between internal shear $V(x)$, moment $M(x)$, and external distributed force $w(x)$:

$$\frac{dV}{dx} = w(x); \quad V(x) = V_0 + \int_0^x w(x)\,dx$$

$$\frac{dM}{dx} = V; \quad M(x) = M_0 + \int_0^x V(x)\,dx$$

Area method for shear and moment diagrams:

1. Change in V in an interval = Area of $w(x)$ curve in that interval
2. Change in M in an interval = Area of V curve in that interval

Effect of external downward point force (F_0) at x on internal shear:

$$V_{x+\delta x} = V_x + F_0$$

Effect of external clockwise point moment (M_0) at x on internal moment:

$$M_x = M_{x+\delta x} + M_0$$

Mirror image method for shear and moment diagrams:

1. Take mirror image of beam of length L (i.e., $x \to L-x$) and determine V and M from left end
2. Change back $x \to L-x$ in the result and reverse sign of V only

Bending strain formula (general):

$$\varepsilon = -\frac{y}{\rho}$$

or

$$\varepsilon = -\frac{y}{c}\varepsilon_{max}$$

ρ is the radius of curvature
y is the height from neutral (centroidal) axis (NA)

Bending stress formula (linear elastic):

$$\sigma = -\frac{y}{c}\sigma_{max}$$

Flexure formula:

$\sigma = -\frac{My}{I}$ or $\sigma_{max} = \frac{Mc}{I}$; c is the absolute max of y

Second moment of area about NA: $I = \int_A y^2 dA$

Parallel axis theorem: $I_D = I_G + A \cdot d^2$
  $I_G = I$ about centroidal axis (NA)
  $I_D = I$ about an axis at distance d from centroidal axis
  A is the area of cross section

Area removal method:
Given: Area O = Area A − Area B
Then: Area moment of O = Area moment of A − Area moment of B
And: Second moment of area O = Second moment of area A − Second moment of area of B
(about the same axis)

Transformed section method (for composite beams):
Given: $n = E_B/E_A$; widen the area of material B by n and analyze as a single material A; multiply resulting bending stress values for material B by n

Shear formula: Shear stress at y from NA

$$\tau = \frac{VQ}{It}$$

V is the internal shear force on X-section
Q is the first moment of area of the X-section above y, about NA
I is the second moment of area of entire X-section about NA
t is the width of X-section at y

Shear flow (q): Shear force per unit length, $q = dF/dx = \tau \cdot t$ or $q = VQ/I$

Elastic curve: Centroidal curve along beam axis

Beam deflection formulas:

$\frac{1}{\rho} = \frac{d^2v}{dx^2}$; $M = EI\frac{d^2v}{dx^2}$; $V = \frac{d}{dx}EI\frac{d^2v}{dx^2}$

Rotation (change in slope) of the beam segment from A to B:

$\theta_{B/A} = \int_A^B (M/EI)dx$ = Area under the M/EI curve from A to B

Compare: $\delta = \int_A^B (P/EA)dx$ for a tensile member; $\phi = \int_A^B (T/GJ)dz$ for a torsion member

PROBLEMS

8.1 A beam is held horizontally by clamping at one end (A). A uniform distributed load w_0 per unit length is applied along its total length L. A bending moment M_o is applied at the free end (see Figure P8.1).
Determine
  a. The reaction loading (force and moment) at the clamped end
  b. The shear and moment diagrams

8.2 A beam of length $2L$ is pinned at end A and supported on rollers at the other end (D). It is loaded as shown in Figure P8.2 (with a uniformly distributed force of w/length from A up to midspan B, and a point force P at C, at distance $3L/2$ from A). Determine the shear diagram and the moment diagram for the beam. Assume $P > wL$.

8.3 A structural member consists of two continuous beams ABD and $CDEF$ joined together and loaded as shown in Figure P8.3. The beams are maintained horizontally using a hinge (smooth) at A, a pin-jointed (smooth) short vertical member across BC, a roller support (smooth) at D, and another hinge (smooth) at F. An external downward force P is applied

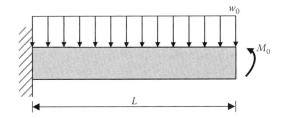

FIGURE P8.1 A cantilever with a distributed force and an end moment.

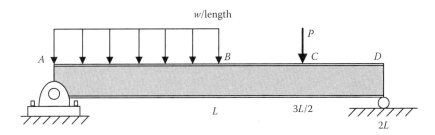

FIGURE P8.2 A beam with a distributed force and a concentrated force.

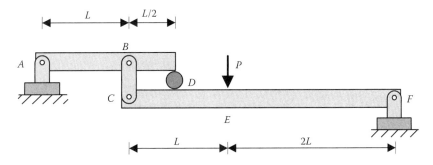

FIGURE P8.3 A structure with two beam segments and cross link.

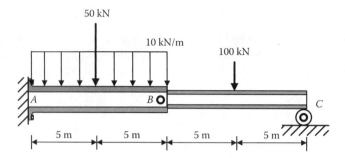

FIGURE P8.4 A cantilever with pin-joined and roller-supported segment.

at E. The dimensions are $AB=L$, $BC=CD=L/2$, $DE=L/2$, and $EF=2L$. Use the graphical method to obtain the shear diagrams and the moment diagrams of the two beams.

8.4 A structure consists two horizontal beam segments of equal length 10 m pinned (smooth) together, clamped at one end, and supported on rollers (smooth) at the other end, as shown in Figure P8.4. Concentrated loads of 50 kN and 100 kN are applied at the midspans of the two beam segments. Also a distributed load of 10 kN/m is applied uniformly on the first beam segment.

Using the graphical method, construct the shear and moment diagrams for the structure. Indicate the numerical values at important points of the two curves. Also, clearly identify the linear segments and the curved segments of the two curves.

8.5 A J-shaped beam $DABC$ is horizontally supported by a smooth roller at A and a smooth hinge at D (see Figure P8.5). A concentrated downward force P is applied at the free end C. Also, uniformly distributed forces of w/length are applied along AD and BC. Take $AD=2L$, $BC=L$, and $w=2P/L$.

Using the graphical method, determine the shear curve and the moment curve of the beam (including the segment AB). Indicate the key values on these curves.

8.6 A cantilever beam of length L is subjected to a linearly varying distributed force (varying from w_0 per unit length to zero) along its length, as shown in Figure P8.6. Using the coordinate reversal (mirror-image) method, determine the shear and moment diagrams for the beam.

8.7 A beam AD of length $2L=8$ m is hinged at end A and free at end D. The midspan C of the beam is supported on a smooth roller (Figure P8.7). A concentrated downward force $P_B=100$ kN is applied at the midpoint B of the beam segment AC. A linearly varying distributed force starting from $w=w_0=60$ kN/m at C and ending at $w=0$ is applied over the beam segment CD.

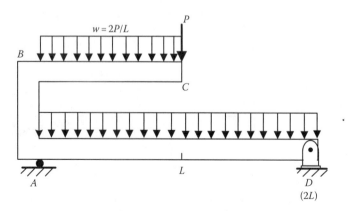

FIGURE P8.5 A J-shaped beam with distributed and concentrated forces.

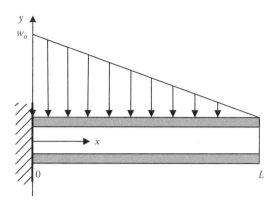

FIGURE P8.6 A cantilever with a distributed force.

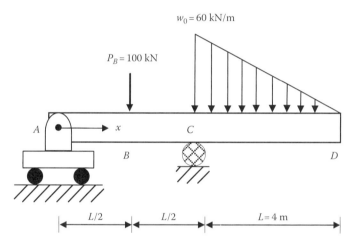

FIGURE P8.7 A beam with concentrated and distributed loading.

Use the method of coordinate transformation (i.e., mirror image) to determine the shear diagram and the moment diagram of the beam.

8.8 A uniform straight strip of brass having rectangular cross section of thickness h is tightly bent around a rigid cylinder of radius r using the bending moment M at its two ends (Figure P8.8). Determine the maximum possible h such that the strip will not yield.
Given
$r = 0.5$ m = radius of the cylinder
$\sigma_Y = 120$ MPa = yield stress of brass
$E = 100$ GPa = Young's modulus of brass

8.9 A beam of length 10 m and made of an I-section is supported at its ends and loaded at the midspan with force 20 kN (Figure P8.9a). Determine the maximum bending stress (magnitude) in the beam and indicate its location. The dimensions of the beam cross section are given in Figure P8.9b.

8.10 In Problem 8.9, suppose that the yield strength of the beam is $\sigma_Y = 250$ MPa. Using a safety factor of 3 for yield, determine the allowable maximum external force at the midspan of the beam.

8.11 A uniform beam AB of length $L = 1$ m is clamped horizontally at end A, and a vertical downward force $P = 5$ kN is applied at the free end B (Figure P8.11a). The beam is formed by

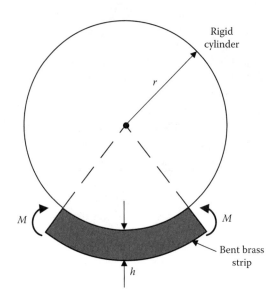

FIGURE P8.8 A metal strip bent around a cylindrical object.

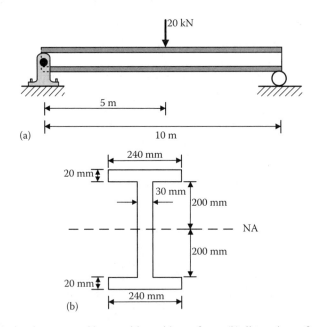

FIGURE P8.9 (a) A simply supported beam with a midspan force; (b) dimensions of the X-section.

drilling an axial circular bore of radius $a = 20$ mm in a solid beam of rectangular cross section of dimensions

Width of the X-section, $b = 60$ mm
Height of the X-section, $h = 100$ mm

Note: The hole is centrally located on the rectangular cross section (see Figure P8.11b). Determine the maximum bending stress on the beam cross section and its location.

Hint (see Figure P.11c and d). Also, see Appendix A:
Second moment of area of a rectangle about centroidal axis Z–Z, $I_{rectangle} = (1/12)bh^3$
Second moment of area of a circle about centroidal diametric axis, $I = (1/4)\pi a^4$

Bending in Beams

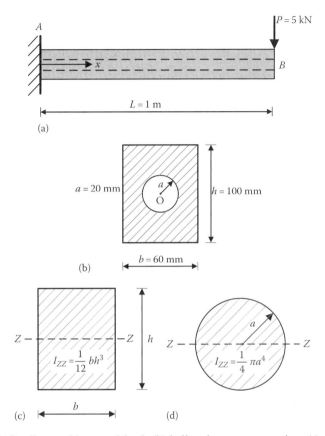

FIGURE P8.11 (a) Cantilever with an end load; (b) hollow beam cross section; (c) rectangular section; (d) circular section.

8.12 A uniform beam AB of length $L=5$ m is clamped at A, and a downward force $P=10$ kN is applied at the free end B (Figure P8.12a). The geometry of the beam cross section is shown in Figure P8.12b. Determine the maximum bending stress (magnitude, direction, and location) of the beam.

8.13 A beam of uniform cross section is made by welding a solid steel beam of square cross section with side $2a$ on a solid steel beam of circular cross section with radius a (Figure P8.13a). The beam is horizontally clamped at one end and a downward force P is applied at the free end of the resulting cantilever (Figure P8.13b).
Given
  Length of beam $L=5$ m
  Radius of circular section $a=0.1$ m
  Yield stress of steel $\sigma_Y=250$ MPa
Using a factor of safety of 2, determine the maximum allowable P.

8.14 A beam AB of length L, uniform thickness h, and a symmetric triangular (i.e., wedge-shaped) plan view is clamped at end A, and a downward force P is applied at the other end (B), as shown in Figure P8.14. The root width of the beam is a_0. Determine the maximum bending stress on a beam cross section at distance x from end A.

8.15 A reinforced concrete beam of rectangular cross section is designed to carry a maximum positive bending moment (causing tension at the bottom and compression at the top) of 20 kN·m.
  Width of the beam section = 0.2 m

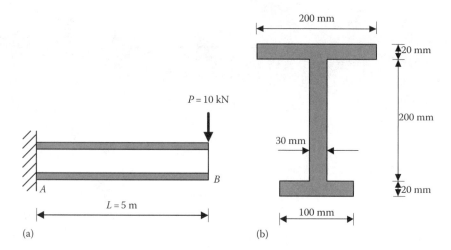

FIGURE P8.12 (a) A cantilever with an end force; (b) geometry of the beam X-section.

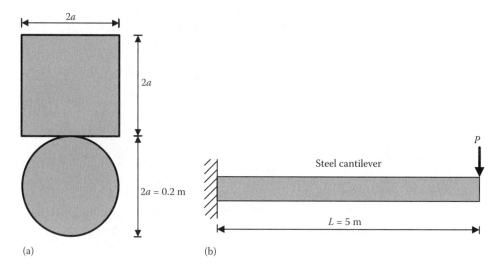

FIGURE P8.13 (a) Cross section of the fabricated beam; (b) loading on the beam.

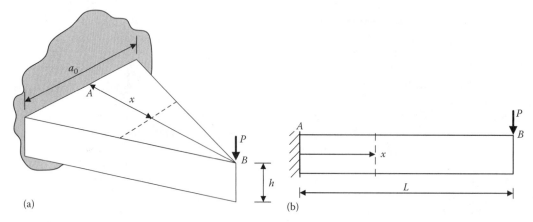

FIGURE P8.14 A cantilever with nonuniform X-section; (a) 3-D view; (b) side view.

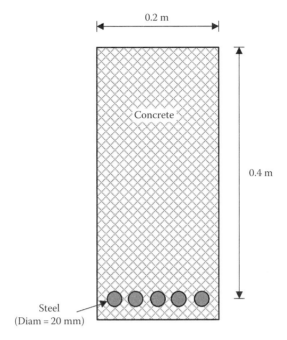

FIGURE P8.15 Cross section of a reinforced concrete beam.

Five steel reinforcement bars each of diameter 20 mm are placed at a depth of 0.4 m from the top surface of the beam (Figure P8.15). Determine:
  The maximum compressive stress in the concrete
  The maximum tensile stress in the steel
  Young's modulus of concrete $E_c = 20$ GPa
  Young's modulus of steel $E_s = 200$ GPa

8.16 A steel girder of I-section is fixed horizontally to a rigid structure at one end and a vertical load of $P = 100$ kN is applied at the other end (Figure P8.16a). Determine the value and the

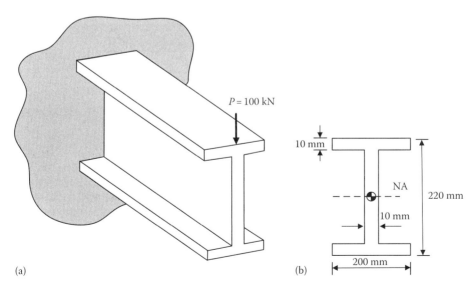

FIGURE P8.16 (a) A clamped girder with an end load; (b) dimensions of the cross section.

location of the maximum shear stress in the cross section. The dimensions of the section are as shown in Figure P8.16b.

8.17 A uniform beam having a square box section is pinned (smooth) at one end and supported on a roller (smooth) at the other end. A concentrated downward force of 200 kN is applied at the midspan of the beam (Figure P8.17a). The dimensions of the beam cross section are given in Figure P8.17b.
  a. Sketch the shear diagram. Determine the maximum shear force, and its location along the beam.
  b. For the location determined in part (a), determine the shear stresses at point X and point Y on the beam cross section.
  c. What is the shear flow at point X on the beam cross section?

8.18 A horizontal cantilever beam AB of length $L=2$ m is rigidly clamped at end A and a vertical force $P=10$ kN is applied at the free end B (Figure P8.18a). The beam has an I-section which is fabricated by seam welding two identical rectangular steel plates of cross-sectional width 100 mm and thickness 10 mm to the top and bottom sides of another rectangular steel plate of cross-sectional height 100 mm and thickness 20 mm (Figure P8.18b). Determine
  a. The complete location (i.e., along the beam and on the X-section) and the value of the maximum normal stress σ_{max} due to bending
  b. The location and the value of the maximum shear stress τ_{max} on the beam cross section
  c. The combined axial force supported by the welded seam joints of a side plate (top plate or bottom plate) per unit length of the beam

8.19 A cantilever beam is made by riveting two identical flat bars on to the top and the bottom flanges of an I-beam. The cantilever is loaded with a downward force $P=100$ kN at its free end (Figure P8.19a). The dimensions of the beam cross section are shown in Figure P8.19b. The diameter of a steel rivet is 10 mm. The allowable shear stress for a rivet is 100 MPa.

Determine the required number of rivets per 1 m length on each side of the beam.

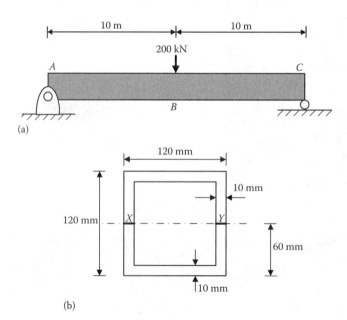

FIGURE P8.17 (a) A simply supported beam with a midspan force; (b) dimensions of the cross section of the beam.

Bending in Beams

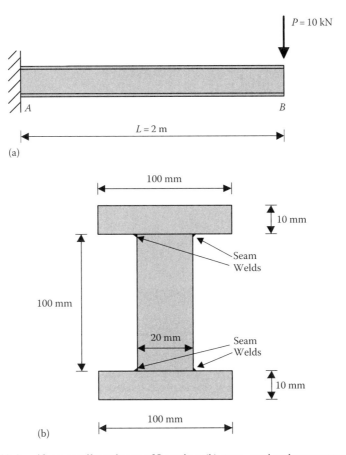

FIGURE P8.18 (a) A uniform cantilever beam of I-section; (b) cross-sectional geometry of the beam.

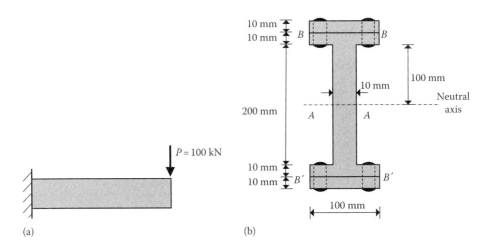

FIGURE P8.19 (a) Cantilever with an end force; (b) details of the cross section.

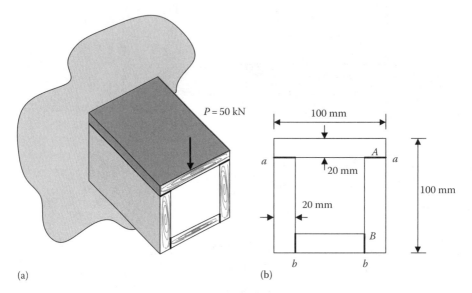

FIGURE P8.20 (a) A beam made of wooden planks; (b) dimensions of the X-section.

8.20 A wooden beam of box section is made by gluing four planks together (Figure P8.20a). The dimensions of the cross section are as shown in Figure P8.20b. A vertical force of $P = 50$ kN is applied to one end of the beam, while the other end is fixed and maintained horizontal. Determine the shear flow q_A and q_B at the glued joints A and B, respectively (which represent the resisting force supported by the glued joint per unit axial length of the beam).

8.21 A uniform beam ABC of length $4L$, second moment of area about the NA (of the cross-section) I, and Young's modulus E is pinned (smooth) at end A and supported on smooth rollers at end C (Figure P8.21). A downward force P is applied at B where $AB = L$. Determine the transverse deflection v of the beam at any general location x from end A, due to the applied force.

8.22 A uniform beam ABC of length $2L$ is supported by a smooth pin at end A and on smooth rollers at end C. A uniformly distributed force w per unit length is applied downward along the entire beam. Also a point moment M_o is applied in the clockwise direction at the midpoint B of the beam (Figure P8.22).

  a. Write expressions for the internal bending moment on a cross section of the beam at a distance x from A for the two segments AB and BC. By successively integrating these expressions and using the boundary conditions, determine a complete expression for the lateral deflection of the elastic curve of the beam.
  b. Determine the unknown reactions at the support points of the beam. Check your results using the static equilibrium method.

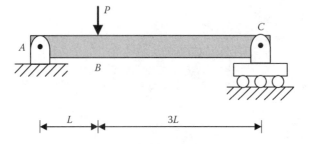

FIGURE P8.21 A simply supported beam with a concentrated force.

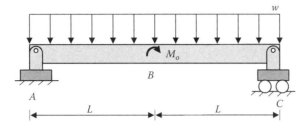

FIGURE P8.22 A simply supported beam with a distributed force and a concentrated moment.

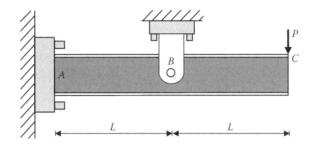

FIGURE P8.23 A clamped girder with pin support.

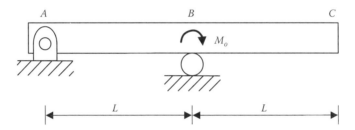

FIGURE P8.24 A uniform beam supported by a hinge and a smooth roller.

8.23 A horizontal girder of length $2L$ is rigidly clamped at one end, rigidly supported by a smooth pin joint at midspan, and experiences a vertical downward force P at the other end (Figure P8.23). Determine
   a. The reaction loads at the supports
   b. The lateral deflection of the elastic curve of the girder due to bending

8.24 A uniform beam ABC of length $2L$ is hinged (smooth) at end A and is supported on a smooth roller at its midpoint B. End C is free (overhung), as shown in Figure P8.24. A concentrated moment M_o is applied to the beam at B, in the clockwise direction.
   a. Determine the support reactions at A and B.
   b. Determine a complete expression for the lateral deflection v of the beam (of its elastic curve) from the original horizontal position, when the external load M_o is applied. The deflection v should be expressed as a function of x, where x is the distance to any general location along the beam from end A. *Note*: Use the method of direct integration.
   c. From the result of part (b), determine the lateral deflection at the free end C of the beam. Verify this result by using the slope of the beam (elastic curve) at B.
   Given
   I is the second moment of area of the beam X-section about its NA
   E is Young's modulus of the beam material

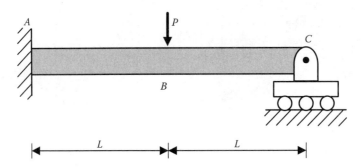

FIGURE P8.25 A beam clamped at one end and supported on rollers at the other.

8.25 A uniform beam ABC of length $2L$, second moment of area I about the cross-sectional NA, and Young's modulus E is clamped at end A and supported on smooth rollers at end C (Figure P8.25). A downward force P is applied at midpoint B of the beam. Determine the lateral deflection v of the elastic curve of the beam at a general axial location x from end A.

9 Stress and Strain Transformations

CHAPTER OBJECTIVES

- Problem of plane stress
- Stress transformation: finding stresses on different planes at the same location
- Use of Mohr's circle for stress transformation
- Determination of principal stresses, maximum in-plane shear stress, and absolute maximum shear stress
- Analysis of thin-walled pressure vessels
- Problem of plane strain
- Strain transformation: finding strains in different directions and corners at the same location
- Use of Mohr's circle for strain transformation
- Determination of principal strains, maximum in-plane shear strain, and absolute maximum shear strain
- Theory of strain measurement using strain gauges; strain-gauge rosettes
- Theories of failure for ductile material and brittle material

9.1 INTRODUCTION

A component may not have the same strength across all planes (and along all directions) at a particular location of it. Also, the strengths in tension, compression, and shear may be different at given location, even with respect to a specified local plane or direction. Since failure in a component occurs when a stress exceeds its limiting value (the ultimate strength), in the design and failure studies of an object it is important to know the magnitudes and directions of the stresses (particularly, the maximum normal and maximum shear stresses) at a location of the object, and it is important as well to know the planes on which these stresses act. Similarly, since the deformations in a body depend on the distribution of the strains (both normal and shear) throughout the body in all directions, in deformation studies it is important to know the strains along different directions and with respect to different corners, at a given location of the body (particularly, the maximum normal and shear strains). Stress transformation and strain transformation concern the determination of local stresses and strains in different planes and directions in a body.

9.1.1 STRESS TRANSFORMATION

Local stress is defined on a local plane at a point. The representation of stress involves the representation of the "plane" on which the stress acts and the specification of the "direction" in which the stress acts on that plane, in addition to giving the magnitude of the stress. This follows from the fact that stress is a two-dimensional (2-D) tensor, not a vector. Given the stresses on the infinitesimal or elemental (i.e., extremely small) orthogonal planes at a point, stress transformation concerns determination of the stress (magnitude and direction) on a different infinitesimal plane at the same point.

9.1.2 Strain Transformation

Local normal strain is defined along a specific direction (or axis) at a given point. Shear strain at a point is defined with respect to a right-angled corner at the point. Given the normal strains along two orthogonal axes and the shear strain of the corner formed by these two axes at the point, strain transformation concerns determining the normal strain along another axis and the shear strain of a right-angled corner at the same point, formed by this axis and an axis orthogonal to it.

We will consider the stress transformation first. In particular, we will consider the *plane-stress* problem where the stresses act on a single plane, which is the 2-D problem. We will present the Mohr's circle for plane stress, which is a graphical method to carry out stress transformation. We will derive the principal stresses, the maximum shear stress (magnitude and the plane of action) for the 2-D case, and the absolute maximum shear stress. Using the associated principles and results, we will then study the practical problem of analysis of thin-walled pressure vessels.

Next, we will consider the problem of strain transformation, particularly for the *plane-strain* case. We will demonstrate an interesting and convenient analogy between stress transformation and strain transformation. We will use this analogy to present the Mohr's circle for plane strain, and derive the principal strains, the maximum shear strain (magnitude and direction) in the case of plane strain, and the absolute maximum shear strain. We will use the associated principles to present a theory for strain measurement using strain gauges. Finally, we will present some theories of failure, which will make use of the concepts of stress transformation.

9.1.3 Coordinate System

A coordinate system with an associated sign convention is needed in the formulation of stress transformation and strain transformation. As usual, we will use a right-handed Cartesian coordinate system x–y–z where the axes x, y, and z are orthogonal to each other. Also, in the applicable sign convention, a rotation from one coordinate axis to the next coordinate axis in the cyclic sequence x–y–z–x is considered a +ve rotation about the third axis in the sequence because a right-handed corkscrew will move in the +ve direction of that axis, in such a rotation (e.g., a rotation from the +ve x-axis toward the +ve y-axis is considered a +ve rotation about the z-axis; similarly, a rotation from the +ve y-axis toward the +ve z-axis is considered a +ve rotation about the x-axis; a rotation in the opposite direction is considered a −ve rotation). The coordinate system and the sign convention are shown in Figure 9.1.

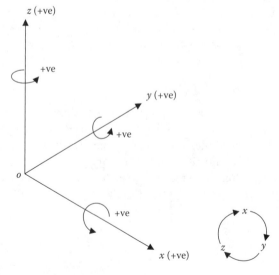

FIGURE 9.1 A Cartesian coordinate system and the sign convention.

9.2 STRESS TRANSFORMATION

Stress is specified by defining the plane on which the stress acts at a point, and the direction and magnitude of that stress.

9.2.1 SPECIFICATION OF STRESS

The plane on which the stress acts (at a given point) is defined by the "outer" normal to the plane (i.e., the normal axis coming out of the plane). Then, three orthogonal components of stress (at the specific point) can be expressed on that plane as normal stress, which acts along the normal axis, and two shear stress components, which act on the plane along two other orthogonal axes. See Figure 9.2, where normal stress σ_x, shear stress τ_{xy}, and shear stress τ_{xz} are along the directions x, y, and z, at point O, on the plane whose outer normal is the x-axis.

9.2.2 SIGN CONVENTION

The convention for the positive direction for stress is shown in Figure 9.3, for the 2-D case (case of plane stress). Consider an elemental rectangle at point O with its sides normal to either the x-axis or the y-axis of a Cartesian coordinate system O–x–y. On the plane where x-axis is the outer normal (i.e., the right vertical plane), the +ve normal stress (σ_x) is in the +ve x direction and the +ve shear

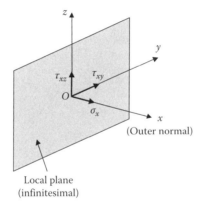

FIGURE 9.2 Specification of three orthogonal components of stress at a point on a plane.

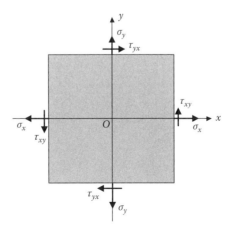

FIGURE 9.3 State of plane stress at a point.

stress (τ_{xy}) is in the +ve y direction. On the plane where x-axis is the inner normal (i.e., the left vertical plane), the +ve normal stress (σ_x) is in the −ve x direction and +ve shear stress (τ_{xy}) is in the −ve y direction. The directions of the positive stresses on the remaining two planes (which are normal to the y-axis) may be similarly defined. Also, from equilibrium of the infinitesimal rectangular element at O, we have (the complementarity property of shear stresses; see Chapter 3)

$$\tau_{yx} = \tau_{xy} \tag{9.1}$$

9.2.3 General State of Stress

The general state of stress at a point O, with respect to an infinitesimal cuboid element defined in a Cartesian coordinate frame ($O-x-y-z$), is illustrated in Figure 9.4 (also see Chapter 3). On each of the 6 planes, there are 3 stress components (1 normal stress and 2 shear stresses) totaling 18 components. However, since the element is infinitesimal, the stress magnitudes of the opposite planes are identical, resulting in just nine components. Furthermore, due to the complementarity property (resulting from static equilibrium) of shear stresses, only three of the six shear stress components will be independent. It is clear that the six stress components σ_x, σ_y, σ_z, τ_{xy}, τ_{yz}, and τ_{zx} specify the general state of stress at point O.

9.2.4 Plane-Stress Problem

The problem of plane stress is the 2-D problem of stresses. Here, the directions of all the stresses form a single plane (there are no stress components normal to that plane). Hence, for the state of plane stress at point O on the plane $O-x-y$, in Figure 9.4, the stress components σ_z, τ_{yz}, and τ_{xz} (and their complementary shear stresses τ_{zy}, and τ_{zx}) will be zero. This state of plane stress is shown in Figure 9.3.

9.2.5 Plane-Stress Transformation

Consider the problem of plane stress where the lines of action of all the stresses form a single plane. Given the local (i.e., at a point) stresses (normal and shear) on two elemental orthogonal planes that

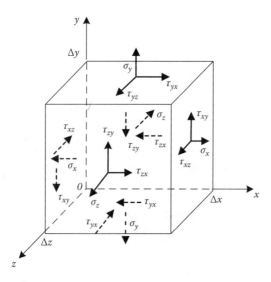

FIGURE 9.4 General state of stress.

Stress and Strain Transformations

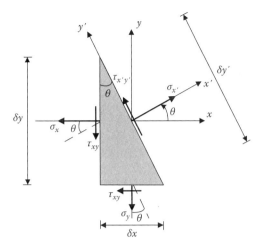

FIGURE 9.5 Description of the problem of stress transformation.

intersect at (and orthogonal to) the plane of stresses, the stress transformation concerns determining the stresses (normal and shear) on some other plane, at the same local point.

Specifically suppose that the local stresses are known on the planes that are defined by the x-axis and the y-axis as their outer normals (i.e., σ_x, σ_y, and $\tau_{xy} = \tau_{yx}$ are known) at a point. We are required to determine the stresses $\sigma_{x'}$ and $\tau_{x'y'}$, which act on a different plane whose outer normal is the axis x', at the same point. The y'-axis is along this plane and is orthogonal to the x'-axis. This situation is shown in Figure 9.5.

Consider the infinitesimal triangular element shown in Figure 9.5, where the side normal to the x-axis has length δy and the side normal to the y-axis has length δx. Suppose that the x'-axis forms angle θ in the counterclockwise (ccw) sense from the x-axis (i.e., about the +ve z-axis in the right-handed cork-screw sense). If the side of the triangle normal to the x'-axis has length $\delta y'$, we have

$$\delta x = \delta y' \sin\theta \qquad (i)$$

$$\delta y = \delta y' \cos\theta \qquad (ii)$$

Suppose that the triangle has a unit depth. We have
  Area of the horizontal side $= \delta x$
  Area of the vertical side $= \delta y$
  Area of the inclined side $= \delta y'$
Equations of equilibrium are now written for the force balance of the triangular element along the x' and y' directions:

$$\sum F_{x'} = 0 \Rightarrow \sigma_{x'} \cdot \delta y' - (\sigma_x \cdot \delta y)\cos\theta - (\tau_{xy} \cdot \delta y)\sin\theta - (\sigma_y \cdot \delta x)\sin\theta - (\tau_{xy} \cdot \delta x)\cos\theta = 0$$

$$\sum F_{y'} = 0 \Rightarrow \tau_{x'y'} \cdot \delta y' + (\sigma_x \cdot \delta y)\sin\theta - (\tau_{xy} \cdot \delta y)\cos\theta - (\sigma_y \cdot \delta x)\cos\theta + (\tau_{xy} \cdot \delta x)\sin\theta = 0$$

Substitute (i) and (ii) and cancel the common term $\delta y' \rightarrow$

$$\sigma_{x'} = \sigma_x \cos^2\theta + \sigma_y \sin^2\theta + 2\tau_{xy}\sin\theta\cos\theta \qquad (9.2)^*$$

$$\tau_{x'y'} = -\sigma_x \sin\theta\cos\theta + \sigma_y \cos\theta\sin\theta + \tau_{xy}\cos^2\theta - \tau_{xy}\sin^2\theta \qquad (9.3)^*$$

Note: Clearly, $\sigma_{y'}$ is obtained by simply changing θ to $\theta + \pi/2$ in (iii). We get

$$\sigma_{y'} = \sigma_x \sin^2\theta + \sigma_y \cos^2\theta - 2\tau_{xy}\sin\theta\cos\theta \tag{9.4}*$$

Substitute the trigonometric identities

$$\cos^2\theta = \frac{1+\cos 2\theta}{2}; \quad \sin^2\theta = \frac{1-\cos 2\theta}{2}; \quad 2\sin\theta\cos\theta = \sin 2\theta$$

We get

$$\sigma_{x'} = \frac{\sigma_x + \sigma_y}{2} + \frac{\sigma_x - \sigma_y}{2}\cos 2\theta + \tau_{xy}\sin 2\theta \tag{9.2}$$

$$\tau_{x'y'} = -\frac{\sigma_x - \sigma_y}{2}\sin 2\theta + \tau_{xy}\cos 2\theta \tag{9.3}$$

$$\sigma_{y'} = \frac{\sigma_x + \sigma_y}{2} - \frac{\sigma_x - \sigma_y}{2}\cos 2\theta - \tau_{xy}\sin 2\theta \tag{9.4}$$

Direct calculation can be done using the formulas (9.2) through (9.4), to carry out stress transformation.

Example 9.1

A handle of a drum is exerted a distributed force as shown in Figure 9.6a. An infinitesimal rectangular element at point O close to the distributed force of the handle experiences the state of plane stress shown in Figure 9.6b, with respect to the Cartesian coordinate frame xOy, where $\sigma_x = 10$ MPa, $\sigma_y = -50$ MPa, and $\tau_{xy} = -20$ MPa. By direct calculation, determine the state of stress at O on an infinitesimal rectangle, which is in the first quadrant of the coordinate frame x'Oy' where the x'-axis is at 30° from the original x-axis, in the negative direction.

Solution

In this problem, $\sigma_x = 10$ MPa, $\sigma_y = -50$ MPa, $\tau_{xy} = -20$ MPa, and $2\theta = -2 \times 30° = -60°$
Apply (9.2)

$$\sigma_{x'} = \frac{\sigma_x + \sigma_y}{2} + \frac{\sigma_x - \sigma_y}{2}\cos 2\theta + \tau_{xy}\sin 2\theta$$

$$= \frac{10-50}{2} + \frac{10-(-50)}{2}\cos(-60°) - 20\sin(-60°)$$

$$= -20 + \frac{60}{2}\times\frac{1}{2} + 20\times\frac{\sqrt{3}}{2} = -20 + 15 + 17.32 = 12.32 \text{ MPa}$$

Apply (9.3)

$$\tau_{x'y'} = -\frac{\sigma_x - \sigma_y}{2}\sin 2\theta + \tau_{xy}\cos 2\theta = -\frac{10-(-50)}{2}\sin(-60°) + (-20)\cos(-60°)$$

Stress and Strain Transformations

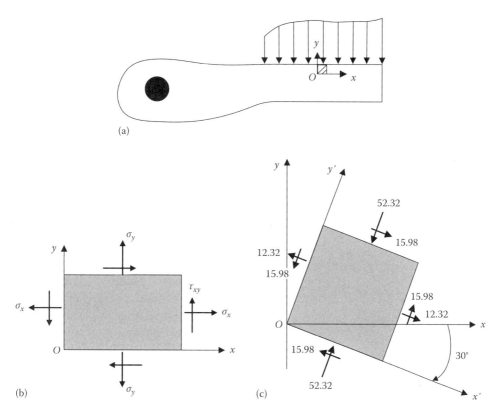

FIGURE 9.6 (a) A handle subjected to a distributed force; (b) state of plane stress on an infinitesimal element at O; (c) state of stress on the rotated element.

$$= -30 \times \left(-\frac{\sqrt{3}}{2}\right) - 20 \times \frac{1}{2} = 25.98 - 10 = 15.98 \text{ MPa}$$

Apply (9.4)

$$\sigma_{y'} = \frac{\sigma_x + \sigma_y}{2} - \frac{\sigma_x - \sigma_y}{2} \cos 2\theta - \tau_{xy} \sin 2\theta = -20 - 15 - 17.32 = -52.32 \text{ MPa}$$

The state of stress on the rotated element is shown in Figure 9.6c.

Primary Learning Objectives
1. Stress transformation using direct computation
2. Identification of the proper planes and directions of action of stresses
3. Familiarization with the sign convention for stresses and stress transformation

■ **End of Solution**

9.2.6 Principal Stresses

At any point of a body, under a given loading condition, there will be an infinitesimal plane on which the maximum normal stress occurs. On a plane normal to that plane at the same point, the minimum normal stress will occur. These are called the principal stresses.

Note: In the 3-D case, there will be three such orthogonal planes at a point, and correspondingly there are three principal stresses. In the 2-D (or plane-stress) case, there are only two principal stresses.

On the planes of principal stress, the shear stress will be zero. Also, the maximum shear stress will occur on a plane at $\pi/4$ to a plane of principal stress. These facts will be proved now, for the case of plane stress.

To determine the principal stresses (i.e., the magnitude of the maximum or minimum normal stress and the orientation of the plane normal to which it acts), differentiate Equation 9.2 with respect to θ and equate to zero. We get

$$\frac{d\sigma_{x'}}{d\theta} = 0 = \frac{\sigma_x - \sigma_y}{2}(-2\sin 2\theta) + \tau_{xy}(2\cos 2\theta)$$

or

$$\tan 2\theta_p = \frac{\tau_{xy}}{(\sigma_x - \sigma_y)/2} \tag{9.5}$$

In (9.5) θ_p denotes the angle of the principal stress plane as measured from the plane of σ_x (i.e., the angle of the outer normal of the principal stress plane, measured from the x-axis). *Note*: +ve direction of the angle is about the +ve z-axis.

The following observations can be made:

1. Since $\tan 2\theta = \tan(2\theta + \pi)$, there will be two distinct solutions for θ_p in (9.5), which are $\pi/2$ apart. Hence, there will be two principal planes, which are orthogonal.
2. From (9.5), we have

$$\sin 2\theta_p = \frac{\tau_{xy}}{\sqrt{((\sigma_x - \sigma_y)/2)^2 + \tau_{xy}^2}} \quad \text{and} \quad \cos 2\theta_p = \frac{(\sigma_x - \sigma_y)/2}{\sqrt{((\sigma_x - \sigma_y)/2)^2 + \tau_{xy}^2}};$$

$$\sin(2\theta_p + \pi) = -\frac{\tau_{xy}}{\sqrt{((\sigma_x - \sigma_y)/2)^2 + \tau_{xy}^2}} \quad \text{and} \quad \cos(2\theta_p + \pi) = -\frac{(\sigma_x - \sigma_y)/2}{\sqrt{((\sigma_x - \sigma_y)/2)^2 + \tau_{xy}^2}}$$

Substitute these angles into (9.2). We get the two principal stresses:

$$\sigma_{1,2} = \frac{\sigma_x + \sigma_y}{2} \pm \sqrt{\left(\frac{\sigma_x - \sigma_y}{2}\right)^2 + \tau_{xy}^2} \tag{9.6}$$

3. Substitute (9.5) in (9.3). On a principal plane

$$\tau_{xy} = 0 \tag{9.7}$$

Example 9.2

A torque is applied to a uniform, solid circular shaft in the direction shown in Figure 9.7a. On a cross section, the maximum shear stress τ_{max} occurs in the radial direction near the outer edge.

Stress and Strain Transformations

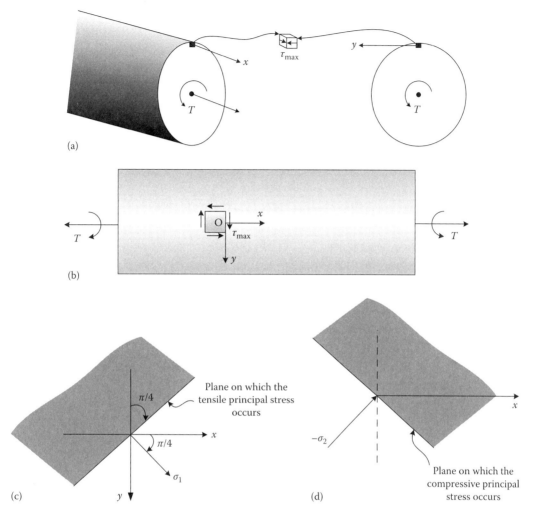

FIGURE 9.7 (a) A circular shaft in pure torsion. (b) State of stress in an X-section near the outer surface of the shaft. (c) The plane of action of the tensile principal stress. (d) The plane of action of the compressive principal stress.

Determine the magnitudes and directions of the principal stresses that result, and the orientations of the planes on which they act.

Solution

The state of plane stress of an infinitesimal rectangular element at a point at the outer surface of the shaft is shown in Figure 9.7b.
  In this example, $\sigma_x = 0$, $\sigma_y = 0$, and $\tau_{xy} = \tau_{max}$
  Apply (9.5)

$$\tan 2\theta_p = \frac{\tau_{xy}}{(\sigma_x - \sigma_y)/2} = \frac{\tau_{max}}{0} \to +\infty$$

$$\to 2\theta_p = 90° \text{ or } 270° \to \theta_p = 45° \text{ or } 135°$$

For $2\theta_p = 90°, \sigma_{x'} = \dfrac{\sigma_x+\sigma_y}{2} + \dfrac{\sigma_x-\sigma_y}{2}\cos 2\theta + \tau_{xy}\sin 2\theta = \tau_{xy}\sin 90° = \tau_{xy} = \tau_{max}$

For $2\theta_p = 270°, \sigma_{x'} = -\tau_{xy} = -\tau_{max}$

Alternatively, from (9.6)

$$\sigma_{1,2} = \dfrac{\sigma_x+\sigma_y}{2} \pm \sqrt{\left(\dfrac{\sigma_x-\sigma_y}{2}\right)^2 + \tau_{xy}^2} = 0 \pm \sqrt{0 \pm \tau_{xy}^2} = \pm\tau_{xy} = \pm\tau_{max}$$

In particular, the tensile principal stress is $\sigma_1 = \tau_{max}$. It occurs on a plane that is oriented at $\pi/4$ (measured in the +ve sense of the z-axis) from the plane whose outer normal is the x-axis, as shown in Figure 9.7c.

The compressive principal stress is $\sigma_2 = -\tau_{max}$. It occurs on a plane orthogonal to the plane of tensile principal stress, as shown in Figure 9.7d.

Primary Learning Objectives

1. Determination of the principal stresses using direct computation
2. Identification of the proper planes and directions of action of principal stresses
3. Familiarization with the sign convention for stresses and stress transformation

■ **End of Solution**

9.2.7 MAXIMUM IN-PLANE SHEAR STRESS

To determine the maximum (or minimum) shear stress (i.e., its magnitude, and the orientation of the plane along which it acts) corresponding to the state of plane stress, differentiate (9.3) with respect to θ and equate to zero. We get

$$\dfrac{d\tau_{x'y'}}{d\theta} = 0 = -\dfrac{\sigma_x-\sigma_y}{2}(2\cos 2\theta) + \tau_{xy}(-2\sin 2\theta)$$

or

$$\tan 2\theta_s = -\dfrac{(\sigma_x-\sigma_y)/2}{\tau_{xy}} \qquad (9.8)$$

In (9.8), θ_s denotes the angle of the plane on which the maximum/minimum shear stress occurs (i.e., the angle of the outer normal of the max shear plane) measured from the plane of σ_x (i.e., x-axis), in the +ve direction (i.e., about the +ve z-axis).

From (9.5) and (9.8), it is clear that there are two solutions such that

$$2\theta_s = 2\theta_p + \dfrac{\pi}{2} \quad \text{or} \quad 2\theta_p - \dfrac{\pi}{2}$$

Hence,

$$\theta_s = \theta_p + \dfrac{\pi}{4} \quad \text{or} \quad \theta_p - \dfrac{\pi}{4} \qquad (9.9)$$

Stress and Strain Transformations

The following observations can be made:

1. From (9.8), we have two solutions:

$$\sin 2\theta_s = \frac{(\sigma_x - \sigma_y)/2}{\sqrt{\left((\sigma_x - \sigma_y)/2\right)^2 + \tau_{xy}^2}} \quad \text{and} \quad \cos 2\theta_s = -\frac{\tau_{xy}}{\sqrt{\left((\sigma_x - \sigma_y)/2\right)^2 + \tau_{xy}^2}};$$

$$\sin(2\theta_s - \pi) = -\frac{(\sigma_x - \sigma_y)/2}{\sqrt{\left((\sigma_x - \sigma_y)/2\right)^2 + \tau_{xy}^2}} \quad \text{and} \quad \cos(2\theta_s - \pi) = \frac{\tau_{xy}}{\sqrt{\left((\sigma_x - \sigma_y)/2\right)^2 + \tau_{xy}^2}}$$

Substitute these angles in (9.3). We get the two shear stresses. The first solution corresponds to the minimum shear stress (–ve) and the second solution corresponds to the maximum shear stress (+ve) with the same magnitude. A simpler form of the result may be expressed in terms of the principal stresses, in view of (9.6). We get

$$\tau_{\min}, \tau_{\max} = \mp \sqrt{\left(\frac{\sigma_x - \sigma_y}{2}\right)^2 + \tau_{xy}^2} = \mp \frac{1}{2}(\sigma_1 - \sigma_2) \tag{9.10}$$

2. From (9.9), it is seen that the maximum "in-plane" shear stress (+ve value) occurs on a plane that is oriented at $\pi/4$ in the –ve sense from the principal plane of maximum normal stress. Its complementary shear stress (having equal magnitude but –ve) occurs on a plane that is oriented at $\pi/4$ in the +ve sense from the principal plane (or at $\pi/2$ from the plane of maximum shear stress).
3. Substitute the results presented earlier for $\sin 2\theta_s$ and $\cos 2\theta_s$ corresponding to the maximum and minimum in-plane shear stresses into (9.2) and (9.4). In both cases, we get the same expression for normal stress. The result may be expressed as well in terms of the principal stresses, in view of (9.6). We get the normal stress on the plane where the minimum or maximum shear stress (9.10) occurs as

$$\sigma\big|_{\tau_{\max}} = \sigma_o = \frac{\sigma_x + \sigma_y}{2} = \frac{\sigma_1 + \sigma_2}{2} \tag{9.11}$$

4. The "absolute maximum shear stress" is not necessarily the maximum "in-plane" shear stress given by (9.10). The third orthogonal direction z, along which the stresses are zero (*Note*: In particular, in the z direction, the normal stress is indeed a principal stress, which is also zero: $\sigma_3 = 0$), has to be considered to determine the absolute maximum shear stress and its plane of action. This will be discussed in a subsequent section.

Example 9.3

A cross section of a beam in bending is considered in this example, as shown in Figure 9.8a. At a point below the neutral axis on the cross section the state of stress was found to be (σ_o, τ_o), where the normal stress is $\sigma_o = 50$ MPa, and the shear stress $\tau_o = 20$ MPa.
  By direct computation, determine

a. The principal stresses σ_1 and σ_2 and the planes on which they occur
b. The maximum in-plane shear stress and the plane on which it occurs
c. The normal stress on the plane where the maximum shear stress occurs

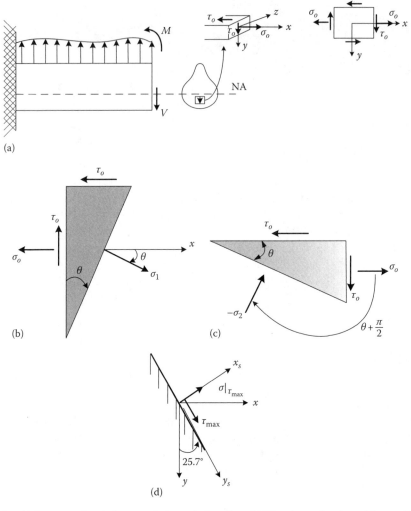

FIGURE 9.8 (a) Cross-sectional element of a beam in bending. (b) The plane of action of the tensile principal stress. (c) The plane of action of the compressive principal stress. (d) The plane of action of the maximum shear stress.

Solution

In this example, $\sigma_x = 50$ MPa, $\sigma_y = 0$, and $\tau_{xy} = 20$ MPa.

a.

$$(9.5): \tan 2\theta_p = \frac{\tau_{xy}}{(\sigma_x - \sigma_y)/2} = \frac{20}{(50-0)/2} = \frac{20}{25} = 0.8$$

$$\rightarrow 2\theta_p = 38.7° \rightarrow \theta_p = 19.3°$$

$$(9.6): \sigma_{1,2} = \frac{\sigma_x + \sigma_y}{2} \pm \sqrt{\left(\frac{\sigma_x - \sigma_y}{2}\right)^2 + \tau_{xy}^2} = \frac{50}{2} \pm \sqrt{\left(\frac{50}{2}\right)^2 + 20^2} = 25 \pm \sqrt{25^2 + 20^2}$$

$$= 25 \pm 32 = 57, -7 \text{ MPa}$$

Stress and Strain Transformations

→ The plane of the principal stress σ_1 makes an angle $\theta = 19.3°$ with the cross-sectional plane (whose outer normal is the x-axis) in the +ve sense about the z-axis (i.e., in the cw direction, in the present example). This is shown in Figure 9.8b.

The principal stress σ_2 occurs on a plane orthogonal to the σ_1 plane, as shown in Figure 9.8c.

b.

$$(9.8): \tan 2\theta_s = -\frac{50/2}{20} = -\frac{25}{20} = -1.25 \rightarrow 2\theta_s = -51.3° \rightarrow \theta_s = -25.7°$$

→25.7° in the ccw direction from the x-axis, in the physical coordinate system.

$$(9.10): \tau_{max} = \sqrt{\left(\frac{\sigma_x - \sigma_y}{2}\right)^2 + \tau_{xy}^2} = \sqrt{\left(\frac{50}{2}\right)^2 + 20^2} = 32 \text{ MPa}$$

The plane of action of the maximum shear stress (whose outer normal is the x_s-axis) is shown in Figure 9.8d.

c.

The normal stress on the plane of maximum shear stress is computed now.

$$(9.11): \sigma\big|_{\tau_{max}} = \frac{\sigma_x + \sigma_y}{2} = \frac{50}{2} = 25 \text{ MPa}$$

Its line of action is shown in Figure 9.8d.

Primary Learning Objectives
1. Determination of the principal stresses using direct computation
2. Determination of the maximum in-plane shear stress using direct computation
3. Identification of the proper planes and directions of action of the principal stresses and the maximum shear stress
4. Familiarization with the sign convention for stresses and stress transformation

■ **End of Solution**

9.3 MOHR'S CIRCLE OF PLANE STRESS

As we studied in the previous section, in the case of plane stress, once the state of stress on two orthogonal planes (whose outer normals are x-axis and y-axis) at a point is given, Equations 9.2 through 9.4 may be used to directly determine the state of stress on a plane (whose outer normal is x'-axis) that is oriented at angle θ in the +ve sense of rotation (i.e., in the ccw sense about the z-axis) from the first plane (whose outer normal is the x-axis) at the same point. This is the method of "direct calculation" for stress transformation. The same transformation may be accomplished "graphically" by using the Mohr's circle. The associated steps are as follows:

1. Mark the horizontal axis as the normal stress axis (σ) and the downward vertical axis as the shear stress axis (τ).
2. Given the stresses σ_x, σ_y, and τ_{xy} (which occur on the orthogonal planes whose outer normals are the x-axis and the y-axis), draw the Mohr's circle as follows. Its center (point O) is on the horizontal axis at $(\sigma_x + \sigma_y)/2$. Draw the radius OA where the coordinates of point A are (σ_x, τ_{xy}). Mohr's circle is drawn with this radius and center O, as shown in Figure 9.9.
3. To determine the state of stress on a plane oriented at angle θ in the +ve sense about the z-axis (i.e., ccw sense in the present case) from the plane whose state of stress is (σ_x, τ_{xy}), mark the corresponding radius OA' which is at angle 2θ in the ccw sense from radius OA.

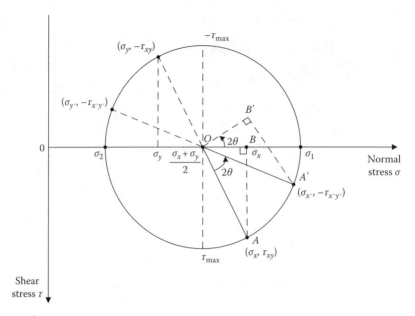

FIGURE 9.9 Proof of applicability of the Mohr's circle for plane stress.

The coordinates of point A' give the state of stress $(\sigma_{x'}, \tau_{x'y'})$ on the new local plane, at the same point of the object.

4. From Step 3, it is clear that the physical angle of orientation of a plane is half the angle of the corresponding radius of the Mohr's circles, measured in the same sense (+ve rotation of a plane—about the z-axis—corresponds to a ccw rotation in the Mohr's circle) from the same reference plane (and from the corresponding reference radius).
5. The two points where the Mohr's circle intersects the horizontal axis (σ-axis) give the principal stresses. The angle of orientation of the plane of the principal stress σ_1 is half the angle from the radius OA to the horizontal radius at σ_1 (as clear from the property of doubled angle, as indicated in Step 4 previously).
6. The maximum shear stress and the plane on which it occurs are given by the maximum vertical coordinate point of the Mohr's circle.

To prove these statements concerning the Mohr's circle, it is only necessary to prove that the state of stress given by point A' on the Mohr's circle satisfies Equations 9.2 and 9.3. This proof is given now.

Proof

1. From point A on the Mohr's circle of Figure 9.9 (i.e., from the given state of stress), draw a vertical line to point B on the horizontal axis (normal stress axis).
2. Rotate the right-angled triangle OAB ccw through angle 2θ about the center O. Point A will move to A' and point B will move to B'.
3. The triangle $OA'B'$ is congruent (identical) to the triangle OAB. Since $OB = \sigma_x - \dfrac{\sigma_x + \sigma_y}{2} = \dfrac{\sigma_x - \sigma_y}{2}$; $AB = \tau_{xy}$

we have

$$\sigma_{x'} = \frac{\sigma_x + \sigma_y}{2} + OB'\cos 2\theta + A'B'\sin 2\theta = \frac{\sigma_x + \sigma_y}{2} + \frac{\sigma_x - \sigma_y}{2}\cos 2\theta + \tau_{xy}\sin 2\theta$$

Stress and Strain Transformations

This is identical to the transformation result (9.2), which was obtained earlier, analytically. Also,

$$\tau_{x'y'} = -OB' \sin 2\theta + A'B' \cos 2\theta = -\frac{\sigma_x - \sigma_y}{2} \sin 2\theta + \tau_{xy} \cos 2\theta$$

This result is identical to the transformation result (9.3), which was obtained analytically.
Hence, the proof is complete.

9.3.1 Principal Stresses

Using the Mohr's circle in Figure 9.9 or 9.10, the principal stresses are obtained as follows: These correspond to the maximum and minimum values of normal stress (σ_1 and σ_2), which are the extreme points of the Mohr's circle on the σ-axis. Immediately, we notice from Figure 9.10 that the corresponding shear stress is zero (as required by the analytical result obtained in the previous section). Also,

$$\text{Radius of the circle} = R = OA = \sqrt{OB^2 + AB^2} = \sqrt{\left(\frac{\sigma_x - \sigma_y}{2}\right)^2 + \tau_{xy}^2}$$

Hence,

$$\sigma_1 = \frac{\sigma_x + \sigma_y}{2} + R = \frac{\sigma_x + \sigma_y}{2} + \sqrt{\left(\frac{\sigma_x - \sigma_y}{2}\right)^2 + \tau_{xy}^2}$$

Similarly,

$$\sigma_2 = \frac{\sigma_x + \sigma_y}{2} - R = \frac{\sigma_x + \sigma_y}{2} - \sqrt{\left(\frac{\sigma_x - \sigma_y}{2}\right)^2 + \tau_{xy}^2}$$

These results agree with the analytical results (9.6) that were obtained in the previous section.

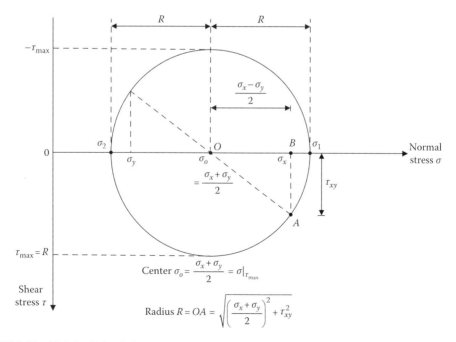

FIGURE 9.10 Mohr's circle of plane stress.

9.3.2 Maximum In-Plane Shear Stress

The maximum (in-plane) shear stress is

$$\tau_{max} = R = \sqrt{\left(\frac{\sigma_x - \sigma_y}{2}\right)^2 + \tau_{xy}^2} = \frac{\sigma_1 - \sigma_2}{2}$$

It is clear from Figure 9.10 that the normal stress on the plane where the minimum or maximum shear stress occurs is given by

$$\sigma\big|_{\tau_{max}} = \sigma_o = \frac{\sigma_x + \sigma_y}{2} = \frac{\sigma_1 + \sigma_2}{2}$$

These results from the Mohr's circle confirm the analytical results as obtained in the previous section.

Example 9.4

A shaft of circular cross section is subjected at torque T and an axial force P at its ends, as shown in Figure 9.11a. Under equilibrium, an infinitesimal rectangular element at point O near the surface of the shaft experiences the state of stress shown in Figure 9.11b, with respect to the Cartesian coordinate frame xOy, where $\sigma_x = 80$ MPa and $\tau_{xy} = -30$ MPa. Determine the principal stresses σ_1 and σ_2; the maximum shear stress τ_{max} at O; the normal stress on the plane of maximum shear stress; and the orientations of the planes on which they act.

Solution

Since there is no lateral loading in this problem, we have $\sigma_y = 0$. Also, this is a case of plane stress.
We will sketch the Mohr's circle using the given state of stress.
Circle center

$$\sigma_o = \frac{1}{2}(\sigma_x + \sigma_y) = \frac{1}{2}(80 + 0) = 40 \text{ MPa}$$

Circle radius

$$R = \sqrt{(\sigma_x - \sigma_o)^2 + \tau_{xy}^2} = \sqrt{(80 - 40)^2 + 30^2} = 50 \text{ MPa}$$

The given two points on the circle are

$$X = (80, -30); \ Y = (0, 30)$$

Using these values, the Mohr's circle for stress is sketched in Figure 9.11c.
The principal stresses are (see Mohr's circle)

$$\sigma_1 = \sigma_o + R = 40 + 50 = 90 \text{ MPa}$$

$$\sigma_2 = \sigma_o - R = 40 - 50 = -10 \text{ MPa}$$

On the Mohr's circle, σ_1 is at an angle $2\theta_1$ from X, where

$$\tan 2\theta_1 = \frac{30}{40} \rightarrow 2\theta_1 = 36.86° \text{ in the cw direction} \leftarrow \text{-ve direction}$$

Stress and Strain Transformations

→ $\theta_1 = 18.43°$ in the −ve sense

→ x_p-axis (the axis normal to the plane of σ_1) is at 18.43° from the x-axis, in the cw direction (i.e., −ve sense).

The corresponding state of stress is shown in Figure 9.11d.

Note: On the principal planes, the shear stress is zero.

The maximum shear stress is (see Mohr's circle in Figure 9.11c)

$$\tau_{max} = R = 30 \text{ MPa}$$

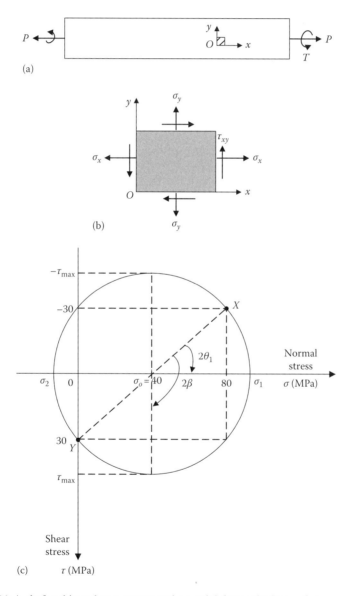

FIGURE 9.11 (a) A shaft subjected to a torque and an axial force; (b) State of stress on an infinitesimal element at point O. (c) Mohr's circle for stress.

(continued)

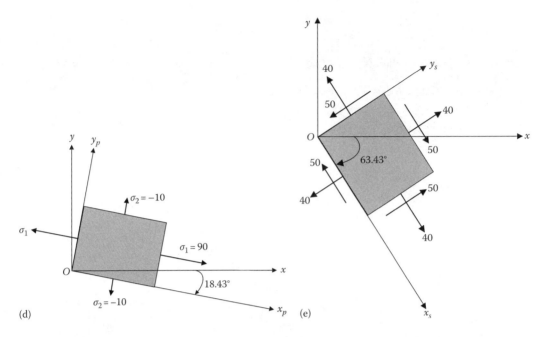

FIGURE 9.11 (continued)   (d) State of principal stress at point O. (e) State of stress on the plane of maximum shear stress at O.

This is at an angle 2β from point X on the Mohr's circle, where

$$2\beta = 2\theta_1 + 90° = 36.86° + 90° \text{ in the cw direction} \leftarrow \text{-ve direction}$$

$$\rightarrow \beta = \theta_1 + 45° = 18.43° + 45° = 63.43° \text{ in the cw direction (i.e., -ve sense)}$$

The normal stresses on the planes of maximum/minimum shear stress are $\sigma_o = 40$ MPa, as clear from the Mohr's circle.

The corresponding state of stress is shown in Figure 9.11e.

Primary Learning Objectives
1. Determination of the principal stresses using Mohr's circle
2. Determination of the maximum in-plane shear stress using Mohr's circle
3. Identification of the proper planes and directions of action of the principal stresses and maximum shear stress, using Mohr's circle
4. Familiarization with the sign convention for stresses and stress transformation when using Mohr's circle

■ **End of Solution**

9.4 THREE-DIMENSIONAL STATE OF STRESS

As noted earlier, a general state of stress at a point may be represented by the components of stress on the faces of an infinitesimal cuboid at the point (see Figure 9.4). On each face, there will be one normal stress component (normal to the face) and two orthogonal components of shear stress (along the face). Hence, in total there will be 18 components of stress (on the 6 faces of the infinitesimal cuboid). Out of them, we noted that only 6 components of stress are independent. In an x–y–z Cartesian coordinate frame at the point, with the axes falling along three edges of the cuboid meeting at that point, only the normal stress components σ_x, σ_y, and σ_z, and the shear stress components τ_{xy}, τ_{yz}, and τ_{zx} will be independent. Any other component will be equal to one of these six components (in view of the zero size of the cuboid and the fact that the cuboid is in static equilibrium).

9.4.1 Stress Transformation in 3-D

Suppose that the six stress components σ_x, σ_y, σ_z, τ_{xy}, τ_{yz}, and τ_{zx} at point P on planes defined in a Cartesian coordinate frame P–x–y–z are known. Then, any general stress component on a plane at P that is different from the planes of the original cuboid (say the stress components on a plane normal to axis x' in a coordinate frame P–x'–y'–z' which is oriented in an arbitrary way at P) is to be determined. For example, $\sigma_{x'}$ will depend on all six stress components σ_x, σ_y, σ_z, τ_{xy}, τ_{yz}, and τ_{zx}, in general. Hence, a Mohr's circle cannot be used to determine $\sigma_{x'}$. Actual transformation analysis in this 3-D case is beyond the scope of the present study.

In the case of plane stress, there will be just three independent stress components (one normal and two shear), which can be represented by a 2-D coordinate frame (say P–x–y; in which case only σ_x, σ_y, and τ_{xy} are nonzero and all the stress components in the z direction, σ_z, τ_{yz}, τ_{zx}, will be zero). This is the situation of plane stress, which can be analyzed using a Mohr's circle, as we have seen.

9.4.2 Principal Stresses in 3-D

In a general (3-D) state of stress at point P, there will be three principal stresses σ_1, σ_2, and σ_3, which are normal stresses on planes perpendicular to three orthogonal Cartesian coordinate axes x_P, y_P, and z_P. The principal coordinate frame P–x_P–y_P–z_P will have a specific orientation from a given frame P–x–y–z (the six stress components corresponding to which are given). Of course, by definition, the shear stress components corresponding to the principal coordinate frame will be zero (see Figure 9.12).

Again, each principal stress σ_i will be determined by a given general state of stress σ_x, σ_y, σ_z, τ_{xy}, τ_{yz}, and τ_{zx} and the orientation of the corresponding coordinate frame P–x–y–z. Transformation relations can be developed by considering the equilibrium of an infinitesimal wedge with one of its sides having the stress component that needs to be determined, and the remaining sides having the given stress components. Clearly, these relations cannot be represented by a Mohr's circle (in 2-D).

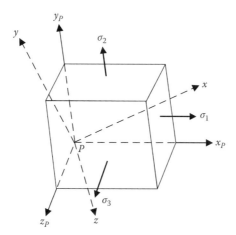

FIGURE 9.12 State of principal stress at a point in 3-D.

9.4.3 Absolute Maximum Shear Stress in Plane Stress

The "absolute maximum" normal stress at point P is one of the principal stresses σ_1, σ_2, or σ_3 there. The "absolute minimum" normal stress is also one of these three principal stresses. General 3-D transformation relations are needed to determine these absolute maximum and absolute minimum normal stresses. This analysis is beyond our scope.

Similarly, to determine the absolute maximum shear stress τ_{max} at P corresponding to a general (3-D) state of stress there, we will have to use the general 3-D transformation relations (analogously to what we did in the case of plane stress, to determine the maximum shear stress corresponding to that plane, which is the maximum in-plane shear stress). This analysis is also beyond our scope.

In the present study, we will consider only a state of plane stress. The absolute maximum normal stress will be one of the two principal stresses σ_1 or σ_2, as determined from the analytical relations or the Mohr's circle (*Note*: $\sigma_3 = 0$). The absolute maximum shear stress, however (in this case of plane stress), is not necessarily the maximum in-plane shear stress, which we have determined by the procedures (using the formula or Mohr's circle) given previously.

It is rather easy to determine the absolute maximum shear stress for the case of plane stress using concepts that we are already familiar with (e.g., Mohr's circle). However, we need to look beyond 2-D and consider 3-D (even though the state of stress is 2-D) in order to determine the absolute maximum shear stress.

Consider a state of plane stress at point P, represented by the Cartesian coordinate frame P–x–y–z (see Figure 9.13). All the stress components in the z direction (σ_z, τ_{yz}, τ_{zx}) will be zero. There will be two principal stresses σ_1 and σ_2, which will not be zero in general, and the third principal stress $\sigma_3 = 0$. The principal axis x_P will be at angle θ_P from x and similarly the principal axis y_P will be at angle θ_P from y (in the +ve sense, as shown in Figure 9.13). The z_P-axis will coincide with the z-axis, and will remain unchanged in the plane stress transformation.

The plane P–x–y is the same as the plane P–x_P–y_P and the plane-stress transformation is carried out on this plane. The state of plane stress at point P is defined on that plane. Mohr's circle can be drawn for this plane, which will determine the state of plane stress on an arbitrary element at P oriented at a general angle θ from the original element, and represented by a corresponding coordinate frame, P–x'–y' (with the z' axis in common with z and z_P) where the x'-axis is at angle θ from the x-axis. The Mohr's circle for the P–x–y plane will give a maximum shear stress τ_{xymax}. This is what is known as the "maximum in-plane shear stress" $\tau_{max\text{-}plane}$. Specifically,

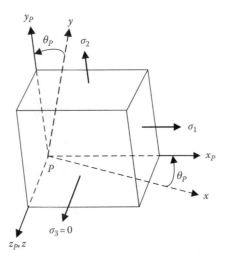

FIGURE 9.13 Principal stresses in a state of plane stress.

Stress and Strain Transformations

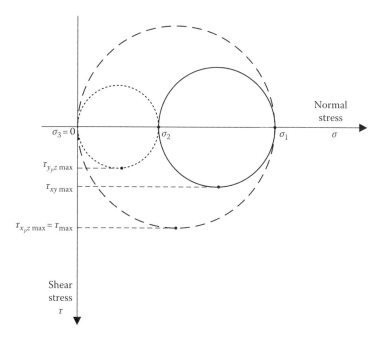

FIGURE 9.14 Case where principal stresses are of the same sign.

$$\tau_{xy\,max} = \tau_{max\text{-plane}} = \frac{1}{2}|\sigma_1 - \sigma_2| \qquad (9.10)*$$

as obtained from the Mohr's circle (see Figure 9.14). This is not necessarily the absolute maximum shear stress, as we can now explain in a simple way.

Consider the plane $P–x_p–z$. Since $\sigma_3 = 0$, the principal stresses on this plane are σ_1 and 0. Correspondingly, there will be a maximum shear stress $\tau_{zxp\,max}$ corresponding to this plane, which is given by

$$\tau_{zxp\,max} = \frac{1}{2}|\sigma_1| \qquad (i)$$

as obtained from the Mohr's circle for the plane $P–x_p–z$ (see the large broken-line circle in Figure 9.14).

Similarly, consider the plane $P - y_p - z$. Again, since $\sigma_3 = 0$, the principal stresses on this plane are σ_2 and 0. The corresponding maximum shear stress $\tau_{yz\,max}$ for this plane is given by

$$\tau_{ypz\,max} = \frac{1}{2}|\sigma_2| \qquad (ii)$$

as obtained from the Mohr's circle for the plane $P - y_p - z$ (see the small dotted circle in Figure 9.14).

The absolute maximum shear stress is the maximum of the three shear stresses, given by $\tau_{xy\,max}$, $\tau_{ypz\,max}$, and $\tau_{zxp\,max}$ in Equations 9.10*, (i), and (ii). Specifically,

$$\tau_{max} = \max[\tau_{xy\,max}, \tau_{ypz\,max}, \tau_{zxp\,max}] = \frac{1}{2}\max\left[|\sigma_1 - \sigma_2|, |\sigma_1|, |\sigma_2|\right] \qquad (9.12)*$$

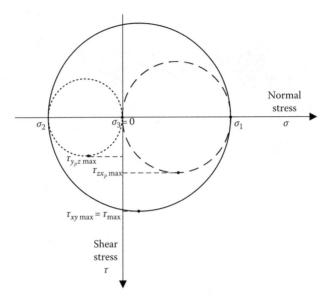

FIGURE 9.15 Case where principal stresses are of opposite signs.

What is shown in Figure 9.14 is the case where both σ_1 and σ_2 are positive (and $\sigma_1 > \sigma_2$). In Figure 9.15 we show the case where $\sigma_1 > 0$ and $\sigma_2 < 0$. From these results it should be clear that the absolute maximum shear stress is given by

$$\tau_{max} = \frac{1}{2} \max\left[|\sigma_1|,|\sigma_2|\right] \text{ if } \sigma_1 \text{ and } \sigma_2 \text{ have same sign}$$

$$= \frac{1}{2}|\sigma_1 - \sigma_2| \text{ if } \sigma_1 \text{ and } \sigma_2 \text{ have opposite signs}$$
(9.12)

Example 9.5

Consider a structural member with the loading condition as shown in Figure 9.16a. An infinitesimal element at point P close to the bottom surface of the member has the state of stress shown in Figure 9.16b, with respect to the Cartesian coordinate frame P–x–y.

a. Determine the principal stresses and the corresponding principal planes.
b. What is the absolute maximum shear stress, and what is the plane on which it occurs at P?

Solution

a. With respect to the Cartesian frame P–x–y, the state of stress at P is given by $X = (\sigma_x, \tau_{xy}) = (90, 40)$ MPa and $Y = (\sigma_y, -\tau_{xy}) = (30, -40)$ MPa.
Radius of the Mohr's circle (for the condition of plane stress on P–x–y) is

$$R = \sqrt{\left(\frac{\sigma_x - \sigma_y}{2}\right)^2 + \tau_{xy}^2} = \sqrt{\left(\frac{90-30}{2}\right)^2 + 40^2} = 50 \text{ MPa}$$

$$\text{Center} = (\sigma_0, 0) = \left(\frac{\sigma_x + \sigma_y}{2}, 0\right) = \left(\frac{90+30}{2}, 0\right) = (60, 0) \text{ MPa}$$

Stress and Strain Transformations

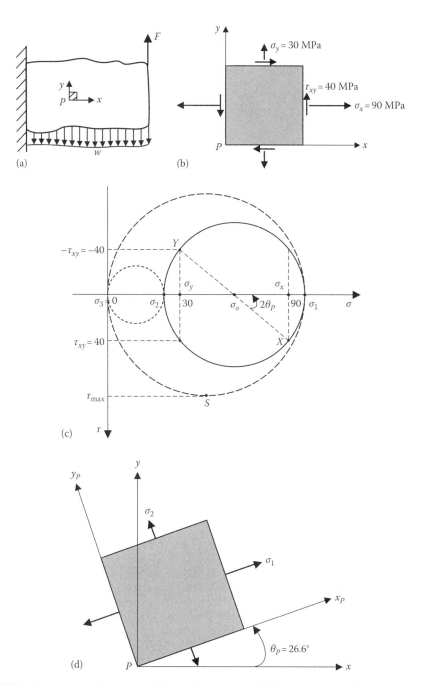

FIGURE 9.16 (a) A structural member. (b) State of stress at P. (c) Mohr's circle for plane stress on x–y plane. (d) State of principal stress on x–y plane.

The Mohr's circle is sketched in Figure 9.16c.
The principal stresses are

$$\sigma_1 = \sigma_o + R = 60 + 50 = 110 \text{ MPa}$$

$$\sigma_2 = \sigma_o - R = 60 - 50 = 10 \text{ MPa}$$

$$\tan 2\theta_P = \frac{\tau_{xy}}{(\sigma_x - \sigma_y)/2} = \frac{40}{(90-30)/2} = \frac{40}{30}$$

$$\rightarrow 2\theta_P = 53.1° \rightarrow \theta_P = 26.6°$$

Hence, the principal axis x_P is at 26.6° from x, in the +ve sense (ccw), as shown in Figure 9.16d.

b. Since σ_1 and σ_2 are of the same sign, from Figure 9.16c, it is clear that the absolute maximum shear stress is given by point S on the Mohr's circle, as

$$\tau_{max} = \frac{1}{2}\sigma_1 = \frac{1}{2} \times 110 = 55 \text{ MPa}$$

Note: This is larger than the maximum in-plane shear stress for the plane P–x–y, which is $R = 50$ MPa.

The plane of the absolute maximum shear stress is obtained as follows.
Point S gives the maximum shear stress corresponding to the plane P–z–x_P where the principal stresses are $\sigma_3 = 0$ and $\sigma_1 = 110$ MPa.

To reach S from σ_1, there is a rotation of 90° in the −ve sense (i.e., cw) on the Mohr's circle (Figure 9.16d, see the large broken-line circle passing through O and σ_1). In the physical domain, this corresponds to a rotation of −45° about the y_P-axis (i.e., a rotation of 45° from x_P toward z). The corresponding axis (say x') to which x_P has rotated gives the plane on which τ_{max} acts (specifically, the plane normal to the x'-axis).

Primary Learning Objectives

1. Determination of the principal stresses using Mohr's circle
2. Determination of the maximum in-plane shear stress using Mohr's circle
3. Determination of the absolute maximum shear stress using Mohr's circle
4. Identification of the plane of action of the absolute maximum shear stress

■ **End of Solution**

9.5 THIN-WALLED PRESSURE VESSELS

Containers subjected to fluid pressure are found in numerous applications. Examples include aircraft fuselages, boilers, gas cylinders, submarines, and chemical storage tanks. Theories governing the analysis of stresses and strains of those pressure vessels are well established. The analysis becomes particularly simple when the wall of the vessel is thin compared to a representative dimension. Thin-walled containers are called *shell structures*. An example of a shell structure is shown in Figure 9.17.

We will now analyze two types of thin-walled pressure vessels for their stresses. Specifically, we will consider cylindrical pressure vessels and spherical pressure vessels. For stress analysis, the thin wall assumption is satisfactory when

$$\frac{\text{Inner radius } (r)}{\text{Wall thickness } (t)} \geq 10$$

In the worst case of $r/t = 10$, it is known that the resulting error in the stress values is no more than 4%. This level of accuracy is quite acceptable in most applications, particularly since a factor of safety is also incorporated into the design.

9.5.1 Cylindrical Pressure Vessels

Consider a uniform cylinder of inner radius r and wall thickness t (Figure 9.18a) closed using end caps (which are typically semispherical in order to minimize the stresses). The internal gauge pressure

Stress and Strain Transformations

FIGURE 9.17 A shell structure. (Courtesy of C.W. de Silva.)

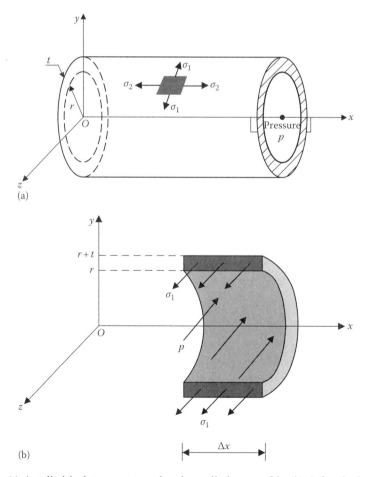

FIGURE 9.18 (a) A cylindrical pressure vessel and a wall element of it. (b) A free-body segment of the cylinder.

p (i.e., with respect to the external atmospheric pressure) is assumed to be uniform throughout the cylinder. An infinitesimal element with its sides along the longitudinal (axial) and transverse (circumferential) directions, as shown in Figure 9.18a, will only experience normal stresses, and no shear stresses, in view of the axis-symmetry of the element (i.e., if there were a shear stress, it is equally likely that it would be directed in either of the two opposite directions, which is a contradiction). The following two normal stresses are shown on the element:

σ_1 is the circumferential stress (or hoop stress or tangential stress)

σ_2 is the longitudinal stress (or axial stress)

In addition, there will be a radial stress σ_3, normal to the face of the element. According to the boundary conditions

$\sigma_3 = p$ (internal gauge pressure) at the inner face and

$\sigma_3 = 0$ (atmospheric, gauge pressure) at the outer face.

Hence, σ_3 will vary from p to 0 across the wall, in the radial direction. However, it will be seen (after obtaining the analytical results) that σ_3 is much smaller than either of σ_1 and σ_2, and can be neglected.

9.5.2 Hoop Stress

First, we determine the circumferential stress σ_1 by simply considering the equilibrium of a specific free-body slice as shown in Figure 9.18b. This virtual segment of the cylindrical is obtained by first cutting (virtually) the cylinder at two cross sections that are Δx apart and then cutting the resulting slice again using an axial plane through the central axis x).

The wall area of the segment normal to the z-axis consists of two regions each of area $t \times \Delta x$. The corresponding force in the z direction is $2 \times \sigma_1 \times t \times \Delta x$.

The projected area of the fluid in the segment, normal to the z-axis, is $2r \times \Delta x$. The corresponding force in the z direction is $-p \times 2r \times \Delta x$ (*Note*: Normal fluid pressure acts opposite to the +ve z direction on the free-body segment.).

Equilibrium of the segment in the z direction gives

$$\sum F_z = 0 : 2\sigma_1 \times t \times \Delta x - p \times 2r \times \Delta x = 0$$

$$\rightarrow \sigma_1 = \frac{p \cdot r}{t} \tag{9.13}$$

9.5.3 Longitudinal Stress

Assume that the cylinder is not restrained in the longitudinal (x-axis) direction, or simply consider an unrestrained end segment. Separate the left-end segment of the cylinder using a virtual cut at a cross section, as shown in Figure 9.19. There is no external, axial force on this segment. The wall area of the cross section is $2\pi r \times t$. The corresponding axial force is $\sigma_2 \times 2\pi r \times t$, in the positive x direction. Area of fluid normal to the x-axis is πr^2. The corresponding fluid force is $-p \times \pi r^2$ (*Note*: The normal fluid pressure acts on the free body in the negative x direction.).

Equilibrium of the free-body segment, in the axial direction gives

$$\sum F_x = 0 : \sigma_2 \times 2\pi r \times t - p \times \pi r^2 = 0$$

$$\rightarrow \sigma_2 = \frac{p \cdot r}{2t} \tag{9.14}$$

Stress and Strain Transformations

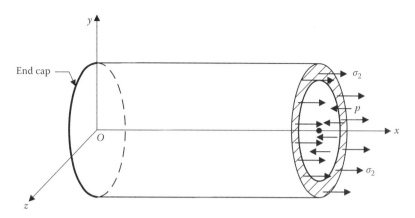

FIGURE 9.19 Free-body diagram of the left end segment of the cylinder.

Note: Since $r/t \geq 10$, we have $\sigma_2 \geq 5p$ and $\sigma_1 \geq 10p$. Hence, σ_3 is at least 5 times smaller than σ_2 and 10 times smaller than σ_1. This is the rationale for neglecting σ_3.

9.5.4 Absolute Maximum Shear Stress

Since there are no shear stresses on the element shown in Figure 9.18a, σ_1 and σ_2 are the principal stresses on the plane of the cylinder wall. The in-plane Mohr's circle will pass through σ_1 and σ_2 on the σ-axis and will have the radius $(1/2)(\sigma_1 - \sigma_2) = p \cdot r/(4t)$, as shown by the solid circle in Figure 9.20a. We note that, maximum in-plane shear stress

$$\tau_{\text{max-plane}} = \frac{p \cdot r}{4t} \tag{9.15}$$

This corresponds to point A on the in-plane Mohr's circle. Hence, its plane of action is at 45° to the plane of σ_1 or σ_2, as obtained by a rotation about a radial axis normal to the surface of the cylinder (see Figure 9.18a).

The absolute maximum shear stress is represented by point B of the (σ_1, σ_3) Mohr's circle—the out-of-plane Mohr's circle—whose diameter is σ_1. (*Note*: Strictly, σ_3 is not zero, but as noted earlier, we take $\sigma_3 = 0$ because it is much smaller than both σ_1 and σ_2.) The absolute maximum shear stress is

$$\tau_{\text{max}} = \frac{\sigma_1}{2} = \frac{p \cdot r}{2t} \tag{9.16}$$

The plane of action of the absolute maximum shear stress is obtained as follows: In Figure 9.18, consider an infinitesimal element of the vessel, such that the hoop stress σ_1 is in the local z direction, and the radial direction of the element is the local y direction where $\sigma_3 = 0$. (*Note*: The longitudinal stress σ_2 of the element is in the local x direction.) From the broken-line Mohr's circle corresponding to the principal stress pair (σ_1, σ_3) in Figure 9.20a, it is seen that the absolute maximum shear stress point (B) is reached by going from σ_1 toward σ_3 through 90° in the negative (cw) sense. Hence, in the physical domain, we rotate the local z-axis of the element toward the local y-axis (*Note*: $z \to y$ is the $-$ve sense) through 45° about the local x-axis. The resulting (rotated) z-axis is the outer normal of the plane of absolute maximum shear stress. This is shown as the z_s-axis in Figure 9.20b. *Note*: It is clear from the Mohr's circle that there will also be a normal stress of $\sigma_1/2$ on this plane.

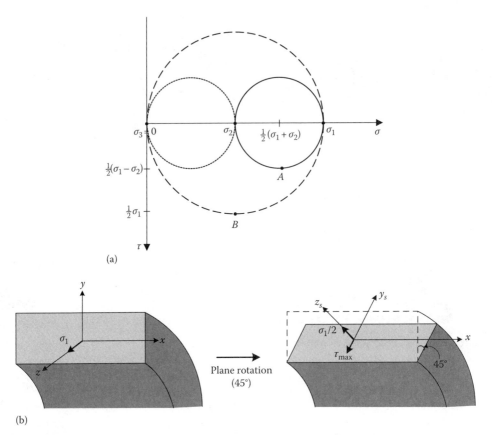

FIGURE 9.20 (a) Mohr's circles and absolute maximum shear stress of cylinder. (b) Plane of action of the absolute maximum shear stress.

Example 9.6

An engine mount uses a gas shock absorber, which is schematically shown in Figure 9.21. The maximum air pressure (gauge) inside the cylinder, during operating conditions, is limited to $p = 2.0 \times 10^6$ Pa. Inside radius of the cylinder is $r = 10$ cm. If the allowable normal stress of the cylinder material is $\sigma_{allow} = 75$ Pa, design a suitable thickness t for the cylinder walls.

Solution

Hoop stress

$$\sigma_1 = \frac{p \cdot r}{t}$$

Longitudinal stress

$$\sigma_1 = \frac{p \cdot r}{2t}$$

Hence, hoop stress is more critical, and should be used in the design. We have

$$75 \times 10^6 \,(\text{Pa}) = \frac{2 \times 10^6 \,(\text{Pa}) \times 0.1 \,(\text{Pa})}{t \,(\text{m})}$$

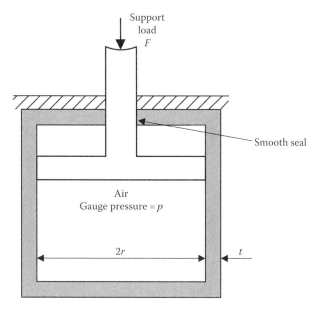

FIGURE 9.21 A gas shock absorber.

$$\rightarrow t = \frac{2 \times 0.1}{75} \text{ m} = 2.67 \text{ mm}$$

We will select $t \sim 3$ mm.

Primary Learning Objectives
1. Determination of the stresses in pressure vessels
2. Design of thin-walled pressure vessels

■ **End of Solution**

Example 9.7

A cylindrical pressure vessel is constructed by forming a flat steel plate into cylindrical shape and riveting the meeting edges to form a butt joint, using outer and inner strips along the joint, in double shear (see Figure 9.22a and b). End caps are welded subsequently. The maximum fluid pressure (gauge) in the cylinder is limited to $p = 2$ MPa.

The following design parameters are given:
  Internal radius of the cylinder, $r = 0.5$ m
  Diameter of a rivet, $d = 8$ mm

Select a suitable thickness for the plate and estimate the required number of rivets per meter length. *Given*:
  For the plate

$$\sigma_{allow} = 50 \text{ MPa}$$

For a rivet

$$\tau_{allow} = 60 \text{ MPa}$$

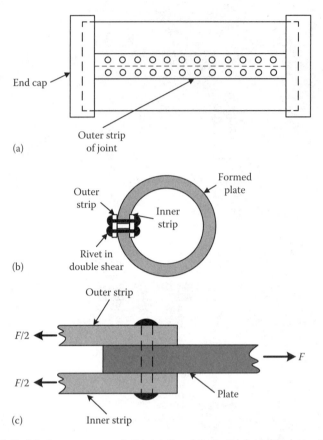

FIGURE 9.22 (a) Cylindrical pressure vessel; (b) details of the butt joint with rivets; (c) loading on a rivet in double shear.

Solution

First, we select the plate thickness such that the hoop stress σ_1 does not exceed σ_{allow}. We have,

$$\text{using } \sigma_1 = \frac{p \cdot r}{t} : 50 \times 10^6 \text{ (Pa)} = \frac{2 \times 10^6 \text{ (Pa)} \times 0.5 \text{ (m)}}{t \text{ (m)}}$$

$$\rightarrow t = \frac{2 \times 0.5}{50} \text{m} = 20 \text{ mm}$$

For a 1 m length of the joint, the hoop force exerted by the cylinder plate is $\sigma_1 \times t \times 1$. This has to be supported by n rivets, in double shear.

Force on each rivet (average) = $(\sigma_1 \times t \times 1)/n$

Since each rivet is in double shear (see Figure 9.22c), the shear force in a rivet cross section = $(\sigma_1 \times t \times 1)/(2 \times n)$.

This must not exceed the allowable shear force for a rivet, $(\pi/4)d^2 \tau_{allow}$.

Hence, for a balanced design, we have

$$\frac{\sigma_{allow} \times t \times 1}{2 \times n} = \frac{\pi}{4} d^2 \tau_{allow}$$

$$\rightarrow \frac{50 \times 10^6 \text{ (Pa)} \times 20 \times 10^{-3} \text{ (m)} \times 1 \text{(m)}}{2 \times n} = \frac{\pi}{4} \times (8 \times 10^{-3})^2 \text{ (m}^2\text{)} \times 60 \times 10^6 \text{ (Pa)}$$

Stress and Strain Transformations

$$\rightarrow n = \frac{50 \times 20 \times 10^{-3}}{2 \times \frac{\pi}{4} \times 64 \times 10^{-6} \times 60} = 165.8 \text{ rivets/m}$$

We will select $n = 170$.

Primary Learning Objectives
1. Determination of the stresses in pressure vessels
2. Design of the joints of thin-walled pressure vessels

■ **End of Solution**

9.5.5 SPHERICAL PRESSURE VESSELS

In view of their symmetry, the analysis of spherical pressure vessels is relatively simpler. Consider a thin-walled spherical pressure vessel of inner radius r and wall thickness t. The internal gauge pressure p is assumed to be uniform throughout the sphere. An infinitesimal element of the vessel will have the same normal stress in any tangential direction. In particular, for the element shown in Figure 9.23a, we have $\sigma_1 = \sigma_2$.

To determine its value, make a virtual cut of the vessel using a plane through its center O, as shown in Figure 9.23b, and consider the equilibrium of one half of the vessel. Assume that there are no external forces on this segment (apart from the atmospheric pressure, which is accounted for as we are using the gauge pressure p).

Wall area of the cross section is $2\pi r \times t$. The corresponding normal force is $\sigma_2 \times 2\pi r \times t$, in the positive x direction. Area of fluid normal to the x-axis is πr^2. The corresponding fluid force is $-p \times \pi r^2$ (*Note*: The normal fluid pressure acts on the free body in the negative x direction).

Equilibrium of the free-body segment in the axial direction gives

$$\sum F_x = 0: \sigma_2 \times 2\pi r \times t - p \times \pi r^2 = 0$$

$$\rightarrow \sigma_1 = \sigma_2 = \frac{p \cdot r}{2t} \tag{9.17}$$

For the corresponding state of stress, the Mohr's circle is a point on the σ-axis at $\sigma = p \cdot r/2t$ (see point A in Figure 9.23c). Hence, the in-plane shear stress (and of course the maximum in-plane shear stress) is zero for a thin-walled spherical pressure vessel (this fact can be confirmed by the symmetry of the vessel).

As for the cylindrical pressure vessel, we can neglect the radial stress of an element of a thin-walled spherical pressure vessel. Hence, the corresponding principal stress $\sigma_3 = 0$. The absolute maximum shear stress is represented by point B of the σ_1, σ_3 Mohr's circle in Figure 9.23c (*Note*: In the present symmetrical problem, the σ_1, σ_3 Mohr's circle and the σ_2, σ_3 Mohr's circle are the same). The absolute maximum shear stress is

$$\tau_{max} = \frac{p \cdot r}{4t} \tag{9.18}$$

The plane of action of the absolute maximum shear stress is obtained as in the case of a cylindrical pressure vessel. However, now (in view of symmetry), either of the two transverse axes of the element may be rotated toward the radial axis of the element, through 45°, to determine the outer normal of the required plane.

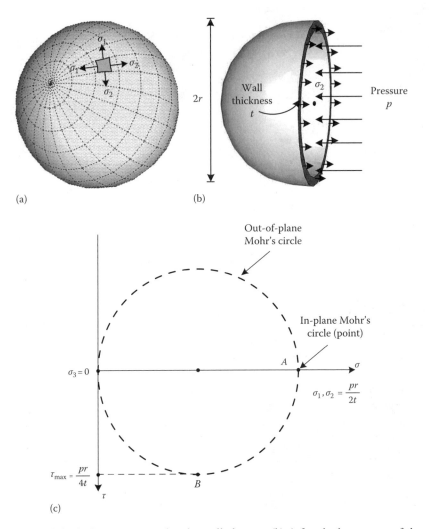

FIGURE 9.23 (a) Spherical pressure vessel and a wall element. (b) A free-body segment of the spherical vessel. (c) Mohr's circle for an element of spherical vessel.

9.6 STRAIN TRANSFORMATION

Local normal strain is defined along a specific direction (or axis) at a given point. Shear strain at a point is defined with respect to two orthogonal planes (infinitesimal) intersecting at the point (i.e., at a right-angled corner). Given the normal strains along two orthogonal axes and the shear strain of the corner at the point, strain transformation concerns determining the normal strain along another axis and the shear strain of the corner at the same point formed by an intersecting pair of orthogonal infinitesimal planes, one of which passes through this new axis.

9.6.1 Sign Convention

For normal strains, "tensile" strain is +ve and "compressive" strain is −ve. Hence, if the length of an infinitesimal element in a particular direction increases due to deformation, the normal strain in that direction is +ve.

For shear strains, if the angle of the defining right-angled corner decreases, then the corresponding shear strain is +ve. If the angle increases, the corresponding shear strain is −ve.

Stress and Strain Transformations

The +ve sense of a Cartesian coordinate system is defined in the usual manner (as in the case of stresses). Specifically, a +ve direction is defined by the +ve direction of the coordinate axis. A +ve rotation about an axis is defined by a right-handed corkscrew rotation, which moves the corkscrew in the +ve direction of that axis (see Figure 9.1).

9.6.2 General State of Strain

Corresponding to the general state of stress at a point O, with respect to an infinitesimal cuboid element defined in a local Cartesian coordinate frame (O–x–y–z), as discussed earlier, there will be a general state of strain. These strains are defined with respected to the edges (for normal strains) and corners (for shear strain) of the six planes of the cuboid. As for the stress, there will be a total of 18 components of strain. However, since the element is infinitesimal, the strain magnitudes of the opposite planes are identical, resulting in just nine components. Furthermore, due to the complementarity property of shear strain, only three of the six shear strain components will be independent. It follows that the six strain components ε_x, ε_y, ε_z, γ_{xy}, γ_{yz}, and γ_{zx} define the general state of strain at point O.

9.6.3 Plane-Strain Problem

The problem of plane strain is the 2-D problem of strains. Here, all the strains form a single plane (there are no strain components normal to that plane). Hence, for a state of plane strain at point O, on the plane O–x–y, the strain components ε_z, γ_{yz}, and γ_{zx} will be zero.

9.6.4 Comparison of Plane-Stress and Plane-Strain Problems

It should be noted that the problems of plane stress and plane strain are not identical. Specifically, a state of plane stress does not imply a state of plane strain, and vice versa. The reason for this is the Poisson effect (see Chapter 5). To confirm this fact, note that for a state of plane stress at point O, on the plane O–x–y, the stress components σ_z, τ_{yz}, and τ_{zx} will be zero. The nonzero normal stress components σ_x and σ_y will generate nonzero normal strain components ε_x and ε_y. Furthermore, due to the Poisson effect, these nonzero normal strain components ε_x and ε_y will generate a nonzero normal strain ε_z as well (in the z direction). The result is not a state of plane strain.

However, a state of plane stress will result in a relatively simpler state of strain. Specifically, consider the case of plane stress where the stress components σ_z, τ_{yz}, and τ_{zx} are zero. Then, the shear strain components γ_{yz} and γ_{zx}, which are proportional to τ_{yz} and τ_{zx}, through Hooke's law (with the shear modulus G as the constant of proportionality) will also be zero. Even though the normal strain component ε_z is not zero in general, it will be a principal strain and will be along the z-axis. This principal strain (ε_z) will not have any contribution to any other orthogonal components of strain. This fact is particularly useful in strain measurement since the strain gauges are mounted on a free surface (with no stresses), which corresponds to a situation of plane stress, even when it is not a situation of plane strain.

9.6.5 Plane-Strain Transformation

In a problem of plane strain, the directions of all the normal strains form a single plane, and the two orthogonal axes that represent the corner with respect to which a shear strain is defined, also fall on the same plane. Given the local normal strains along two orthogonal axes that intersect at the point of interest and the shear strain with respect to the right-angled corner defined by these two axes, on the plane of strains, the strain transformation concerns determining the normal strain along some other direction at the same point and on the same plane of strains, and also determining the shear

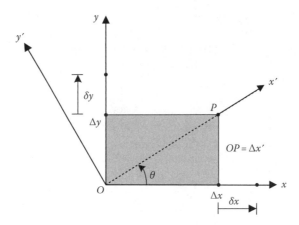

FIGURE 9.24   Definition of the problem of plane-strain transformation.

strain defined by a right-angled corner whose first side is along this new axis and the second side is orthogonal to it (in the +ve sense of coordinate rotation).

Specifically, suppose that the local normal strains ε_x, ε_y are known along the x-axis and the y-axis, respectively, which lie on the plane of strains and intersect at the point of interest; and the shear strain $\gamma_{xy} = \gamma_{yx}$ defined by the corner whose sides fall on the x-axis and the y-axis, is also known (see Figure 9.24). First, we will obtain an expression for the normal strain $\varepsilon_{x'}$ in the +ve direction of the axis x' and making an angle θ in the +ve sense from the x-axis (i.e., a +ve rotation about the z-axis—ccw direction in Figure 9.24). Next, we will obtain an expression for the shear strain $\gamma_{x'y'}$ defined by a right-angled corner at the same point whose first side is along the x'-axis and the other side is orthogonal to it, in the +ve sense (i.e., 90° ccw from the x'-axis), shown as the y'-axis in Figure 9.24.

a. Determination of Transformed Normal Strains $\varepsilon_{x'}$ and $\varepsilon_{y'}$

Note that $\varepsilon_{x'}$ has contributions from all three strain components ε_x, ε_y, and γ_{xy}. We will consider them separately and then together to obtain the overall strain.

Contribution to Normal Strain from Normal Strains

Consider an infinitesimal rectangular element at the point of interest (O in Figure 9.24) of length Δx along the x-axis, height Δy along the y-axis, and diagonal $OP = \Delta x'$ along the x'-axis.

The side Δx will extend through $\varepsilon_x \Delta x$ due to ε_x. Its contribution to the extension of OP along the x'-axis is $\varepsilon_x \Delta x \cos \theta$.

Similarly, the side Δy will extend through $\varepsilon_y \Delta y$ due to ε_y. Its contribution to the extension of OP along the x'-axis is $\varepsilon_y \Delta y \sin \theta$.

Contribution to Normal Strain from Shear Strain

Suppose that the side Δx is fixed and the side Δy rotates (deforms) in the cw direction through angle γ_{xy} due to shear (see Figure 9.25). The corresponding deformation of P along the x direction is $\gamma_{xy} \Delta y$. Its contribution to the extension of OP along the x'-axis is $\gamma_{xy} \Delta y \cos \theta$.

Note: We could have fixed Δy and rotated Δx in the ccw direction through angle γ_{xy} to obtain the same result, without loss of generality.

Hence, the overall extension of OP is

$$\delta x' = \varepsilon_x \Delta x \cos \theta + \varepsilon_y \Delta y \sin \theta + \gamma_{xy} \Delta y \cos \theta \qquad (i)$$

Stress and Strain Transformations

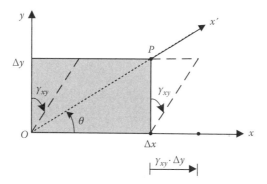

FIGURE 9.25 Contribution to normal strain from original shear strain.

Divide throughout by the original length $OP = \Delta x'$ of the element in the x' direction. This gives the normal strain along the x' direction at O

$$\frac{\delta x'}{\Delta x'} = \varepsilon_{x'} = \varepsilon_x \frac{\Delta x}{\Delta x'}\cos\theta + \varepsilon_y \frac{\Delta y}{\Delta x'}\sin\theta + \gamma_{xy}\frac{\Delta y}{\Delta x'}\cos\theta$$

Substitute the trigonometric facts (see Figure 9.24):

$$\frac{\Delta x}{\Delta x'} = \cos\theta \quad \text{and} \quad \frac{\Delta y}{\Delta x'} = \sin\theta$$

We get

$$\varepsilon_{x'} = \varepsilon_x\cos^2\theta + \varepsilon_y\sin^2\theta + \gamma_{xy}\sin\theta\cos\theta \qquad (9.19)*$$

The transformation equation for the normal strain in the y'-direction is obtained from (9.19) simply by changing θ to $\pi/2 + \theta$ in there. We get

$$\varepsilon_{y'} = \varepsilon_x\sin^2\theta + \varepsilon_y\cos^2\theta - \gamma_{xy}\cos\theta\sin\theta \qquad (9.20)*$$

These are the results of strain transformation for normal strain. Next, we obtain the result for shear strain transformation.

b. Determination of Shear Strain $\gamma_{x'y'}$

Consider a rectangular element $OPRQ$ at the point of interest (O) of length $OP = \Delta x'$ along the x'-axis, and height $OQ = \Delta y'$ along the y'-axis, as shown in Figure 9.26. With reference to point O, deformation of this element is shown by dotted lines as $OP'R'Q'$ (in the +ve sense). Note that the deformation includes both normal deformation and shear deformation. Specifically, the shear deformation (and shear strain) is determined by the amount of shrinking of the corner O (*Note:* A shrinking of the angle is +ve according to our sign convention). We have

$$\gamma_{x'y'} = \angle QOP - \angle Q'OP'$$

To determine this angle of shrink, we first find $\angle P'OP$. Then, we obtain $\angle QOQ'$ simply by changing θ to $\pi/2 + \theta$ in this result, as we will see (because OQ is 90° away from OP).

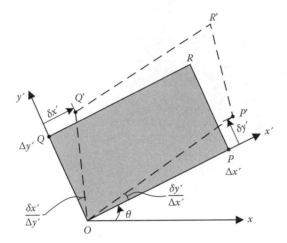

FIGURE 9.26 Geometry of deformation of an element in the new coordinate frame.

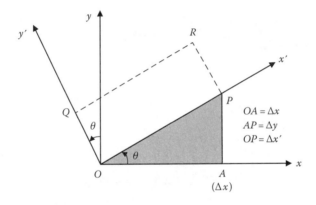

FIGURE 9.27 Triangular element.

Specifically, first we need to determine the rotation of the side OP about O (due to shear deformation of the plane). For this, consider the triangular element OAP (see Figure 9.27), which is integral with the original rectangular element $OPQR$. Here note that P moves (deforms) to P' due to three effects, as we noted earlier when determining the normal strain transformations.

1. Extension of OP (or OA) in the x direction (due to normal strain ε_x). This is equal to $\varepsilon_x \Delta x$.
2. Extension of OP (or AP) in the y direction (due to normal strain ε_y). This is equal to $\varepsilon_y \Delta y$.
3. Sliding of P in the x direction, with respect to side OA (due to shear strain γ_{xy}). This is equal to $\gamma_{xy} \Delta y$.

An exaggerated representation of these three (very small) movements is given in Figure 9.28.

The movement from P to P' may also be resolved into the following two orthogonal components (see Figure 9.28):

1. Movement through $\delta x'$ in the x' direction. This is given by (also see the derivation of normal strain transformation)

$$\delta x' = (\varepsilon_x \Delta x + \gamma_{xy} \Delta y) \cos \theta + \varepsilon_y \Delta y \sin \theta \tag{i}$$

Stress and Strain Transformations

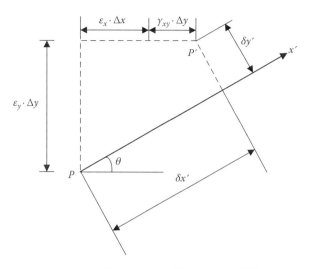

FIGURE 9.28 Orthogonal components of deformation with respect to different coordinate frames.

2. Movement through $\delta y'$ in the y' direction. This is given by

$$\delta y' = -(\varepsilon_x \Delta x + \gamma_{xy} \Delta y) \sin\theta + \varepsilon_y \Delta y \cos\theta \qquad \text{(ii)}$$

The component $\delta x'$ does not contribute to the rotation of OP. Only the component $\delta y'$ does. Hence, we have

$$\angle P'OP = \frac{\delta y'}{OP'} \approx \frac{\delta y'}{OP} = \frac{\delta y'}{\Delta x'}$$

Substitute (ii)

$$\angle P'OP \approx \frac{-(\varepsilon_x \Delta x + \gamma_{xy} \Delta y)\sin\theta + \varepsilon_y \Delta y \cos\theta}{\Delta x'}$$

Now, substitute

$$\frac{\Delta x}{\Delta x'} = \cos\theta \quad \text{and} \quad \frac{\Delta y}{\Delta x'} = \sin\theta$$

We get

$$\angle P'OP \approx -\varepsilon_x \cos\theta \sin\theta - \gamma_{xy} \sin^2\theta + \varepsilon_y \sin\theta \cos\theta \qquad \text{(iii)}$$

Now, we obtain $\angle QOQ'$ simply by changing θ to $\pi/2 + \theta$ in this result (because OQ is 90° away from OP).

Note: What we get then is the ccw rotation of OQ (according to the sign convention), which is the −ve of $\angle QOQ'$ as shown in Figure 9.26.

We use $\sin(\pi/2+\theta)=\cos\theta$ and $\cos(\pi/2+\theta)=-\sin\theta$ with the result (iii) to get

$$\angle QOQ' \approx -(\varepsilon_x \cos\theta \sin\theta - \gamma_{xy} \cos^2\theta - \varepsilon_y \sin\theta \cos\theta) \qquad \text{(iv)}$$

From (iii) and (iv), in view of the fact that, $\gamma_{x'y'} = \angle P'OP - \angle QOQ'$ we get

$$\gamma_{x'y'} = -\varepsilon_x\cos\theta\sin\theta - \gamma_{xy}\sin^2\theta + \varepsilon_y\sin\theta\cos\theta - (\varepsilon_x\cos\theta\sin\theta - \gamma_{xy}\cos^2\theta - \varepsilon_y\sin\theta\cos\theta)$$

or

$$\gamma_{x'y'} = -2\varepsilon_x\sin\theta\cos\theta + 2\varepsilon_y\cos\theta\sin\theta + \gamma_{xy}(\cos^2\theta - \sin^2\theta) \qquad (9.21)^*$$

This is the transformation equation for shear strain.

Now, substitute the trigonometric identities into (9.19)* through (9.21)*:

$$\sin\theta\cos\theta = \frac{1}{2}\sin 2\theta; \quad \cos^2\theta = \frac{1}{2}(1+\cos 2\theta); \quad \sin^2\theta = \frac{1}{2}(1-\cos 2\theta)$$

We can express the transformation equations for plane strain as

$$\varepsilon_{x'} = \frac{\varepsilon_x + \varepsilon_y}{2} + \frac{\varepsilon_x - \varepsilon_y}{2}\cos 2\theta + \frac{\gamma_{xy}}{2}\sin 2\theta \qquad (9.19)$$

$$\varepsilon_{y'} = \frac{\varepsilon_x + \varepsilon_y}{2} - \frac{\varepsilon_x - \varepsilon_y}{2}\cos 2\theta - \frac{\gamma_{xy}}{2}\sin 2\theta \qquad (9.20)$$

$$\frac{\gamma_{x'y'}}{2} = -\frac{(\varepsilon_x - \varepsilon_y)}{2}\sin 2\theta + \frac{\gamma_{xy}}{2}\cos 2\theta \qquad (9.21)$$

These equations can be used to determine strain transformations by the so-called method of "direct computation."

Example 9.8

The state of strain of an infinitesimal element at point O, which was rectangular originally as shown in Figure 9.29a, is $\varepsilon_x = -200\ \mu$, $\varepsilon_y = 600\ \mu$, and $\gamma_{xy} = -600\ \mu$, expressed with respect to the Cartesian coordinate frame xOy shown in the figure.

a. Give an exaggerated sketch to indicate the nature of deformation of the original rectangular element.
b. Consider an infinitesimal unstrained rectangular element at O with its corresponding sides along another Cartesian coordinate frame $x'Oy'$, which is obtained by rotating the frame xOy through $-15°$ (i.e., clockwise). By direct computation, determine the state of strain of this element when the state of strain given in the example is present. Also, give an exaggerated sketch indicating the nature of its deformation.

Solution

In this example, $\varepsilon_x = -200\ \mu$, $\varepsilon_y = 600\ \mu$, $\gamma_{xy} = -600\ \mu$, $\theta = -15° \rightarrow 2\theta = -30°$.

Stress and Strain Transformations

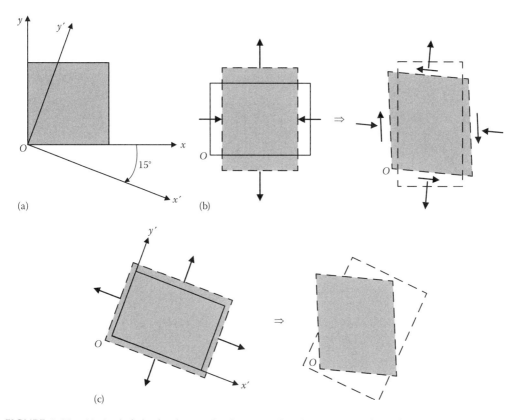

FIGURE 9.29 (a) An infinitesimal unstrained rectangular element at O. (b) Deformation of the original element. (c) Deformation of the new element.

a.
With ε_x (–ve) and ε_y (+ve) alone are present, the element would shrink in the x direction and extend in the y direction. Next, due to γ_{xy} (–ve), the corner angle at O will increase. This sequence of deformations is schematically shown in Figure 9.29b.

b.

$$\varepsilon_{x'} = \frac{\varepsilon_x + \varepsilon_y}{2} + \frac{\varepsilon_x - \varepsilon_y}{2}\cos 2\theta + \frac{\gamma_{xy}}{2}\sin 2\theta$$

$$= \frac{-200 + 600}{2} + \frac{(-200 - 600)}{2}\cos(-30°) - \frac{600}{2}\sin(-30°)$$

$$= 200 - 400\cos 30° + 300\sin 30°$$

$$= 200 - 346.41 + 150 = 3.59\ \mu$$

$$\varepsilon_{y'} = \frac{\varepsilon_x + \varepsilon_y}{2} - \frac{\varepsilon_x - \varepsilon_y}{2}\cos 2\theta - \frac{\gamma_{xy}}{2}\sin 2\theta$$

$$= 200 + 346.41 - 150 = 396.41\ \mu$$

$$\gamma_{x'y'/2} = -\frac{\varepsilon_x - \varepsilon_y}{2}\sin 2\theta + \frac{\gamma_{xy}}{2}\cos 2\theta = -\frac{(-200-600)}{2}\sin(-30°) + \frac{(-600)}{2}\cos(-30°)$$

$$= -400\sin 30° - 300\cos 30° = -200 - 259.81 = -459.81\ \mu$$

$$\gamma_{x'y'} = -919.62\ \mu$$

With $\varepsilon_{x'}$ (+ve) and $\varepsilon_{y'}$ (+ve) alone are present, the new element would extend in both x' and y' directions. Next, due to $\gamma_{x'y'}$ (−ve), the corner at O will enlarge. This sequence of deformation is schematically shown in Figure 9.29c.

Primary Learning Objectives

1. Strain transformation using direct computation
2. Identification of the proper orientations of strains
3. Recognition of the nature of local deformation due to strains
4. Familiarization with the sign convention for strains and strain transformation

■ **End of Solution**

9.6.6 PRINCIPAL STRAINS

As in the determination of principal stresses, Equation 9.19 is differentiated with respect to θ and set to zero, to determine the principal strains and their angles of orientation at a point. The following results are obtained:

$$\varepsilon_{1,2} = \frac{\varepsilon_x + \varepsilon_y}{2} \pm \sqrt{\left[\frac{\varepsilon_x - \varepsilon_y}{2}\right]^2 + \left[\frac{\gamma_{xy}}{2}\right]^2} \qquad (9.22)$$

The angle of the direction of the principal strain ε_1 is θ_p in the +ve sense from the axis of ε_x, and is given by

$$\tan 2\theta_p = \frac{\gamma_{xy}}{(\varepsilon_x - \varepsilon_y)} \qquad (9.23)$$

Also, it can be shown that in a right-angled corner with the principal axes falling on its sides, the shear strain $\gamma_{x\gamma} = 0$.

Example 9.9

The state of strain of an infinitesimal element at point O, which was rectangular originally as shown in Figure 9.30a, is $\varepsilon_x = 800\ \mu$, $\varepsilon_y = -200\ \mu$, and $\gamma_{xy} = -600\ \mu$.

  a. Give an exaggerated sketch of the nature of the deformation of the element.
  b. Determine the principal strains at O and the corresponding rectangular element to which these strains are subjected. Also, give an exaggerated sketch to show the corresponding deformation of the element.

Solution

In this example, $\varepsilon_x = 800\ \mu$, $\varepsilon_y = -200\ \mu$, and $\gamma_{xy} = -600\ \mu$.

  a.
   With ε_x (+ve) and ε_y (−ve) alone are present, the element will extend in the x direction and shrink in the y direction. Next, due to γ_{xy} (−ve), the corner angle at O will increase. This sequence of deformations is schematically shown in Figure 9.30b.

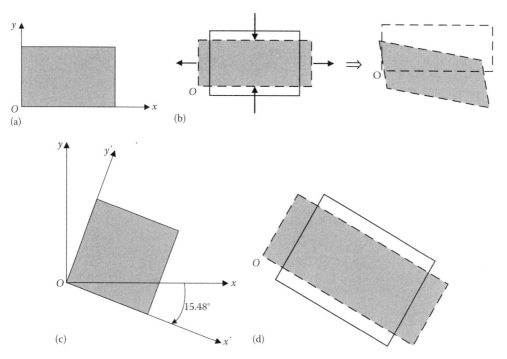

FIGURE 9.30 (a) An infinitesimal unstrained rectangular element. (b) Deformation of the original element. (c) Coordinate frame corresponding to principal strains. (d) Deformation due to principal strains.

b.

$$\tan 2\theta_p = \frac{(-600)/2}{(800-(-200))/2} = -\frac{600}{1000} = -0.6$$

$$\rightarrow 2\theta_p = -30.96° \rightarrow \theta_p = -15.48°$$

$$\varepsilon_{1,2} = \frac{800+(-2000)}{2} \pm \sqrt{\left(\frac{800-(-200)}{2}\right)^2 + \left(-\frac{600}{2}\right)^2}$$

$$= 300 \pm \sqrt{500^2 + 300^2} = 300 \pm 583.1 = 883.1\,\mu \quad \text{and} \quad -283.1\,\mu$$

It follows that the coordinate frame $x'Oy'$, which corresponds to the planes of principal strain, is obtained by rotating the original coordinate frame xOy through 15.48° in the cw direction (i.e., −ve direction). This is shown in Figure 9.30c.

In the state of principal strain, there are no shear strains. Since ε_1 extends the element in the x' direction and ε_2 shrinks the element in the y' direction, the corresponding deformation is exaggerated in Figure 9.30d.

Primary Learning Objectives
1. Determination of principal strains by using direct computation
2. Identification of the proper orientations of principal strains
3. Recognition of the nature of local deformation due to strains
4. Familiarization with the sign convention for strains and strain transformation

■ **End of Solution**

9.6.7 Maximum In-Plane Shear Strain

By differentiating Equation 9.21 with respect to θ and setting the result to zero, it can be shown that the maximum (in-plane) shear strain will occur in a right-angled corner whose first side is at $\pi/4$ to the direction of the first principal strain, in the −ve sense (cw in the present case). The minimum shear strain will occur in a right-angled corner that is placed at $\pi/2$ from the corner of maximum shear strain, in the +ve sense. We get

$$\gamma_{max} = 2\sqrt{\left[\frac{\varepsilon_x - \varepsilon_y}{2}\right]^2 + \left[\frac{\gamma_{xy}}{2}\right]^2} = \varepsilon_1 - \varepsilon_2 \qquad (9.24)$$

The corresponding angle is given by

$$\tan 2\theta_s = -\frac{(\varepsilon_x - \varepsilon_y)}{\gamma_{xy}} \qquad (9.25)$$

By substituting (9.25) into (9.19) (or (9.20)), the normal strain where the shear strain is maximum (or minimum) is given by (*Note*: The normal strain is the same for both orthogonal directions that define this maximum [or minimum] shear strain)

$$\varepsilon\big|_{\gamma_{max}} = \varepsilon_o = \frac{\varepsilon_x + \varepsilon_y}{2} = \frac{\varepsilon_1 + \varepsilon_2}{2} \qquad (9.26)$$

Example 9.10

A state of plane strain exists at a location P in a member with respect to a Cartesian coordinate system (right-handed) x–y–z. The normal strains along the x and y axes are found to be $\varepsilon_x = 150 \times 10^{-6}$ and $\varepsilon_y = -50 \times 10^{-6}$.

For a right-angled corner at P with its sides along the +ve x-axis and +ve y-axis, the shear strain (the angle of shrink of the corner) is found to be $\gamma_{xy} = 150 \times 10^{-6}$ rad. By direct calculation

a. Determine the principal strains ε_1 and ε_2 and their directions at P
b. Determine the maximum in-plane shear strain γ_{max}, the right-angled corner with respect to which the strain is defined at P, and the normal strain ε_o along the corresponding edges

Solution

In this example, $\varepsilon_x = 150\,\mu$, $\varepsilon_y = -50\,\mu$, $\gamma_{xy} = 150\,\mu$.

a.

$$\tan 2\theta_p = \frac{\gamma_{xy}/2}{(\varepsilon_x - \varepsilon_y)/2} = \frac{150/2}{[150-(-50)]/2} = \frac{75}{100} = 0.75$$

$$\rightarrow 2\theta_p = 36.87° \rightarrow \theta_p = 18.43°$$

$$\varepsilon_{1,2} = \frac{\varepsilon_x + \varepsilon_y}{2} \pm \sqrt{\left(\frac{\varepsilon_x - \varepsilon_y}{2}\right)^2 + \left(\frac{\gamma_{xy}}{2}\right)^2} = \frac{150+(-50)}{2} \pm \sqrt{\left(\frac{150-(-50)}{2}\right)^2 + \left(\frac{150}{2}\right)^2}$$

$$= 50 \pm \sqrt{100^2 + 75^2} = 50 \pm 125 = 175\,\mu \quad \text{and} -75\,\mu$$

Stress and Strain Transformations

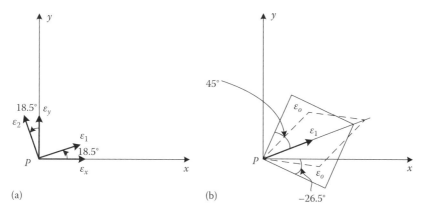

FIGURE 9.31 (a) Directions of the principal strains. (b) Corner and deformation corresponding to maximum in-plane shear strain.

The directions of the principal strains at P with respect to the x–y–z coordinate system are shown in Figure 9.31a.

b.

$$\tan 2\theta_s = -\frac{(\varepsilon_x - \varepsilon_y)/2}{\gamma_{xy}/2} = -\frac{(150-(-50))/2}{150/2} = \frac{-100}{75}$$

$\rightarrow 2\theta_s = -53.13° \rightarrow \theta_s = -26.57° \rightarrow x'$ is at 26.57° in the –ve (cw) sense.

$$\frac{\gamma_{max}}{2} = \sqrt{\left(\frac{\varepsilon_x - \varepsilon_y}{2}\right)^2 + \left(\frac{\gamma_{xy}}{2}\right)^2} = \sqrt{\left(\frac{150-(-50)}{2}\right)^2 + \left(\frac{150}{2}\right)^2} = \sqrt{100^2 + 75^2} = 125\,\mu$$

$\rightarrow \gamma_{max} = 250\,\mu$

One side of the right-angled corner corresponding to this shear strain is at angle $-(1/2)(90°-2\theta)$ from the x-axis. The other side is normal to this in the +ve sense.

This corner segment is sketched in Figure 9.31b (solid line) with the possible shape of its deformation (broken lines).

The normal strain corresponding to the maximum in-plane shear strain is given by

$$\varepsilon\big|_{\gamma_{max}} = \varepsilon_o = \frac{\varepsilon_x + \varepsilon_y}{2} = \frac{\varepsilon_1 + \varepsilon_2}{2} = \frac{175 + (-75)}{2}\,\mu$$

We have

$$\varepsilon_o = 50 \times 10^{-6}\,\mu$$

Primary Learning Objectives
1. Determination of the maximum in-plane shear strain by using direct computation
2. Identification of the corner of maximum in-plane shear strain
3. Recognition of the nature of local deformation due to shear strain
4. Determination of the normal strain along the edges that define the corner of maximum in-plane shear strain

■ **End of Solution**

9.7 MOHR'S CIRCLE OF PLANE STRAIN

In the case of plane strain, given the normal strains along two orthogonal axes (say, ε_x and ε_y, along x-axis and y-axis) at a point on the plane defined by these axes, and also given the shear strain of a right-angled corner at that point with its sides coinciding with the two axes (say, γ_{xy}), Equation 9.19 may be used to determine the normal strain along an axis ($\varepsilon_{x'}$ along x'-axis) that is at an angle θ in the +ve sense (i.e., ccw direction about the z-axis in the present case) from the first axis (x-axis); and Equation 9.21 may be used to determine the shear strain of a right-angled corner with x'-axis coinciding with its first side and y'-axis (which is orthogonal to the x'-axis, in the +ve sense) coinciding with the second side, at the same point and on the same plane. This is known as the "direct method" of strain transformation. In the previous section, we studied the derivation of the applicable equations, and the use of those equations for the "direct computation" of the transformation results.

The same transformation of strains may be accomplished graphically by means of the Mohr's circle, as in the case of stress transformation. The associated steps are quite analogous to the case of stress transformation. The relevant results can be obtained without further derivation, simply by noticing that the equations of strain transformation are analogous to those of stress transformation. Specifically, by comparing the strain transformation Equations 9.19 through 9.21 with the stress transformation Equations 9.2 through 9.4, we establish the following analogies

Normal stress → Normal strain

Shear stress → ½ × Shear strain

Specifically, the Mohr's circle for plane strain may be drawn as in Figure 9.32, using the two diametrically opposite points on it as given by $(\varepsilon_x, \gamma_{xy}/2)$ and $(\varepsilon_y, -(\gamma_{xy}/2))$.

From the Mohr's circle for plane strain, the results we obtained in the previous section by using the direct approach may be easily confirmed, in a manner analogous to that of plane stress. These results are given in the following sections.

9.7.1 PRINCIPAL STRAINS

At any point of a body, under a given loading condition, there will be a direction along which the maximum normal strain occurs. In a direction orthogonal to this, at the same point, the minimum normal strain will occur. *Note*: In the 3-D case, there will be three such orthogonal principal strains at a point. In the 2-D (i.e., plane-strain) case, there will be two principal strains. In a right-angled corner at the point, with its sides falling along the directions of two principal strains, the shear strain will be zero.

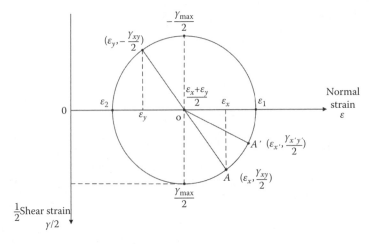

FIGURE 9.32 Mohr's circle of plane strain.

Stress and Strain Transformations

The radius of the Mohr's circle is

$$R = \sqrt{\left[\frac{\varepsilon_x - \varepsilon_y}{2}\right]^2 + \left[\frac{\tau_{xy}}{2}\right]^2} \qquad (9.27a)$$

Hence, the principal strains are given by

$$\varepsilon_{1,2} = \frac{\varepsilon_x + \varepsilon_y}{2} \pm \sqrt{\left[\frac{\varepsilon_x - \varepsilon_y}{2}\right]^2 + \left[\frac{\gamma_{xy}}{2}\right]^2} \qquad (9.22)$$

Also from the Mohr's circle in Figure 9.32, in a right-angled corner with the principal axes falling on its sides, shear strain $\gamma_{xy} = 0$.

The angle of the direction of the principal strain ε_1 is θ_p in the +ve sense from the axis of ε_x, and is given by

$$\tan 2\theta_p = \frac{\gamma_{xy}}{(\varepsilon_x - \varepsilon_y)} \qquad (9.23)$$

Note: The angles on the Mohr's circle are twice the physical angles. Also, the +ve sense is the ccw sense on the Mohr's circle.

9.7.2 Maximum In-Plane Shear Strain

The maximum shear strain will occur in a right-angled corner whose first side is at $\pi/4$ to the direction of the first principal strain, in the −ve sense (cw in the present case). The minimum shear strain will occur in a right-angled corner that is placed at $\pi/2$ from the corner of maximum shear strain, in the +ve sense. Its magnitude is twice the radius R of the Mohr's circle. We have

$$\gamma_{max} = 2R = 2\sqrt{\left[\frac{\varepsilon_x - \varepsilon_y}{2}\right]^2 + \left[\frac{\gamma_{xy}}{2}\right]^2} = \varepsilon_1 - \varepsilon_2 \qquad (9.24)$$

The corresponding angle is given by

$$\tan 2\theta_s = -\frac{(\varepsilon_x - \varepsilon_y)}{\gamma_{xy}} \qquad (9.25)$$

The normal strain where the shear strain is maximum is given by (see Mohr's circle in Figure 9.32)

$$\varepsilon\big|_{\gamma_{max}} = \varepsilon_o = \frac{\varepsilon_x + \varepsilon_y}{2} = \frac{\varepsilon_1 + \varepsilon_2}{2} \qquad (9.26)$$

Example 9.11

A member under plane strain has the following state of strain at a point:
Normal strain in the x direction, $\varepsilon_x = 0$
Normal strain in the y direction, $\varepsilon_y = 0$
Shear strain in a rectangular corner with sides along the x and y axes, $\gamma_{xy} = 160 \times 10^{-6}$

a. Sketch the Mohr's circle of plane strain at the given point (P).
b. Determine the principal strains ε_1 and ε_2 and their directions at P.

c. Determine the maximum in-plane shear strain γ_{max} and the orientation of the rectangular corner at the given point P to which this maximum shear strain is referred.
d. Sketch the shape to which the rectangular corner in part (c) is deformed.
e. Using the same sketch as in part (d), show that the shear strain of a corner that is oriented at 90° (in the +ve direction) from that of part (c), at point P, is negative and has the same maximum magnitude as in part (c). Confirm this observation using the Mohr's circle.

Solution

a. Given $\varepsilon_x = 0$, $\varepsilon_y = 0$, $\gamma_{xy} = 160 \times 10^{-6}$

Hence, the two diametrically opposite points on the Mohr's circle of strain are $(0, 80 \times 10^{-6})$ and $(0, -80 \times 10^{-6})$, which correspond to a state of pure shear strain.

The Mohr's circle is drawn as in Figure 9.33a.

b. From the Mohr's circle, the +ve principal strain is

$$\varepsilon_1 = 80 \times 10^{-6}$$

This is along an axis that is oriented at $1/2 \times \pi/2 = \pi/4$ from the x-axis in the +ve sense (about the z-axis of a right-handed x–y–z Cartesian coordinate system).

The −ve principal strain is

$$\varepsilon_2 = -80 \times 10^{-6}$$

This is along an axis that is oriented at $1/2 \times ((\pi/2) + \pi) = 3\pi/4$ from the x-axis in the +ve sense.

The direction of the principal strains and the elliptical shape of a small circular element at P due to the strains at the location are shown in Figure 9.33b.

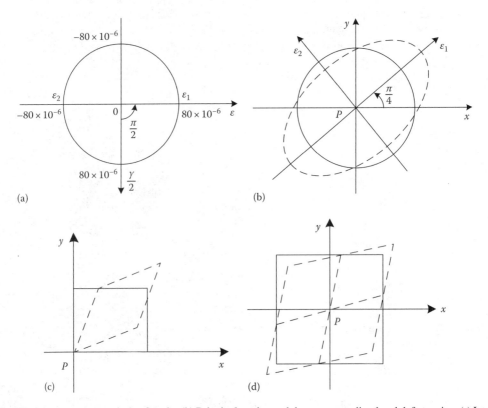

FIGURE 9.33 (a) Mohr's circle of strain. (b) Principal strains and the corresponding local deformation. (c) Local deformation due to maximum in-plane shear strain. (d) Local deformation of a square element centered at P.

c. Clearly, from the Mohr's circle in Figure 9.33a, maximum shear strain, $\gamma_{max} = 160 \times 10^{-6}$. It corresponds to a right-angled corner with sides along the +ve x-axis and the +ve y-axis.
d. The deformed shape of the corner of part (c) is sketched in Figure 9.33c.
Note the shrinking of the angle of the corner ⇒ +ve shear strain.
e. The local deformation of a square element at P is sketched in Figure 9.33d.
Note how the angle of the right-angled corner formed by the y-axis and the −ve x-axis has enlarged by the same angle as the corner in part (d) has shrunk.
⇒ It corresponds to a −ve shear of the same magnitude as γ_{max}.

Primary Learning Objectives

1. Determination of the principal strains using Mohr's circle
2. Determination of the maximum in-plane shear strain using Mohr's circle
3. Identification of the proper directions and the corner of principal strains and maximum shear strain, using Mohr's circle
4. Recognition of the nature of local deformation due to principal strains and due to maximum shear strain

■ **End of Solution**

9.8 THREE-DIMENSIONAL STATE OF STRAIN

A general state of strain at a point P in a body may be represented by the deformation of an infinitesimal cuboid with its corner located at P. Suppose that a Cartesian coordinate frame is located at P, with the orthogonal axes x, y, and z falling along the three edges of the cuboid at P. Positive normal strains along the edges of the cuboid correspond to extensions of these edges. A positive shear strain corresponds to the angle of shrink of a right-angled corner formed by two intersecting sides of the cuboid. Out of the 18 components of strain that are possible for the cuboid, there will be just 6 independent strain components: the 3 normal strains ε_x, ε_y, and ε_z, and 3 shear strains γ_{xy}, γ_{yz}, and γ_{zx} (see Figure 9.34). The remaining 12 strain components will be equal to 1 of these 6 components.

9.8.1 STRAIN TRANSFORMATION IN 3-D

Suppose that the six strain components ε_x, ε_y, ε_z, γ_{xy}, γ_{yz}, and γ_{zx} at point P, defined in a Cartesian frame P–x–y–z (Figure 9.34), are known. Consider another infinitesimal cuboid that is oriented arbitrarily at location P with its own Cartesian frame P–x'–y'–z'. If the angles of orientation (say,

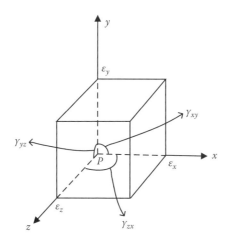

FIGURE 9.34 Three-dimensional state of strain at P.

the direction cosines) of the coordinate axes of frame $P-x'-y'-z'$ are known with respect to the original coordinate frame $P-x-y-z$, then any strain component ($\varepsilon_{x'}$, $\varepsilon_{y'}$, $\varepsilon_{z'}$, $\gamma_{x'y'}$, $\gamma_{y'z'}$, or $\gamma_{z'x'}$) corresponding to the new cuboid can be determined in terms of the six strain components of the original cuboid. These are the relations of general (3-D) strain transformation. A special case of this is the problem of plane strain (2-D), which has been analyzed in the previous sections. Unlike in the case of plane strain, a single Mohr's circle cannot represent the 3-D problem of strain transformation. The general analysis of the problem of 3-D strain transformation is beyond the scope of the present study.

In the case of plane strain, there will be just three independent components of strain (normal strains ε_x and ε_y, and the shear strain γ_{xy}) and all the remaining strain components (corresponding to the z-axis) will be zero.

9.8.2 Principal Strains in 3-D

In a general (3-D) state of strain at location P, there will be three principal strains ε_1, ε_2, and ε_3, which are the normal strains in the principal directions with respect to a Cartesian coordinate frame $P-x_P-y_P-z_P$. From the general relations of strain transformation, these principal strains and their orientation (i.e., the orientation of the principal Cartesian frame $P-x_P-y_P-z_P$) can be determined once the six strain components of an infinitesimal cuboid at P in its coordinate frame $P-x-y-z$ are given. By definition, the shear strains will be zero in the principal cuboid. The special case of plane strain has been analyzed in the previous sections.

9.8.3 Absolute Maximum Shear Strain in Plane Strain

The absolute maximum normal strain (at P) is the largest of the three principal strains ε_1, ε_2, or ε_3 at P. Also, the absolute minimum normal strain is the smallest of the three principal strains. The absolute maximum shear strain at P in the 3-D case can be determined analytically from the general transformation relation for general strain (by setting the derivatives with respect to the angles of orientation zero, as for the case of plane strain). This analysis is beyond the present scope. In the present study, we will limit our focus to the problem of plane strain.

The in-plane maximum shear strain is not necessarily the absolute maximum shear strain, even in the case of plane strain. The reason for this is, there is the third orthogonal direction (z_P) in which the principal strain $\varepsilon_3 = 0$. We have to consider this third dimension as well, when determining the absolute maximum shear strain in the problem of plane strain.

The necessary arguments and the procedure of analysis of the absolute maximum shear strain are similar to those used in determining the absolute maximum shear stress. They are not repeated here. Only the final result is given. (*Note*: There is a factor of 2 in the results of shear strain in comparison to the corresponding results for shear stress.) Specifically, in view of the result (9.12) for shear stress, we have the absolute maximum shear strain (in plane strain)

$$\gamma_{\max} = \max\left[\left|\varepsilon_1 - \varepsilon_2\right|, \left|\varepsilon_1\right|, \left|\varepsilon_2\right|\right] \tag{9.27)*}$$

In other words,

$$\gamma_{\max} = \max[|\varepsilon_1|, |\varepsilon_2|] \text{ if } \varepsilon_1 \text{ and } \varepsilon_2 \text{ have the same sign}$$

$$= \max\left[\left|\varepsilon_1 - \varepsilon_2\right|\right] \text{ if } \varepsilon_1 \text{ and } \varepsilon_2 \text{ have opposite signs} \tag{9.27b}$$

The corresponding Mohr's circles are shown in Figure 9.35.

Stress and Strain Transformations

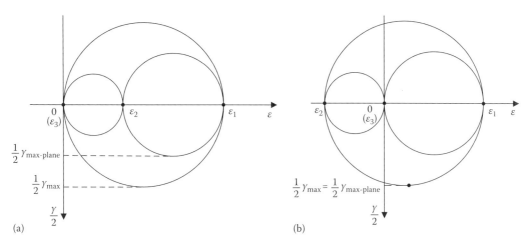

FIGURE 9.35 Absolute maximum shear strain: (a) Principal strains are of the same sign; (b) principal strains are of opposite signs.

Example 9.12

The state of plane strain at point P in a body is given by $\varepsilon_x = 900 \times 10^{-6}$ m/m; $\varepsilon_y = 300 \times 10^{-6}$ m/m; and $\gamma_{xy} = 800 \times 10^{-6}$ rad. Determine the maximum in-plane shear strain and the absolute maximum shear strain at P.

Solution

For the given problem of plane strain,

Center of Mohr's circle, $\varepsilon_o = \dfrac{1}{2}(\varepsilon_x + \varepsilon_y) = \dfrac{1}{2}(900 + 300)\,\mu = 600\,\mu$

Radius of Mohr's circle, $R = \sqrt{\left(\dfrac{\varepsilon_x - \varepsilon_y}{2}\right)^2 + \left(\dfrac{\tau_{xy}}{2}\right)^2}$

$$\rightarrow R = \sqrt{\left(\dfrac{900-300}{2}\right)^2 + \left(\dfrac{800}{2}\right)^2}\,\mu = 500\,\mu$$

The Mohr's circle is shown by a solid line in Figure 9.36.
The principal strains are

$$\varepsilon_1 = \varepsilon_o + R = 600 + 500\,\mu = 1100\,\mu$$

$$\varepsilon_2 = \varepsilon_o - R = 600 - 500\,\mu = 100\,\mu$$

Maximum in-plane shear strain

$$\gamma_{\text{max-plane}} = 2R = 2 \times 500\,\mu = 1000\,\mu$$

The Mohr's circle corresponding to the two principal strains (ε_1, ε_3) is shown by the broken line in Figure 9.36. This circle gives the absolute maximum shear strain. Absolute maximum shear strain

$$\gamma_{\text{max}} = \varepsilon_1 = 1100\,\mu$$

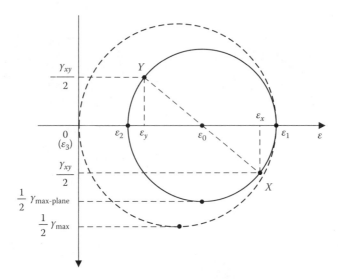

FIGURE 9.36 In-plane Mohr's circle and Mohr's circle for absolute maximum shear strain.

Note: Absolute maximum shear strain γ_{max} occurs on a local cuboid formed by rotating the x_p-axis of the principal cuboid (i.e., the principal axis of ε_1) through 45° about the y_p-axis (i.e., principal axis of ε_2) toward the z-axis (which is also the z_p-axis, the principal axis of ε_3). This is a rotation in the negative sense, which is consistent with the −ve rotation on the Mohr's circle when moving from ε_1 to γ_{max}. Maximum in-plane shear strain occurs on a local cuboid formed by rotating the original cuboid (defined in the original coordinate frame P–x–y–z) about the original z-axis.

Primary Learning Objectives

1. Determination of the maximum in-plane shear strain using Mohr's circle
2. Determination of the absolute maximum shear strain using Mohr's circle
3. Identification of the proper corners (cuboids) of maximum in-plane shear strain and absolute maximum shear strain, using Mohr's circle

■ **End of Solution**

9.9 STRAIN MEASUREMENT

Strain is measured using strain gauges. In Chapter 4, we discussed the associated theory and procedures. Multigauge rosettes are commonly used to measure the state of strain (all the strain components simultaneously) at a given location. We will present and illustrate the underlying theory and procedure of strain-gauge rosettes now.

For the problem of plane strain, we derived the transformation relation

$$\varepsilon_{x'} = \varepsilon_x \cos^2\theta + \varepsilon_y \sin^2\theta + \gamma_{xy} \sin\theta \cdot \cos\theta \qquad (9.19)^*$$

As indicated in Figure 9.37, using this relation we can determine the normal strain in an arbitrary x' direction, once we are given the strain components ε_x, ε_y, and γ_{xy}, which are defined with respect to an infinitesimal rectangular element with its edges along the x-axis and the y-axis of a local Cartesian coordinate frame. Angle θ is the angle of the x'-axis, measured from the x-axis, in the positive sense (i.e., from x to y or ccw sense in the figure).

Relation (9.19)* is valid for the case of "plane strain" where ε_z, γ_{yz}, and γ_{zx} are zero. However, consider the case of "plane stress" where σ_z, τ_{yz}, and τ_{zx} are zero. Then, γ_{yz} and γ_{zx}, which are proportional to τ_{yz} and τ_{zx} in view of Hooke's law (with the shear modulus G as the constant of proportionality),

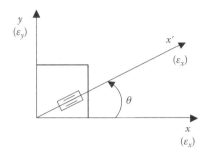

FIGURE 9.37 Strain transformation in plane strain or plane stress.

will also be zero, even when ε_z is not zero. Then, ε_z will be a principal strain and it will act along the z-axis. This principal strain (ε_z) will not have any contribution to $\varepsilon_{x'}$. Hence, even if the condition of plane strain is not satisfied, Equation 9.19* will be valid if the z direction is a direction of principal strain (which is the case in a problem of plane stress). In particular, if the x–y plane in Figure 9.1 is a free surface, then σ_z, τ_{yz}, and τ_{zx} will be zero and as a result z will be the principal direction. Under these conditions, Equation 9.19* will be valid even though $\varepsilon_z \neq 0$. This is typically the case in strain measurement because strain gauges are usually mounted on free surfaces.

Consider the "inverse problem" of determining ε_x, ε_y, and γ_{xy}. Three strain gauges (which measure normal strains only, in their direction of sensitivity—see Chapter 4) may be used to estimate these three strain components. Specifically, a strain gauge mounted in the x' direction as shown in Figure 9.37, will measure the normal strain $\varepsilon_{x'}$. If three such measurement are made using strain gauges mounted at angles $\theta = \theta_1$, θ_2, and θ_3 on the common surface, then we will have the three equations

$$\varepsilon_1 = \varepsilon_x \cos^2 \theta_1 + \varepsilon_y \sin^2 \theta_1 + \gamma_{xy} \sin \theta_1 \cdot \cos \theta_1$$

$$\varepsilon_2 = \varepsilon_x \cos^2 \theta_2 + \varepsilon_y \sin^2 \theta_2 + \gamma_{xy} \sin \theta_2 \cdot \cos \theta_2 \qquad (9.28)$$

$$\varepsilon_3 = \varepsilon_x \cos^2 \theta_3 + \varepsilon_y \sin^2 \theta_3 + \gamma_{xy} \sin \theta_3 \cdot \cos \theta_3$$

With proper choice of the angles of orientation θ_1, θ_2, and θ_3, we will have three independent simultaneous equations in (9.28) from which the unknowns ε_x, ε_y, and γ_{xy} can be computed. This is the principle of *strain rosettes*, which are integral units of multiple strain gauges based on the same film, but mounted in three different directions. They are commonly used in strain measurement (see Chapter 4 under strain measurement).

In the case of plane stress (where the stress components in the z direction are all zero), it can be shown that

$$\varepsilon_z = -\frac{\nu}{(1-\nu)} (\varepsilon_x + \varepsilon_y) \qquad (9.29)$$

where ν is the Poisson's ratio.

Hence, with the knowledge of the measured (and computed) ε_x and ε_y, we can determine ε_z as well, in a problem of plane stress (i.e., when the surface on which the strain gauges are mounted is free, which is the typical situation).

Example 9.13

Consider a strain-gauge rosette, which is mounted as in Figure 9.38. The following readings were obtained from the rosette: $\varepsilon_1 = 500 \times 10^{-6}$ m/m, $\varepsilon_2 = 100 \times 10^{-6}$ m/m, $\varepsilon_3 = -200 \times 10^{-6}$ m/m. Estimate the state of plane strain in the neighborhood where the rosette is mounted.

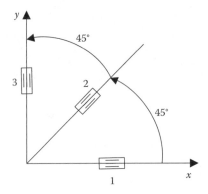

FIGURE 9.38 Strain measurement using a three-gauge rosette.

Solution

In this example we have

$$\theta_1 = 0°;\ \theta_2 = 45°;\ \theta_3 = 90°$$

Hence, from Equations 9.28 we get

$$\varepsilon_1 = \varepsilon_x;\ \varepsilon_2 = \varepsilon_x \times \frac{1}{2} + \varepsilon_y \times \frac{1}{2} + \gamma_{xy} \times \frac{1}{2};\ \varepsilon_3 = \varepsilon_y$$

Hence,

$$\varepsilon_x = 500 \times 10^{-6}\ \text{m/m}$$

$$\varepsilon_y = -200 \times 10^{-6}\ \text{m/m}$$

and

$$\gamma_{xy} = 2\varepsilon_2 - \varepsilon_x - \varepsilon_y = [2 \times 100 - 500 - (-200)] \times 10^{-6}\ \text{rad}$$

$$\rightarrow \gamma_{xy} = -100 \times 10^{-6}\ \text{rad}$$

Primary Learning Objectives

1. Application of the theory of strain measurement
2. Understanding the principle of strain-gauge rosettes
3. Measurement of a state of plane strain or the state of strain under plane stress
4. Measurement of shear strain using strain gauges (which measure normal strain)

■ **End of Solution**

9.10 THEORIES OF FAILURE

Material strength, deformation, and failure conditions are primary considerations in engineering design. A theory of failure is used to predict when a material (or engineering component) would fail (under a given arrangement of external loading). In general, the condition of failure will depend on

1. Type of material
2. Nature of external loading

Also, the following general observations can be made with regard to failure

1. A ductile material tends to be weakest in shear.
2. A brittle material tends to be weakest in tension.
3. Since a ductile material yields before failure, its yield strength and strain energy may be related to the failure criterion.
4. A brittle material fails without undergoing noteworthy yielding. Hence, its failure may be determined by the ultimate strength.
5. A brittle material is typically weaker in tension than in compression.

Based on these concepts, four theories of failure are given now, two for ductile materials and two for brittle materials. No single theory is uniformly applicable for all materials under all conditions—hence, the need for multiple criteria of failure.

9.10.1 Failure Theories for Ductile Material

Two theories of failure are given now for ductile materials. One is based on maximum shear stress and the other is based on strain energy density.

9.10.1.1 Maximum Shear Stress Theory

This theory seems pertinent since ductile material is generally weakest in shear. Also known as the *Tresca yield criterion*, this theory states that yielding (under general loading) begins when the absolute maximum shear stress reaches the maximum shear stress when the same material yields in pure tension.

Suppose that the yield strength (in tension) of the material in tension is σ_Y. (*Note*: This is also the principal stress of yielding since the specimen is in pure tension.) From Mohr's circle, the corresponding maximum shear stress is $\sigma_Y/2$.

For the case of plane stress, the absolute maximum shear stress is

$$\tau_{max} = \frac{1}{2} \max\left[|\sigma_1 - \sigma_2|, |\sigma_1|, |\sigma_3| \right] \tag{9.30}$$

It follows that, under plane stress (or biaxial state of stress), yielding begins when the principal stresses satisfy

$$\max\left[|\sigma_1 - \sigma_2|, |\sigma_1|, |\sigma_3| \right] = \sigma_Y \tag{9.31}$$

Equation 9.31 is plotted in Figure 9.39. According to the present criterion, the material yields when its principal stresses σ_1 and σ_2 fall outside the boundary of Figure 9.39.

9.10.1.2 Maximum Distortion Energy Theory

As noted in Chapter 5, the strain energy density is given by

$$u = \frac{1}{2}\sigma\varepsilon \tag{9.32}$$

For a general (triaxial) state of stress with principal stresses σ_1, σ_2, and σ_3, and principal strains ε_1, ε_2, and ε_3, the strain energy density is given by

$$u = \frac{1}{2}\left(\sigma_1\varepsilon_1 + \sigma_2\varepsilon_2 + \sigma_3\varepsilon_3\right) \tag{9.33}$$

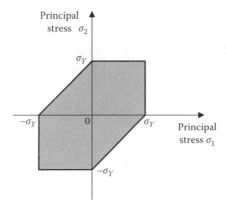

FIGURE 9.39 Maximum shear stress theory of yielding.

The state of principal stress may be decomposed into an average normal stress $\sigma_{av} = (\sigma_1 + \sigma_2 + \sigma_3)/3$ and three remaining components of normal stress $(\sigma_1 - \sigma_{av})$, $(\sigma_2 - \sigma_{av})$, and $(\sigma_3 - \sigma_{av})$ in the directions of principal stress. It is known that the average normal stress, an example of which is *hydrostatic pressure*, does not cause yielding in the material regardless of the stress magnitude. Only the remaining three components of stress will be responsible for yielding. The strain energy associated with these three "excessive" components of principal stress is responsible for distortion of a principal element, and will have the strain energy density (under condition of linear elasticity)

$$u_d = \frac{(1+\nu)}{6E}\left[(\sigma_1 - \sigma_2)^2 + (\sigma_2 - \sigma_3)^2 + (\sigma_3 - \sigma_1)^2\right] \qquad (9.34)$$

where
E is the Young's modulus
ν is the Poisson's ratio

Equation 9.34 gives the "distortion energy density" for a general condition of loading. Maximum distortion energy theory of yielding (due to Huber, von Mises, and Hencky) states that the material yields when its distortion energy density exceeds that of a tensile member of the same material when the tensile member begins to yield.

In Equation 9.34, when a tensile member begins to yield, we have $\sigma_1 = \sigma_Y$, $\sigma_2 = 0$, and $\sigma_3 = 0$, and the corresponding distortion energy density is

$$u_{dY} = \frac{(1+\nu)}{3E}\sigma_Y^2 \qquad (9.35)$$

Hence, from Equations 9.34 and 9.35, maximum distortion energy theory of yielding is

$$(\sigma_1 - \sigma_2)^2 + (\sigma_2 - \sigma_3)^2 + (\sigma_3 - \sigma_1)^2 = 2\sigma_Y^2 \qquad (9.36)$$

In the case of plane stress, $\sigma_3 = 0$, and we have

$$\sigma_1^2 + \sigma_2^2 - \sigma_1\sigma_2 = \sigma_Y^2 \qquad (9.37)$$

The relation (9.37) is sketched in Figure 9.40, which is an ellipse.

For comparison, the yield boundary corresponding to the maximum shear stress theory is also sketched in Figure 9.40, as a broken line. It is seen that the maximum distortion energy

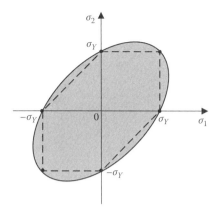

FIGURE 9.40 Maximum distortion energy theory of yielding.

theory is less conservative. However, it is known that this theory is more accurate than the maximum shear stress theory.

Note: From Figure 9.40, it is observed that both theories give the same results when either one of the principal stresses is zero or when the two principal stresses are equal.

9.10.2 Failure Theories for Brittle Material

Two theories of failure are now given for brittle material. One theory is based on the maximum normal (principal) stress. The other theory is based on the limiting Mohr's circles under pure tensile loading, pure compressive loading, and pure torsional loading.

9.10.2.1 Maximum Normal Stress Theory

This theory states that the material will fail when its maximum principal stress exceeds the ultimate stress corresponding to tensile loading (σ_U). Specifically, the failure boundary is given by

$$\max\left[|\sigma_1|, |\sigma_2|, |\sigma_3|\right] = \sigma_U \tag{9.38}$$

Note: This criterion assumes that the material fails in a similar manner under compressive principal stress as with tensile principal stress.

For the case of plane stress, the relation (9.38) is sketched in Figure 9.41.

9.10.2.2 Mohr's Failure Criterion

In some brittle materials, the strengths in tension and compression are different (an extreme example is concrete, which is much weaker in tension than in compression). For such materials, Mohr's failure criterion is applicable.

In this criterion, the failure boundary is given on the Mohr's circle plane (σ–τ) for plane stress. Specifically, three tests are conducted to obtain the failure boundary: a tensile test until failure (with the corresponding normal stress at failure, σ_{ut}), a compressive test until failure (with the corresponding normal stress at failure, σ_{uc}), and a pure torsion test until failure (with the corresponding shear stress at failure, τ_u). The three Mohr's circles are drawn corresponding to these three conditions (Figure 9.42). The envelope of these three circles is the failure boundary. If the Mohr's circle corresponding to a state of stress does not fall entirely within this envelope, the material will fail under that state of stress.

In plane stress, a Mohr's circle has a specific pair of principal stresses (σ_1 and σ_2). The Mohr's failure criterion may be represented on the (σ_1, σ_2) plane as well, as shown in Figure 9.43. If a principal stress falls outside the boundary of Figure 9.43, the material will fail.

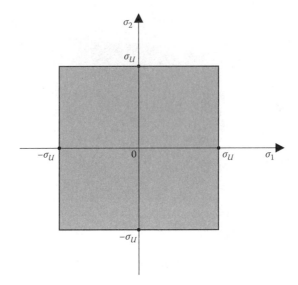

FIGURE 9.41 Maximum normal stress theory of failure.

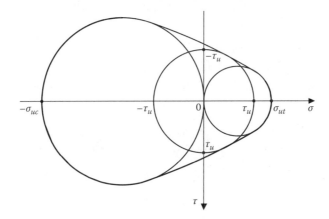

FIGURE 9.42 Mohr's failure criterion.

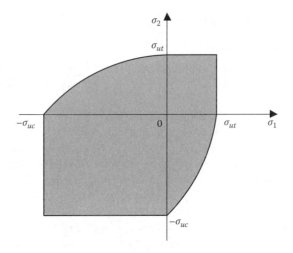

FIGURE 9.43 Mohr's failure criterion on the principal stress plane.

SUMMARY SHEET

Stress transformation: Finding stresses on different planes at the same location

Strain transformation: Finding strains in different directions and corners at the same location

Coordinate system and sign convention: Cartesian frame with orthogonal axes and +ve cyclic sequence x–y–z–x; right-handed corkscrew rule applies: a positive rotation moves the corkscrew in the +ve direction of the axis; rotation from x-axis to y-axis corresponds to a +ve rotation about z-axis, and so on

General state of stress: Requires six stress components σ_x, σ_y, σ_z, τ_{xy}, τ_{yz}, and τ_{zx}

Plane stress problem: Requires three stress components σ_x, σ_y, and τ_{xy}

Equations of stress transformation (plane stress):

$$\sigma_{x'} = \sigma_x \cos^2\theta + \sigma_y \sin^2\theta + 2\tau_{xy}\sin\theta\cos\theta$$

$$\sigma_{y'} = \sigma_x \sin^2\theta + \sigma_y \cos^2\theta - 2\tau_{xy}\sin\theta\cos\theta$$

$$\tau_{x'y'} = -\sigma_x \sin\theta\cos\theta + \sigma_y \cos\theta\sin\theta + \tau_{xy}\cos^2\theta - \tau_{xy}\sin^2\theta$$

or

$$\sigma_{x'} = \frac{\sigma_x + \sigma_y}{2} + \frac{\sigma_x - \sigma_y}{2}\cos 2\theta + \tau_{xy}\sin 2\theta$$

$$\sigma_{y'} = \frac{\sigma_x + \sigma_y}{2} - \frac{\sigma_x - \sigma_y}{2}\cos 2\theta - \tau_{xy}\sin 2\theta$$

$$\tau_{x'y'} = -\frac{\sigma_x - \sigma_y}{2}\sin 2\theta + \tau_{xy}\cos 2\theta$$

Principal stresses and orientation:

$$\sigma_{1,2} = \frac{\sigma_x + \sigma_y}{2} \pm \sqrt{\left(\frac{\sigma_x - \sigma_y}{2}\right)^2 + \tau_{xy}^2}; \quad \tan 2\theta_p = \frac{\tau_{xy}}{(\sigma_x - \sigma_y)/2}$$

Maximum/minimum in-plane shear stress and orientation:

$$\tau_{\min}, \tau_{\max} = \mp\sqrt{\left(\frac{\sigma_x - \sigma_y}{2}\right)^2 + \tau_{xy}^2} = \mp\frac{1}{2}(\sigma_1 - \sigma_2); \quad \tan 2\theta_s = -\frac{(\sigma_x - \sigma_y)/2}{\tau_{xy}}$$

Corresponding normal stress: $\sigma\big|_{\tau_{\max}} = \sigma_o = \dfrac{\sigma_x + \sigma_y}{2} = \dfrac{\sigma_1 + \sigma_2}{2}$

Mohr's circle for plane stress:
Axes: Normal stress σ and shear stress τ
  Circle center

$$\sigma_o = \frac{1}{2}(\sigma_x + \sigma_y) = \frac{1}{2}(\sigma_1 + \sigma_2)$$

Circle radius

$$R = \sqrt{(\sigma_x - \sigma_o)^2 + \tau_{xy}^2} = \frac{1}{2}(\sigma_1 - \sigma_2)$$

Physical +ve rotation of $\theta \leftrightarrow$ ccw rotation of 2θ on Mohr's circle
   Observations:

$$\sigma_{1,2} = \sigma_o \pm R;\ \tau_{max} = R;\ \sigma\big|_{\tau_{max}} = \sigma_o$$

Absolute maximum shear stress:

$$\tau_{max} = \frac{1}{2}\max\big[|\sigma_1|,|\sigma_2|\big]\ \text{if } \sigma_1 \text{ and } \sigma_2 \text{ have the same sign}$$

$$= \frac{1}{2}|\sigma_1 - \sigma_2|\ \text{if } \sigma_1 \text{ and } \sigma_2 \text{ have opposite signs}$$

Thin-walled pressure vessels:

$$\frac{\text{Inner radius } (r)}{\text{Wall thickness } (t)} \geq 10$$

Cylindrical pressure vessels (internal gauge pressure p):

$$\text{Hoop stress } \sigma_1 = \frac{p \cdot r}{t};\ \text{longitudinal stress } \sigma_2 = \frac{p \cdot r}{2t};$$

Absolute maximum shear stress $\tau_{max} = \dfrac{\sigma_1}{2} = \dfrac{p \cdot r}{2t}$

Spherical pressure vessels: $\sigma_1 = \sigma_2 = \dfrac{p \cdot r}{2t};\ \tau_{max} = \dfrac{p \cdot r}{4t}$

Equations of strain transformation (plane strain):

$$\varepsilon_{x'} = \varepsilon_x \cos^2\theta + \varepsilon_y \sin^2\theta + \gamma_{xy}\sin\theta\cos\theta$$

$$\varepsilon_{y'} = \varepsilon_x \sin^2\theta + \varepsilon_y \cos^2\theta - \gamma_{xy}\cos\theta\sin\theta$$

$$\gamma_{x'y'} = -2\varepsilon_x\sin\theta\cos\theta + 2\varepsilon_y\cos\theta\sin\theta + \gamma_{xy}(\cos^2\theta - \sin^2\theta)$$

Or

$$\varepsilon_{x'} = \frac{\varepsilon_x + \varepsilon_y}{2} + \frac{\varepsilon_x - \varepsilon_y}{2}\cos 2\theta + \frac{\gamma_{xy}}{2}\sin 2\theta$$

$$\varepsilon_{y'} = \frac{\varepsilon_x + \varepsilon_y}{2} - \frac{\varepsilon_x - \varepsilon_y}{2}\cos 2\theta - \frac{\gamma_{xy}}{2}\sin 2\theta$$

$$\frac{\gamma_{x'y'}}{2} = -\frac{(\varepsilon_x - \varepsilon_y)}{2}\sin 2\theta + \frac{\gamma_{xy}}{2}\cos 2\theta$$

Stress and Strain Transformations

Principal strains and orientation:

$$\varepsilon_{1,2} = \frac{\varepsilon_x + \varepsilon_y}{2} \pm \sqrt{\left[\frac{\varepsilon_x - \varepsilon_y}{2}\right]^2 + \left[\frac{\gamma_{xy}}{2}\right]^2}\,; \quad \tan 2\theta_p = \frac{\gamma_{xy}}{(\varepsilon_x - \varepsilon_y)}$$

Maximum/minimum in-plane shear strain and orientation:

$$\gamma_{max} = 2\sqrt{\left[\frac{\varepsilon_x - \varepsilon_y}{2}\right]^2 + \left[\frac{\gamma_{xy}}{2}\right]^2} = \varepsilon_1 - \varepsilon_2;\quad \tan 2\theta_s = -\frac{(\varepsilon_x - \varepsilon_y)}{\gamma_{xy}}$$

Corresponding normal strain:

$$\varepsilon\big|_{\gamma_{max}} = \varepsilon_o = \frac{\varepsilon_x + \varepsilon_y}{2} = \frac{\varepsilon_1 + \varepsilon_2}{2}$$

Mohr's circle for plane strain:
Axes: normal strain ε and 1/2 shear strain $(1/2)\gamma$
   Circle center

$$\varepsilon_o = \frac{\varepsilon_x + \varepsilon_y}{2} = \frac{\varepsilon_1 + \varepsilon_2}{2}$$

Circle radius

$$R = \sqrt{\left[\frac{\varepsilon_x - \varepsilon_y}{2}\right]^2 + \left[\frac{\gamma_{xy}}{2}\right]^2} = \frac{1}{2}(\varepsilon_1 - \varepsilon_2)$$

Physical +ve rotation of θ $\leftrightarrow$ ccw rotation of 2θ on Mohr's circle
   Observations:

$$\varepsilon_{1,2} = \varepsilon_o \pm R;\quad \gamma_{max} = 2R;\quad \varepsilon\big|_{\gamma_{max}} = \varepsilon_o$$

Absolute maximum shear strain:
$\gamma_{max} = \max[|\varepsilon_1|, |\varepsilon_2|]$ if ε_1 and ε_2 have the same sign
       $= \max[|\varepsilon_1 - \varepsilon_2|]$ if ε_1 and ε_2 have opposite signs

Strain-gauge rosette equations:
Strain gauges mounted at angles: $\theta = \theta_1, \theta_2,$ and θ_3 from x-axis

$$\varepsilon_1 = \varepsilon_x \cos^2\theta_1 + \varepsilon_y \sin^2\theta_1 + \gamma_{xy} \sin\theta_1 \cdot \cos\theta_1$$

$$\varepsilon_2 = \varepsilon_x \cos^2\theta_2 + \varepsilon_y \sin^2\theta_2 + \gamma_{xy} \sin\theta_2 \cdot \cos\theta_2$$

$$\varepsilon_3 = \varepsilon_x \cos^2\theta_3 + \varepsilon_y \sin^2\theta_3 + \gamma_{xy} \sin\theta_3 \cdot \cos\theta_3$$

$$\varepsilon_z = -\frac{\nu}{(1-\nu)}(\varepsilon_x + \varepsilon_y)$$

THEORIES OF FAILURE FOR DUCTILE MATERIAL

Maximum shear stress theory: $\tau_{max} = \dfrac{1}{2}\max\left[|\sigma_1-\sigma_2|,|\sigma_1|,|\sigma_3|\right]; \max\left[|\sigma_1-\sigma_2|,|\sigma_1|,|\sigma_3|\right]=\sigma_Y$

Maximum distortion energy theory: $(\sigma_1-\sigma_2)^2+(\sigma_2-\sigma_3)^2+(\sigma_3-\sigma_1)^2=2\sigma_Y^2$

THEORIES OF FAILURE FOR BRITTLE MATERIAL

Maximum normal stress theory: $\max[|\sigma_1|,|\sigma_2|,|\sigma_3|]=\sigma_U$

Mohr's failure criterion: Draw Mohr's circle for a tensile test until failure (σ_{ut}), a compressive test until failure (σ_{uc}), and a pure torsion test until failure (τ_u). The envelope of these three circles is the failure boundary.

PROBLEMS

9.1 A uniform member is subjected to a tensile force, generating a normal stress σ_{max} on its cross section (Figure P9.1). By direct computation, determine the resulting maximum shear stress τ_o and the orientation of the plane on which it acts. What is the normal stress σ_o on this plane?

9.2 A shaft of circular cross section is subjected to a torque T and an axial force P at its ends, as shown in Figure P9.2. Under equilibrium, a local rectangular element at point O experiences the state of stress: $\sigma_x=80$ MPa and $\tau_{xy}=-30$ MPa. By direct computation, determine the principal stresses σ_1 and σ_2; the maximum shear stress τ_{max} at O; the normal stress on the plane of maximum shear stress; and the orientations of the planes on which they act.

9.3 A cantilevered beam is subjected to a downward force P at the free end, as shown in Figure P9.3a. An infinitesimal rectangular element at point O near the bottom edge of the beam has the state of stress shown in Figure P9.3b with respect to the Cartesian coordinate

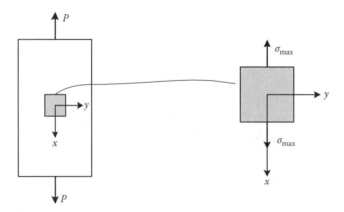

FIGURE P9.1 A tensile member and its state of stress.

FIGURE P9.2 A shaft subjected to a torque and an axial force.

Stress and Strain Transformations 409

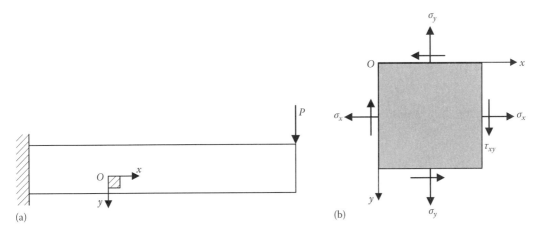

FIGURE P9.3 (a) A cantilever subjected to an end force; (b) state of stress on an infinitesimal element at O.

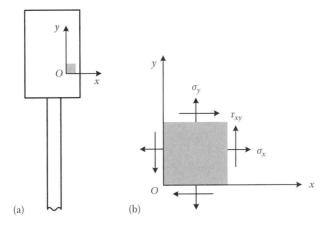

FIGURE P9.4 (a) An element on a billboard. (b) The state of stress of the element.

frame xOy, where $\sigma_x = -160$ MPa and $\tau_{xy} = 60$ MPa. By using the method of direct computation, determine the principal stresses σ_1 and σ_2, and the maximum shear stress τ_{max} at O and the orientations of the planes on which they cut.

9.4 An element at location O in a billboard structure has the state of plane stress given by $\sigma_x = -40$ MPa, $\sigma_y = -10$ MPa, and $\tau_{xy} = 30$ MPa, expressed in the Cartesian coordinate frame O–x–y (see Figure P9.4a and b).
By direct calculation, determine
  a. The principal stresses at O and the planes on which they act
  b. The maximum shear stress and the plane on which it acts
  c. Normal stress on the plane of maximum shear stress

9.5 A uniform member is subjected to a tensile force, generating a normal stress σ_{max} on its cross section (see Problem 9.1). Using the Mohr's circle method, determine the resulting maximum shear stress τ_o and the orientation of the plane on which it acts. What is the normal stress σ_o on this plane?

9.6 A torque is applied to a uniform, solid circular shaft in the direction shown in Figure P9.6. In a cross section, the maximum shear stress τ_{max} occurs in the radial direction near the outer edge. Using the method of Mohr's circle, determine the magnitudes and directions of the principal stresses that result and the orientations of the planes on which they act.

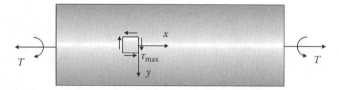

FIGURE P9.6 State of stress on a cross section near the outer surface of the shaft.

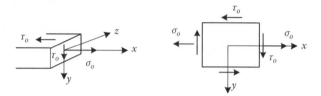

FIGURE P9.7 The state of stress at a location on the X-section of a beam in bending.

9.7 At a point below the neutral axis on a cross section of a beam in bending, there exists the state of stress (σ_o, τ_o) as shown in Figure P9.7:
Normal stress

$$\sigma_0 = 50 \text{ MPa}$$

Shear stress

$$\tau_o = 20 \text{ MPa}$$

Draw the Mohr's circle for determining the state of stress on different planes at the same point. From it, determine
  a. The principal stresses σ_1 and σ_2 and the planes on which they occur
  b. The maximum shear stress τ_{max} and the plane on which it occurs
  c. The normal stress on the plane where the maximum shear stress occurs

9.8 A cantilevered beam is subjected to a downward force P at the free end (see Problem 9.3). An infinitesimal rectangular element at point O on the beam cross section has the state of stress $\sigma_x = -160$ MPa, $\sigma_y = 0$, and $\tau_{xy} = 60$ MPa with respect to the Cartesian coordinate frame xOy there. By using the method of Mohr's circle, determine the principal stresses σ_1 and σ_2, and the maximum shear stress τ_{max} at O and the orientations of the planes on which they act.

9.9 A handle of a drum is exerted a distributed force. As a result, an infinitesimal rectangular element at point O close to the distributed force of the handle experiences the state of stress: $\sigma_x = 10$ MPa, $\sigma_y = -50$ MPa, and $\tau_{xy} = -20$ MPa, with respect to the Cartesian coordinate frame xOy. Using the method of Mohr's circle, determine the state of stress at O on an infinitesimal rectangle, which is in the first quadrant of a new coordinate frame $x'Oy'$ where x'-axis is at 30° from the x-axis, in the negative direction.

9.10 An element at location O of a thick beam (see Figure P9.10a and b) has the state of plane stress given by $\sigma_x = -90$ MPa, $\sigma_y = -30$ MPa, and $\tau = -26.5$ MPa. Determine the principal stresses, the absolute maximum shear stress, and the corresponding planes of action at O.

9.11 A cylindrical pressure vessel has a 60° helical weld as shown in Figure P9.11. The pressure (gauge pressure, with respect to the outside atmospheric pressure) inside the cylinder is $p = 1$ MPa. The cylinder has radius $r = 1.0$ m and wall thickness $t = 20$ mm. Determine the normal stress $\sigma_{x'}$ perpendicular to the weld and the shear stress $\tau_{x'y'}$ along the weld.

Stress and Strain Transformations

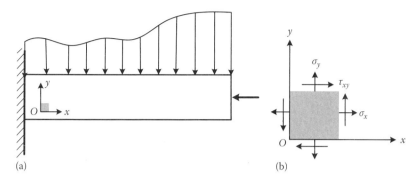

FIGURE P9.10 (a) A thick beam and an element on it; (b) the state of planes stress of the element.

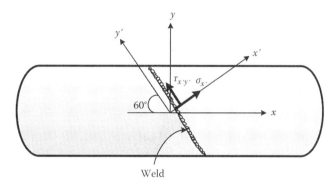

FIGURE P9.11 A pressure vessel with a 60° helical weld.

9.12 A member under the condition of plane strain has the following state of strain at a point:
Normal strain in the x direction, $\varepsilon_x = 0$
Normal strain in the y direction, $\varepsilon_y = 0$
Shear strain in a rectangular corner with sides along the x and y axes, $\gamma_{xy} = 160 \times 10^{-6}$ rad.
  a. By direct computation, determine the principal strains ε_1 and ε_2 and their directions at P.
  b. By direct computation, determine the maximum shear strain γ_{max} and the orientation of the rectangular corner at the given point P to which this maximum shear strain is referred.

9.13 The local state of plane strain of an element at point O of a structure is given by (see Figure P9.13): $\varepsilon_x = -100\ \mu$, $\varepsilon_y = -50\ \mu$, and $\gamma_{xy} = 80\ \mu$.
By direct calculation, determine
  a. The principal shear strains and the corresponding corners
  b. The maximum shear strain and the corresponding corner
  c. The normal strain along the edges of the corner corresponding to the maximum shear strain

9.14 A state of plane strain exists at a location P in a member with respect to a Cartesian coordinate system (right-handed) x–y–z. The normal strains along the x and y axes are found to be $\varepsilon_x = 150 \times 10^{-6}$ and $\varepsilon_y = -50 \times 10^{-6}$. For a right-angled corner at P with its sides along the +ve x-axis and +ve y-axis, the shear strain (the angle of shrink of the corner) is found to be $\gamma_{xy} = 150 \times 10^{-6}$ rad.
  a. Draw the Mohr's circle representing the state of strain at P.
  b. Determine the principal strains ε_1 and ε_2 and their directions at P.
  c. Determine the maximum in-plane shear strain γ_{max}, the right-angled corner with respect to which this strain in defined at P, and the normal strain ε_o along the corresponding edges.

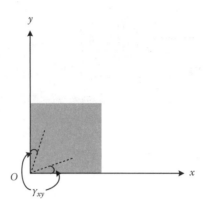

FIGURE P9.13 A state of plane strain.

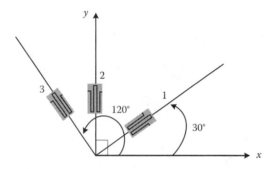

FIGURE P9.18 A strain-gauge rosette.

9.15 The state of strain of an infinitesimal element at point O, which was rectangular originally, is $\varepsilon_x = 800\ \mu$, $\varepsilon_y = -200\ \mu$, and $\gamma_{xy} = -600\ \mu$. Using the method of Mohr's circle, determine the principal strains at O and the corresponding rectangular element to which these strains are subjected.

9.16 The state of strain of an infinitesimal element at point O, which was rectangular originally, is given by $\varepsilon_x = -200\ \mu$, $\varepsilon_y = 600\ \mu$, and $\gamma_{xy} = -600\ \mu$, expressed with respect to the Cartesian coordinate frame xOy. Consider an infinitesimal rectangular element (un-deformed) at O with its corresponding sides along another Cartesian coordinate frame $x'Oy'$, which is obtained by rotating the frame xOy through $-15°$ (i.e., clock-wise). Using the method of Mohr's circle, determine the state of strain of this element.

9.17 The state of plane strain of an element at a point of a structure is given by $\varepsilon_x = 450\ \mu$, $\varepsilon_y = 150\ \mu$, and $\gamma_{xy} = 265\ \mu$.

Determine the principal strains ε_1 and ε_2, the maximum in-plane shear strain $\gamma_{\text{max-plane}}$, and the absolute maximum shear strain γ_{max}.

9.18 In a strain-gauge rosette, the gauges 1, 2, and 3 are oriented at $\theta = 30°$, $\theta_2 = 90°$, and $\theta_3 = 120°$ from the x-axis on a free surface, as shown in Figure P9.18. The readings of the strain gauges were found to be $\varepsilon_1 = 400\ \mu$, $\varepsilon_2 = -300\ \mu$, and $\varepsilon_3 = 200\ \mu$. Estimate the strains ε_x, ε_y, and γ_{xy} with respect to the Cartesian coordinate system x–y.

Appendix A: Geometric Properties of Planar Shapes

Here, G = centroid; A = area; I_a = second moment of area about axis "a"

1. Circle

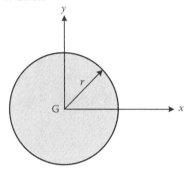

$$A = \pi r^2$$

$$I_x = I_y = \frac{\pi r^4}{4}$$

$$I_z = \frac{\pi r^4}{2}$$

2. Hollow circle

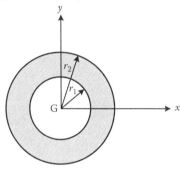

$$A = \pi \left(r_2^2 - r_1^2 \right)$$

$$I_x = I_y = \frac{\pi}{4} \left(r_2^4 - r_1^4 \right)$$

$$I_z = \frac{\pi}{2} \left(r_2^4 - r_1^4 \right)$$

3. Semicircle

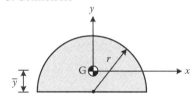

$$A = \frac{\pi}{2} r^2; \quad \bar{y} = \frac{4r}{3\pi}$$

$$I_x = \left(\frac{\pi}{8} - \frac{8}{9\pi} \right) r^4; \quad I_y = \frac{\pi r^4}{8}$$

4. Rectangle

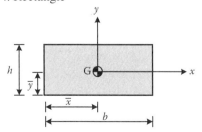

$$A = bh; \quad \bar{x} = \frac{b}{2}; \quad \bar{y} = \frac{h}{2}$$

$$I_x = \frac{bh^3}{12}$$

$$I_y = \frac{hb^3}{12}$$

5. Triangle

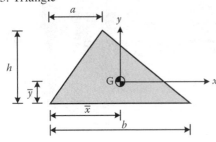

$$A = \frac{bh}{2}; \quad \bar{x} = \frac{a+b}{3}; \quad \bar{y} = \frac{h}{3}$$

$$I_x = \frac{bh^3}{36}$$

$$I_y = \frac{bh^3}{36}(a^2 - ab + b^2)$$

Appendix B: Deflections and Slopes of Beams in Bending

Appendix B: Deflections and Slopes of Beams in Bending

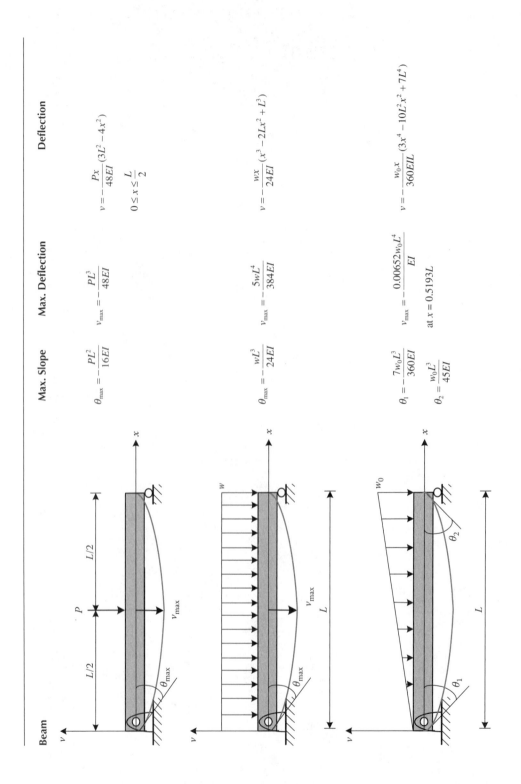

Beam	Max. Slope	Max. Deflection	Deflection
	$\theta_{max} = -\dfrac{PL^2}{16EI}$	$v_{max} = -\dfrac{PL^3}{48EI}$	$v = -\dfrac{Px}{48EI}(3L^2 - 4x^2)$ $0 \le x \le \dfrac{L}{2}$
	$\theta_{max} = -\dfrac{wL^3}{24EI}$	$v_{max} = -\dfrac{5wL^4}{384EI}$	$v = -\dfrac{wx}{24EI}(x^3 - 2Lx^2 + L^3)$
	$\theta_1 = -\dfrac{7w_0L^3}{360EI}$ $\theta_2 = \dfrac{w_0L^3}{45EI}$	$v_{max} = -\dfrac{0.00652 w_0 L^4}{EI}$ at $x = 0.5193L$	$v = -\dfrac{w_0 x}{360 EIL}(3x^4 - 10L^2x^2 + 7L^4)$

Appendix B: Deflections and Slopes of Beams in Bending

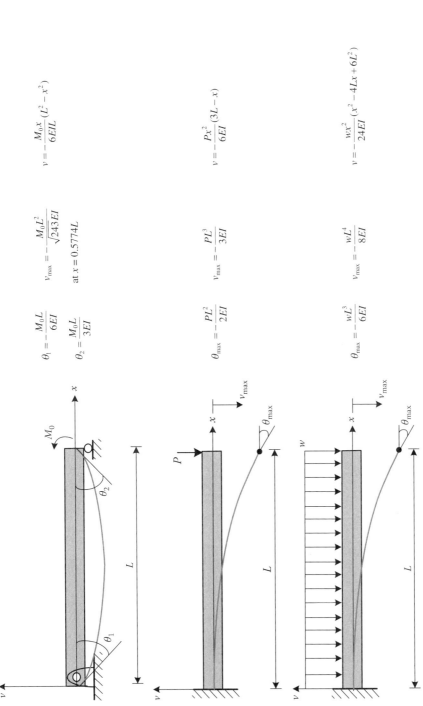

$\theta_1 = -\dfrac{M_0 L}{6EI}$

$\theta_2 = \dfrac{M_0 L}{3EI}$

$v_{\max} = -\dfrac{M_0 L^2}{\sqrt{243}EI}$

at $x = 0.5774L$

$v = -\dfrac{M_0 x}{6EIL}(L^2 - x^2)$

$\theta_{\max} = -\dfrac{PL^2}{2EI}$

$v_{\max} = -\dfrac{PL^3}{3EI}$

$v = -\dfrac{Px^2}{6EI}(3L - x)$

$\theta_{\max} = -\dfrac{wL^3}{6EI}$

$v_{\max} = -\dfrac{wL^4}{8EI}$

$v = -\dfrac{wx^2}{24EI}(x^2 - 4Lx + 6L^2)$

(continued)

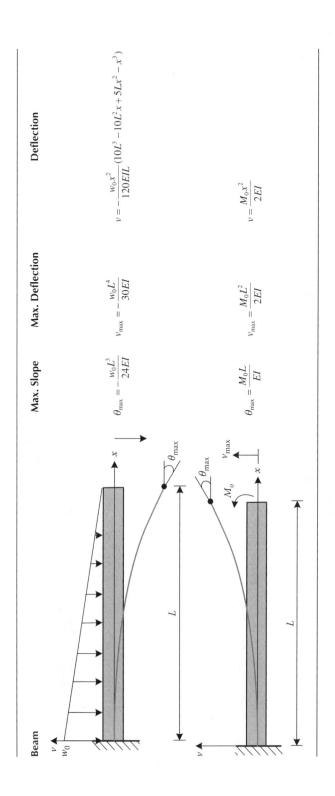

Appendix C: Buckling of Columns

Strength and deflection are two important considerations in the design of structures. Another important issue is stability. Buckling of columns is a type of structural stability.

C.1 BUCKLING

A column is a long and slender axial member subjected to a compressive axial force, as shown in Figure C.1.

Suppose that the compressive force P is gradually increased from zero. Initially, for small P, a slight lateral push will be quickly recovered, with the column returning to the original straight configuration. This corresponds to *stable equilibrium*. When P is increased to a specific, critical value P_{cr}, a slight lateral push will not be restored, and the column will remain in the deflected configuration. This corresponds to *neutral equilibrium*. For $P > P_{cr}$, a slight lateral deflection will continue to grow, causing buckling of the column. This corresponds to *unstable equilibrium*. An analogy of a ball rolling on a surface is shown in Figure C.2.

C.2 CRITICAL LOAD OF PIN-JOINTED COLUMN

In obtaining an analytical expression for the critical load, we make the following assumptions:

1. The member is long and lender.
2. The force is axial along the centroid of the cross-section.
3. The material is linear elastic.
4. The proportional limit is not exceeded.
5. The column bends in a single plane.

Consider a uniform column, one end of which is pinned while the other end moving axially on smooth rollers, as shown in Figure C.3a.

L = length
A = area of cross section
I = second moment of area about the neutral axis
E = Young's modulus

Suppose that the column is deflected under compressive force P.

Consider the segment of length x from the upper end of the column. Its free-body diagram is shown in Figure C.3b.

v = lateral deflection of the column axis (elastic curve) at x
Equilibrium: $M + Pv = 0$
Now we use the fact (see Chapter 8), $M = EI(d^2v/dx^2)$
We have $EI(d^2v/dx^2) + Pv = 0$
or

$$\frac{d^2v}{dx^2} + \lambda^2 v = 0 \tag{C.1}$$

FIGURE C.1 A column subjected to a compressive force.

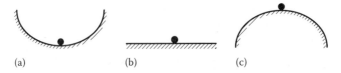

FIGURE C.2 (a) Stable equilibrium; (b) neutral equilibrium; and (c) unstable equilibrium.

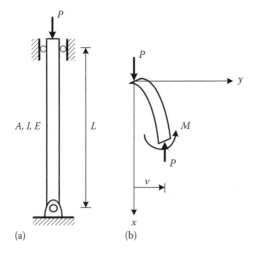

FIGURE C.3 (a) Pin-supported column and (b) segment of the deflected column.

where

$$\lambda = \sqrt{\frac{P}{EI}} \tag{C.2}$$

It is well known that the general solution of the homogeneous ordinary differential Equation C.1 is

$$v = c_1 \sin \lambda x + c_2 \cos \lambda x \tag{C.3}$$

Appendix C: Buckling of Columns

$$\text{Boundary conditions: } v = 0 \text{ at } x = 0 \to C_2 = 0$$
$$v = 0 \text{ at } x = L \to C_1 \sin \lambda L = 0$$

One possible solution for (C.3) is the trivial solution, $C_1 = 0$.

This corresponds to the undeflected, straight configuration of the column. We are interested in the deflected configuration, expressed as $\sin \lambda L = 0$ for any C_1.

The solutions are $\lambda L = n\pi$, $n = 1, 2, 3\ldots$

These solutions correspond to neutral equilibrium and hence the limiting (critical) load.

In particular, the lowest load corresponding to neutral equilibrium is expressed as $n = 1$ or $\lambda L = \pi$, and it gives the critical load.

Substitute in (C.2):

$$\frac{\pi}{L} = \sqrt{\frac{P_{cr}}{EI}}$$

We have

$$P_{cr} = \frac{\pi^2 EI}{L^2} \tag{C.4}$$

The corresponding deflected shape is expressed as

$$v = C_1 \sin \frac{\pi x}{L} \tag{C.5}$$

Here, C_1 is the amplitude of the lateral deflection.

Note: Critical load is also called the *Euler load*.

Observations:

1. Critical load decreases with L and increases with I and E.
2. Critical load does not depend on the strength of the column.

Critical stress,

$$\sigma_{cr} = \frac{P_{cr}}{A} = \frac{\pi^2 EI}{AL^2} \tag{C.6}$$

$$I = r^2 A \tag{C.7}$$

where r is the radius of gyration.

We have

$$\sigma_{cr} = \frac{\pi^2 E}{(L/r)^2} \tag{C.8}$$

Here, L/r is the slenderness ratio.

C.3 COLUMNS WITH GENERAL BOUNDARY CONDITIONS

Equation C.4 gives the critical load for a pin-supported column. For other boundary conditions, the critical load may be expressed as

$$P_{cr} = \frac{\pi^2 EI}{(KL)^2} \tag{C.9}$$

The corresponding critical stress is expressed as

$$\sigma_{cr} = \frac{\pi^2 E}{(KL/r)^2} \tag{C.10}$$

The values of K depend on the boundary conditions.
Some cases are given in Table C.1.
Effective length,

$$L_e = KL \tag{C.11}$$

This is the length of the pinned–pinned column that has the same critical load as the column of the given boundary conditions and length L.

C.4 YIELDING AND BUCKLING

If the yield stress $\sigma_Y < \sigma_{cr}$, which is the case for a column that is not slender, the column will yield before buckling.

If $\sigma_Y < \sigma_{cr}$, which is the case for a column that is slender, the column will buckle before yielding. The curve of σ_{cr} versus $(L/r)^2$ is hyperbolic.

Consider the curves of σ_Y and σ_{cr} as shown in Figure C.4. The yielding region and the buckling region are indicated. Note that the formula (C.10) for the critical stress is valid only in the elastic region (i.e., $\sigma_{cr} < \sigma_Y$). Specifically, it is not valid where $\sigma_{cr} > \sigma_Y$.

C.4.1 DESIGN FOR BUCKLING RESISTANCE

For a column of given material and for a given load P, yielding depends on the area of cross section A whereas buckling depends on the length L and the second moment of area I. It follows that, for a given area A, the buckling resistance can be improved by increasing the radius of gyration r. For example, a box section is better than a solid section of the same area.

TABLE C.1
Buckling Coefficients for Different Support Conditions

End Condition	Coefficient K
Pinned–pinned	1.0
Fixed–free	2.0
Fixed–fixed	0.5
Fixed–pinned	0.7

Appendix C: Buckling of Columns

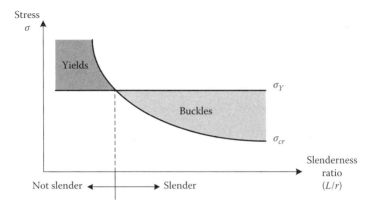

FIGURE C.4 Regions of buckling and yielding.

Appendix D: Advanced Topics

This appendix presents some advanced results in Mechanics of Materials and Theory of Elasticity, which will complement the basic material presented in the book.

D.1 GENERAL PROBLEM OF ELASTICITY

D.1.1 Strain Components

By definition, for deformations u, v, and w in the Cartesian directions x, y, and z, we have the corresponding direct strain (ε) and shear strain (γ) components:

$$\varepsilon_x = \frac{\partial u}{\partial x} \qquad \gamma_{xy} = \frac{\partial v}{\partial x} + \frac{\partial u}{\partial y} = \gamma_{yx}$$

$$\varepsilon_y = \frac{\partial v}{\partial y} \qquad \gamma_{yz} = \frac{\partial w}{\partial y} + \frac{\partial v}{\partial z} = \gamma_{zy} \qquad (D.1)$$

$$\varepsilon_z = \frac{\partial w}{\partial z} \qquad \gamma_{zx} = \frac{\partial u}{\partial z} + \frac{\partial w}{\partial x} = \gamma_{xz}$$

D.1.2 Constitutive Equations

These are stress (direct stress σ or shear stress τ) versus strain (direct strain ε or shear strain γ) equations.

Assumptions:

1. Linear elastic ⇒ stress–strain relations are first degree
2. Homogeneous ⇒ uniform material
3. Isotropic ⇒ material properties are independent of direction

We get

$$\varepsilon_x = \frac{1}{E}\left[\sigma_x - \nu(\sigma_y + \sigma_z)\right] \qquad \gamma_{xy} = \frac{2(1+\nu)}{E}\tau_{xy}$$

$$\varepsilon_y = \frac{1}{E}\left[\sigma_y - \nu(\sigma_z + \sigma_x)\right] \qquad \gamma_{yz} = \frac{2(1+\nu)}{E}\tau_{yz} \qquad (D.2)$$

$$\varepsilon_z = \frac{1}{E}\left[\sigma_z - \nu(\sigma_x + \sigma_y)\right] \qquad \gamma_{zx} = \frac{2(1+\nu)}{E}\tau_{zx}$$

where
  E is the Young's modulus (of elasticity)
  ν is the Poisson's ratio

In addition to these six constitutive equations, we have the following equations.

D.1.3 Equilibrium Equations

From equilibrium of a parallelepiped $\delta x \times \delta y \times \delta z$, we get

$$\sum_i \frac{\partial \sigma_{ij}}{\partial x_i} + X_j = 0 \tag{D.3}$$

We have three equilibrium equations, for $j = 1, 2, 3$.

Note: X_j is the body force in the j direction.

Now taking moment about central axis, canceling $\delta x \times \delta y \times \delta z$, and then neglecting terms of $O(\delta)$ (*Note*: Body forces and derivatives contain $O(\delta)$ multiples. Neglect them), we get

$$\sigma_{ij} = \sigma_{ji} \quad \text{(in the absence of body moments)} \tag{D.4}$$

This is indeed the "complementarity property" for shear strain.

D.1.4 Compatibility Equations

Compatibility means, in addition to continuity, there are no kinks (geometric singularities or discontinuities in higher derivatives). These are equations satisfied by the second derivatives of strains. They are obtained by double-differentiating (D.1) and eliminating the terms containing u, v, and w. There are six compatibility equations.

Principal stresses: Maximum or minimum direct stresses. On the corresponding planes, the shear stress will be zero.

Principal strains: Maximum or minimum direct strains. On the corresponding planes, the shear strains will be zero.

Note: For isotropic solids, principal planes of stress and strain coincide.

D.2 PLANE STRAIN PROBLEM

1. Strains in the z direction are zero.
2. Derivatives wrt z are zero; that is, all stresses are functions of x, y; all existing strains are functions of x, y; body forces are functions of x, y.
3. Displacements in the z direction are zero (this is governed by Item (1)).

Note: (2) and (3) → All properties are independent of z (not functions of z). Still, stresses in the z direction may not be zero.

D.2.1 Constitutive Equations

$$\varepsilon_x = \frac{1}{E}\left[\sigma_x - v(\sigma_y + \sigma_z)\right] \tag{D.5}$$

$$\varepsilon_y = \frac{1}{E}\left[\sigma_y - v(\sigma_z + \sigma_x)\right] \tag{D.6}$$

$$0 = \frac{1}{E}\left[\sigma_z - v(\sigma_x + \sigma_y)\right] \tag{D.7}$$

$$\gamma_{xy} = \frac{2(1+v)}{E}\tau_{xy} \tag{D.8}$$

Appendix D: Advanced Topics

$$\gamma_{yz} = \frac{2(1+v)}{E}\tau_{yz} \tag{D.9}$$

$$\gamma_{zx} = \frac{2(1+v)}{E}\tau_{zx} \tag{D.10}$$

Substitute (D.7) in (D.5) and (D.6):

$$\varepsilon_x = \frac{1}{E}\left[\sigma_x(1-v^2) - v\sigma_y(1+v)\right]$$

$$\varepsilon_y = \frac{1}{E}\left[\sigma_y(1-v^2) - v\sigma_x(1+v)\right]$$

D.2.2 Equilibrium Equations

$$\frac{\partial \sigma_x}{\partial x} + \frac{\partial \tau_{xy}}{\partial y} + X_x = 0 \tag{D.11}$$

$$\frac{\partial \tau_{xy}}{\partial x} + \frac{\partial \sigma_y}{\partial y} + X_y = 0 \tag{D.12}$$

$$X_z = 0 \tag{D.13}$$

Note: Compatibility equations are identically satisfied.

D.3 PLANE STRESS PROBLEM

1. Stresses in the z direction are zero.
2. Properties are independent of z (i.e., $\partial/\partial z = 0$).

D.3.1 Constitutive Equations

$$\varepsilon_x = \frac{1}{E}[\sigma_x - v\sigma_y] \tag{D.14}$$

$$\varepsilon_y = \frac{1}{E}[\sigma_y - v\sigma_x] \tag{D.15}$$

$$\varepsilon_z = -\frac{1}{E}v(\sigma_x + \sigma_y) \tag{D.16}$$

$$\gamma_{xy} = \frac{2(1+v)}{E}\tau_{xy} \tag{D.17}$$

$$\gamma_{yz} = \frac{2(1+v)}{E}\tau_{yz} = 0 \tag{D.18}$$

$$\gamma_{zx} = \frac{2(1+v)}{E}\tau_{zx} = 0 \quad (D.19)$$

D.3.2 Equilibrium Equations

$$\frac{\partial \sigma_x}{\partial x} + \frac{\partial \tau_{xy}}{\partial y} + X_x = 0 \quad (D.20)$$

$$\frac{\partial \tau_{xy}}{\partial x} + \frac{\partial \sigma_y}{\partial y} + X_y = 0 \quad (D.21)$$

$$X_z = 0 \quad (D.22)$$

Note: Unlike in the plane strain problems, compatibility equations are not identically satisfied, in general, in the plane stress problem.

D.3.3 Plane Stress Problem in Polar Coordinates

D.3.3.1 Strain Components

$$\begin{aligned}\varepsilon_r &= \frac{\partial u}{\partial r} \\ \varepsilon_\theta &= \frac{1}{r}\frac{\partial v}{\partial \theta} + \frac{u}{r} \\ \gamma_{r\theta} &= \frac{1}{r}\frac{\partial u}{\partial \theta} + \frac{\partial v}{\partial r} - \frac{v}{r}\end{aligned} \quad (D.23)$$

D.3.3.2 Constitutive Equations

$$\begin{aligned}\varepsilon_r &= \frac{1}{E}(\sigma_r - v\sigma_\theta) \\ \varepsilon_\theta &= \frac{1}{E}(\sigma_\theta - v\sigma_r) \\ \gamma_{r\theta} &= \frac{2(1+v)}{E}\tau_{r\theta}\end{aligned} \quad (D.24)$$

D.3.3.3 Equilibrium Equations

$$\begin{aligned}\frac{\partial \sigma_r}{\partial r} + \frac{1}{r}\frac{\partial \tau_{r\theta}}{\partial \theta} + \frac{\sigma_r - \sigma_\theta}{r} + F_r &= 0 \\ \frac{\partial \tau_{r\theta}}{\partial r} + \frac{1}{r}\frac{\partial \sigma_\theta}{\partial \theta} + \frac{2\tau_{r\theta}}{r} + F_\theta &= 0\end{aligned} \quad (D.25)$$

D.4 ROTATING MEMBERS

Rotates at angular speed ω.

D.4.1 Rotating Discs

The equilibrium equations reduce to (*Note*: axisymmetric member)

$$\frac{d\sigma_r}{dr} + \frac{\sigma_r - \sigma_\theta}{r} + \rho\omega^2 r = 0 \quad (\textit{Note: } \rho = \text{density in absolute units}) \tag{D.26}$$

Also,

$$\varepsilon_\theta = \frac{u}{r} = \frac{1}{E}(\sigma_\theta - \nu\sigma_r)$$

Hence,

$$\frac{du}{dr} = \frac{d}{dr}\frac{r}{E}(\sigma_\theta - \nu\sigma_r) = \frac{1}{E}\frac{d}{dr}\left[r(\sigma_\theta - \nu\sigma_r)\right]$$

Also,

$$\varepsilon_r = \frac{du}{dr} = \frac{1}{E}(\sigma_r - \nu\sigma_\theta)$$

Hence,

$$\sigma_r - \nu\sigma_\theta = \frac{d}{dr}\left[r(\sigma_\theta - \nu\sigma_r)\right] \tag{D.27}$$

Introduce a function ψ such that

$$\sigma_r = \frac{\psi}{r}$$
$$\sigma_\theta = \frac{d\psi}{dr} + \rho\omega^2 r^2 \tag{D.28}$$

Substitute (D.28). Then, (D.26) will be identically satisfied. But (D.27) becomes

$$\frac{d^2\psi}{dr^2} + \frac{1}{r}\frac{d\psi}{dr} - \frac{\psi}{r^2} = -(3+\nu)\rho\omega^2 r$$

or,

$$\frac{d}{dr}\left[\frac{1}{r}\frac{d}{dr}(r\psi)\right] = -(3+\nu)\rho\omega^2 r \tag{D.29}$$

Solution:

$$\psi = Ar + \frac{B}{r} - \frac{(3+\nu)}{8}\rho\omega^2 r^3 \tag{D.30}$$

From (D.28),

$$\sigma_r = A + \frac{B}{r^2} - \frac{(3+\nu)}{8}\rho\omega^2 r^2$$
$$\sigma_\theta = A - \frac{B}{r^2} - \frac{(1+3\nu)}{8}\rho\omega^2 r^2 \tag{D.31}$$

D.4.2 Rotating Thick Cylinders

With similar end conditions, it is assumed that plane sections remain plane, if taken sufficiently distant from ends. Hence, the strain in the direction of the axis is constant.

1. e_z = constant c.
2. By symmetry and similarity of sections remote from ends, all the shear stresses are absent in an element considered.
3. By symmetry, $F_\theta = 0$.
4. $F_r = \rho \omega^2 r$ in the case of rotating cylinder.

D.4.2.1 Strain Equations

$$e_r = \frac{du}{dr}$$
$$e_\theta = \frac{u}{r}$$
(D.32)

Hence,

$$\frac{d}{dr}(re_\theta) = e_r = r\frac{de_\theta}{dr} + e_\theta$$

or

$$\frac{de_\theta}{dr} = \frac{e_r - e_\theta}{r}$$

D.4.2.2 Stress–Strain (Constitutive) Relations

$$e_r = \frac{1}{E}\left[\sigma_r - \nu(\sigma_\theta + \sigma_z)\right]$$
$$e_\theta = \frac{1}{E}\left[\sigma_\theta - \nu(\sigma_r + \sigma_z)\right]$$
$$e_z = \frac{1}{E}\left[\sigma_z - \nu(\sigma_r + \sigma_\theta)\right]$$
(D.33a)

D.4.2.3 Equilibrium Equations

Consider the element shown in Figure D.1.

$$\left(\sigma_r + \frac{d\sigma_r}{dr} \cdot dr\right)(r+dr)d\theta - \sigma_r \cdot r \cdot d\theta - 2\sigma_\theta \cdot dr \frac{d\theta}{2} + \rho\omega^2 r r d\theta \cdot dr = 0$$

We get

$$\frac{d\sigma_r}{dr} + \frac{\sigma_r - \sigma_\theta}{r} + \rho\omega^2 r = 0$$
(D.34)

Appendix D: Advanced Topics

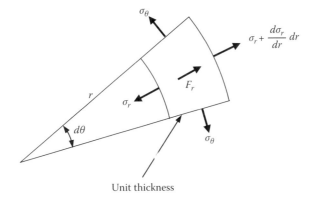

FIGURE D.1 An element of a rotating cylinder.

D.4.2.4 Final Result

$$\sigma_r = A + \frac{B}{r^2} - \frac{(3-2v)}{8(1-v)}\rho\omega^2 r^2$$

$$\sigma_\theta = A - \frac{B}{r^2} - \frac{(1+2v)}{8(1-v)}\rho\omega^2 r^2$$

(D.35a)

D.4.2.5 Thermal Strains and Stresses

Here, Equations D.34 and D.32 are the same. But (D.33a) takes the form

$$e_r = \frac{1}{E}\left[\sigma_r - v(\sigma_\theta + \sigma_z)\right] + \alpha T$$

$$e_\theta = \frac{1}{E}\left[\sigma_\theta - v(\sigma_r + \sigma_z)\right] + \alpha T$$

$$e_z = \frac{1}{E}\left[\sigma_z - v(\sigma_r + \sigma_\theta)\right] + \alpha T$$

(D.33b)

Note: α = coefficient of thermal expansion

D.4.3 Particular Cases of Cylinders

D.4.3.1 Case 1: Axially Restrained Ends

$$\sigma_r = A + \frac{B}{r^2}$$

$$\sigma_\theta = A - \frac{B}{r^2}$$

$$\sigma_z = EC + v(\sigma_r + \sigma_\theta) = \text{constant}$$

(D.35b)

Note: c = axial strain (in the z direction)

Axial force $P = \int_a^b \sigma_z \cdot 2\pi r \cdot dr = \sigma_z \int_a^b 2\pi r \cdot d = \pi(b^2 - a^2)\sigma_z$

Hence,

$$\sigma_z = \frac{P}{\pi(b^2 - a^2)}$$

D.4.3.2 Case 2: Thick Pressure Vessel

Internal radius $= a$
External radius $= b$
Internal pressure $= p$
Hence, end force exerted on the cylinder: $P = \pi a^2 p$

Boundary conditions:

$$\sigma_r = -p \text{ at } r = a \quad \text{and} \quad \sigma_r = 0 \text{ at } r = b$$

Hence,

$$\sigma_r = -\frac{pa^2}{(b^2 - a^2)}\left(\frac{b^2}{r^2} - 1\right)$$

$$\sigma_\theta = \frac{pa^2}{(b^2 - a^2)}\left(\frac{b^2}{r^2} + 1\right) \quad \text{(D.35c)}$$

$$\sigma_z = \frac{pa^2}{(b^2 - a^2)}$$

D.4.3.3 Case 3: Thin Pressure Vessel (Chapter 9)

Here $t/a \ll 1$, where thickness $t = (b - a)$

$$b^2 - a^2 \to 2 \text{ at } b - r = t' \quad \text{where } 0 < t' < t$$

Hence,

$$\sigma_r = -\frac{pt'}{t}$$

$$\sigma_\theta = \frac{pa}{t} \quad \text{(D.35d)}$$

$$\sigma_z = \frac{pa}{2t}$$

Note: σ_θ and σ_z are much larger than σ_r.

The same relations may be obtained by considering equilibrium of the sectioned cylinder.

D.4.3.4 Case 4: Rotating Cylinder with Free Ends

Boundary conditions: $\sigma_r = 0$ at $r = a$ and $r = b$
Hence, from (D.35),

$$A = \frac{(3 - 2v)}{8(1 - v)}(a^2 + b^2)\rho\omega^2$$

$$B = \frac{(3 - 2v)}{8(1 - v)}a^2 b^2 \rho\omega^2$$

Appendix D: Advanced Topics

$$\sigma_z = EC + 2\nu A - \frac{\nu\rho\omega^2 r^2}{2(1-\nu)} = K - \frac{\nu\rho\omega^2 r^2}{2(1-\nu)}$$

$$\text{End force} \quad P = \int_a^b \sigma_z \cdot 2\pi r \cdot dr = \int_a^b \left[K - \frac{\nu\rho\omega^2 r^2}{2(1-\nu)} \right] 2\pi r \cdot dr = 0$$

Hence,

$$K = \frac{\nu\rho\omega^2(a^2 + b^2)}{4(1-\nu)}$$

We get

$$\sigma_z = \frac{\nu}{4(1-\nu)} \left[1 + \frac{a^2}{b^2} - \frac{2r^2}{b^2} \right] \rho\omega^2 b^2$$

Note: $\sigma_{\theta\max}$ and $\sigma_{z\max}$ occur at the lowest value of r, that is, at $r = a$.

D.5 BEAMS IN BENDING

Consider a slender beam with internal bending moment M, internal shear force V, and transverse deflection v (of the elastic curve) at the longitudinal location x (see Chapter 8).

E = Young's modulus of elasticity
I = second moment area of the beam cross section about its neutral axis of bending

D.5.1 Mohr's Theorems

1. For a beam transversely loaded in any manner,

    slope at B – slope at $A = \dfrac{1}{EI}$ [area of bending moment diagram between AB]

2. $\dfrac{1}{EI}$ [moment about a line of bending moment diagram from A to B]

    = Intercept made on the line by the tangents to beam at A and B

D.5.2 Maxwell's Theorem of Reciprocity

Suppose that a force system P_i has corresponding deflection δ_i, and another force system P'_i has δ'_i. Then, $\sum P_i \delta'_i = \sum P'_i \delta_i$.

Note: True for moments as well, provided that slopes are used instead of deflections.

D.5.3 Castigliano's First Theorem

$$\frac{\partial U}{\partial P_r} = \delta_r; \quad \frac{\partial U}{\partial \delta_r} = P; \quad \frac{\partial U}{\partial M_r} = \theta_r; \quad \frac{\partial U}{\partial \theta_r} = M_r$$

D.5.4 Elastic Energy of Bending

For a small beam element of length ds and flexural rigidity EI, bent by moment M, the elastic energy of bending is

$$dU = \frac{1}{2} M \frac{ds}{r} = \frac{1}{2} \frac{M^2}{EI} ds \tag{D.36}$$

where r is the radius of curvature.

D.6 OPEN-COILED HELICAL SPRINGS

D.6.1 Case 1: Axial Load W

Consider element AB of a spring. The moments at A are shown in Figure D.2.
   The bending moment component tends to unwind the coil.
   The twisting couple component twists the wire inwards, causing deflection.

Bending:

$$d\phi = \frac{ds}{R} = \left(\frac{1}{R_2} - \frac{1}{R_1} \right) ds$$

Beam bending result:

$$\frac{M}{EI} = \frac{1}{R} = \left(\frac{1}{R_2} - \frac{1}{R_1} \right) = \frac{d\phi}{ds}$$

Hence,

$$d\phi = \frac{M}{EI} ds$$

Twisting:

$$d\theta = \frac{T}{GJ} ds$$

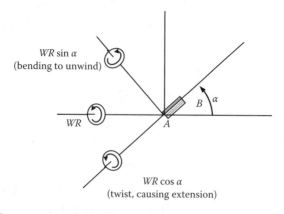

FIGURE D.2 Spring element under axial load.

Corresponding vertical deflection $d\Delta$ is expressed as

$$\frac{1}{2}wd\Delta = \frac{1}{2}Md\phi + \frac{1}{2}Td\theta = \frac{1}{2}\frac{M^2}{EI}ds + \frac{1}{2}\frac{T^2}{GJ}ds = \frac{1}{2}\frac{w^2R^2\sin^2\alpha}{EI}ds + \frac{1}{2}\frac{w^2R^2\cos^2\alpha}{GJ}ds$$

or

$$d\Delta = w^2R^2ds\left[\frac{\cos^2\alpha}{GJ} + \frac{\sin^2\alpha}{EI}\right]$$

For the full spring:

$$\Delta = w^2R^2L\left[\frac{\cos^2\alpha}{GJ} + \frac{\sin^2\alpha}{EI}\right] \quad (D.37)$$

where L is the length $= \dfrac{2\pi Rn}{\cos\alpha}$.

The axial winding due to $T = \dfrac{T}{GJ}ds\sin\alpha = \dfrac{wR}{GJ}\cos\alpha\sin\alpha \cdot ds$.

The axial unwinding due to $M = \dfrac{M}{EI}ds\cos\alpha = \dfrac{wR}{EI}\sin\alpha\cos\alpha \cdot ds$.

Overall winding $= \dfrac{wR}{GJ}\cos\alpha\sin\alpha \cdot ds - \dfrac{wR}{EI}\sin\alpha\cos\alpha \cdot ds$ in an element ds.

Overall winding:

$$\theta = wRL\sin\alpha\cos\alpha\left[\frac{1}{GJ} - \frac{1}{EI}\right] \quad (D.38)$$

D.6.2 Case 2: Axial Couple M

Consider the spring element shown in Figure D.3.
  Axial deflection is $d\theta$.
  We have

$$\frac{1}{2}Md\theta = \frac{1}{2}\frac{T'^2}{GJ}ds + \frac{1}{2}\frac{M'^2}{EI}ds = \frac{1}{2}\frac{M^2\sin^2\alpha}{GJ}ds + \frac{1}{2}\frac{M^2\cos^2\alpha}{EI}ds$$

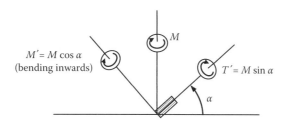

FIGURE D.3 Spring element under axial couple.

Hence,

$$d\theta = Mds\left[\frac{\sin^2\alpha}{GJ} + \frac{\cos^2\alpha}{EI}\right]$$

or

$$\theta = ML\left[\frac{\sin^2\alpha}{GJ} + \frac{\cos^2\alpha}{EI}\right] \tag{D.39}$$

Vertical deflection due to $T' = \dfrac{T'}{GJ}\cos\alpha \cdot R \cdot ds$ downward

Vertical deflection due to $M' = \dfrac{M'}{EI}\sin\alpha \cdot R \cdot ds$ upward

Net vertical deflection downward $d\Delta = \dfrac{T'}{GJ}\cos\alpha \cdot R \cdot ds - \dfrac{M'}{EI}\sin\alpha \cdot R \cdot ds$

$$d\Delta = \frac{MR\sin\alpha\cos\alpha}{GJ}\cdot ds - \frac{MR\sin\alpha\cos\alpha}{EI}\cdot ds$$

$$d\Delta = MR\sin\alpha\cos\alpha \cdot ds\left[\frac{1}{GJ} - \frac{1}{EI}\right]$$

or

$$\Delta = MRL\sin\alpha\cos\alpha\left[\frac{1}{GJ} - \frac{1}{EI}\right] \tag{D.40}$$

D.7 CIRCULAR PLATES WITH AXISYMMETRIC LOADING

Assumptions:

a. Material: homogeneous; isotropic; obeys Hooke's law.
b. Geometry: flat plates; deflections and slopes small; mid-plane is neutral plane; planes normal to mid-plane before deflection remain normal.
c. Stresses: work within elastic limit; normal forces and central shears are negligible.

By symmetry $\partial/\partial\theta = 0$, and $\gamma_{r\theta}$, $\tau_{r\theta}$, $M_{r\theta}$ are absent. From Figure D.4, deflection in the planar (x) direction at depth z from the neutral plane is $u = -z(dw/dr)$.

D.7.1 Strains

$$\varepsilon_r = \frac{du}{dr} = -z\frac{d^2w}{dr^2}$$

$$\varepsilon_\theta = \frac{u}{r} = -\frac{z}{r}\frac{dw}{dr} \tag{D.41}$$

Appendix D: Advanced Topics

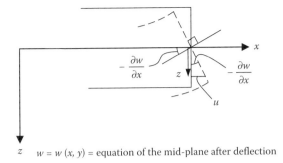

z $w = w(x, y) =$ equation of the mid-plane after deflection

FIGURE D.4 Element of plate in bending.

D.7.2 STRESSES

$$\sigma_r = \frac{E}{(1-v^2)}[\varepsilon_r + v\varepsilon_\theta] = -\frac{Ez}{(1-v^2)}\left[\frac{d^2w}{dr^2} + v\frac{1}{r}\frac{dw}{dr}\right]$$

$$\sigma_\theta = \frac{E}{(1-v^2)}[\varepsilon_\theta + v\varepsilon_r] = -\frac{Ez}{(1-v^2)}\left[\frac{1}{r}\frac{dw}{dr} + v\frac{d^2w}{dr^2}\right]$$
(D.42)

D.7.3 MOMENTS

Considering a unit width in its logical sense and not in its physical sense, we get moments per unit length at a particular point:

$$M_r = \int (\sigma_r \cdot 1 \cdot dz)z = -\frac{E}{(1-v^2)}\left[\frac{d^2w}{dr^2} + v\frac{1}{r}\frac{dw}{dr}\right]\int z^2 \cdot dz$$

$$M_\theta = \int (\sigma_\theta \cdot 1 \cdot dz)z = -\frac{Ez}{(1-v^2)}\left[\frac{1}{r}\frac{dw}{dr} + v\frac{d^2w}{dr^2}\right]\int z^2 \cdot dz$$

or

$$M_r = -D\left[\frac{d^2w}{dr^2} + v\frac{1}{r}\frac{dw}{dr}\right]$$

$$M_\theta = -D\left[\frac{1}{r}\frac{dw}{dr} + v\frac{d^2w}{dr^2}\right]$$
(D.43)

D.7.4 EQUILIBRIUM EQUATIONS

See Figure D.5.

Shear force balance: $-\left[Q_r + \frac{dQ_r}{dr}dr\right][r+dr]d\theta + Q_r \cdot r \cdot d\theta = q \cdot dr \cdot r \cdot d\theta$

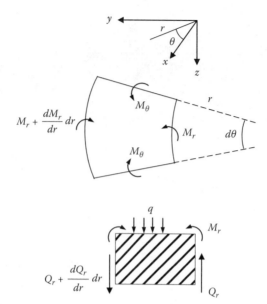

FIGURE D.5 Equilibrium condition of plate element.

Hence,

$$Q_r \cdot dr \cdot d\theta + r\frac{dQ_r}{dr} dr \cdot d\theta = q \cdot dr \cdot r \cdot d\theta$$

or

$$\frac{d(rQ_r)}{dr} = -qr \tag{D.44}$$

Moment balance: $\left[M_r + \dfrac{dM_r}{dr} dr \right][r+dr]d\theta - M_r \cdot r \cdot d\theta - Q_r \cdot r \cdot d\theta \cdot dr - M_\theta \cdot dr \cdot d\theta = 0$

or

$$M_r + r\frac{dM_r}{dr} - M_\theta = Q_r \cdot r$$

Substitute (D.43) and (D.44):

$$-D\left[\frac{d^2w}{dr^2} + \frac{v}{r}\frac{dw}{dr} + r\left(\frac{d^3w}{dr^3} + \frac{v}{r}\frac{d^2w}{dr^2} - \frac{v}{r^2}\frac{dw}{dr} \right) - \frac{1}{r}\frac{dw}{dr} - v\frac{d^2w}{dr^2} \right] = rQ_r$$

We get

$$\frac{d}{dr}\left[\frac{1}{r}\frac{d}{dr}\left(r\frac{dw}{dr} \right) \right] = -\frac{Q}{D} \tag{D.45}$$

where $Q_r = Q$

Substitute (D.44):

$$\frac{1}{r}\frac{d}{dr}\left\{r\frac{d}{dr}\left[\frac{1}{r}\frac{d}{dr}\left(r\frac{dw}{dr}\right)\right]\right\} = \frac{q}{D} \qquad (D.46)$$

D.7.5 BOUNDARY CONDITIONS

D.7.5.1 Fixed Edge

$w = 0$ at $r = a$

$$\frac{dw}{dr} = 0 \text{ at } r = a \text{ and } r = 0$$

D.7.5.2 Simply Supported Edge

$w = 0$ at $r = a$

$$M_r = -D\left[\frac{d^2w}{dr^2} + \frac{v}{r}\frac{dw}{dr}\right] = 0 \text{ at } r = a$$

$$\frac{dw}{dr} = 0 \text{ at } r = 0$$

D.7.5.3 Partially Restrained Edge

$$\frac{dw}{dr} = 0 \text{ at } r = a$$

Edge deflection of plate = deflection of edge beam.
Edge rotation of plate = rotation of edge beam.

Index

A

Absolute maximum shear strain in plane strain
    3-D case, 396
    in-plane Mohr's circle, 397–398
    principal strains, 396, 397, 407
Absolute maximum shear stress in plane stress
    analytical relations/Mohr's circle, 368
    Cartesian coordinate frame P-x-y-z, 368
    3-D transformation relations, 368
    Mohr's circle, 371–372
    principal stresses
        axis, 368, 369
        opposite signs, 370, 406
        same signs, 369, 406
        state of plane stress, 368–369
    state of stress, 370, 371
    structural member, 370, 371
Analysis of trusses
    landing gear, 26–28
    loading, 30
    method of joints, 24–25
    method of sections, 25
    structural testing systems, 29
    symmetric loadings, 30
    two-force members, 25–26
Angle of twist, torsion
    belt-drive unit, 247–248
    conical torsion member, 250–252
    deformation, 244
    description, 244–245, 269
    drilling, hard object, 252, 253
    factor of safety (FoS), 246
    free-body diagram
        drill-bit, 252, 253
        shaft, 247, 248
        wrench stem, 245–246
    internal torque diagram
        drill-bit, 253–254
        shaft, 247, 248
    lug wrench, 245, 246
    polar moment of area, 248
    steel and brass shaft segments, 249–250
Area method (graphical method), shear and moment diagrams, 288–290
Area removal method, 336
Average normal strain
    elemental segment, 106, 107
    extension, rod A, 103, 104
    geometry, beam movement, 103
    joint clearance, 103
    metal bar, 106
    quadratic deflection profile, 106, 108
    reinforcing steel wire, 102
    rigid beam, hanger rods, 103
    state, strain, 102
    strain distribution, 106
    two segments, unequal dimensions, 105
Average shear strain, 120
Axial force diagram
    cable-anchored deck, 66–68
    internal axial force, 66
Axial loading
    bending (flexural) loading, 180, 181, 216
    brass tube, 220–221
    bronze and steel hangers, 220
    cable and pulley shaft, capsule winch, 228
    ceiling and floor, steel post fitted, 226, 227
    collar, 231
    deflection, 221
    description, 8, 180, 216
    force-fitted metal shaft, axial load, 228–229
    hanger rods
        hinged rigid beam, 223, 224, 225
        liftguard deck, 225–226
        rigid beam, 218, 219
        stress $vs.$ strain curve, 218–220
    horizontal shafts, gap, 224–225
    mechanical structure, smooth joints, 229–230
    member, see Axial member
    normal stress under
        axial force diagram, 66–68
        definition, 59
        description, 87
        hanger rods, 62–63
        prismatic rod, 60
        rod, tensile loading, 61
        solution steps, 62–66
        tensile loading scenario, 60, 61
        water tank, 63–66
    PoS, see Principle of superposition (PoS)
    PVC rod, 217–219
    representation, 9
    rigid beam, 223–224, 230
    rigid ceiling
        frame hung, 217, 218
        peg-shaped steel member hangs, 217, 218
    sail post anchoring cable system, 222–223
    Saint-Venant's principle, 181–184
    shear loading, 180, 181, 216
    statically indeterminate structures, 200–204
    steel shaft, 231–232
    stepped flat bar, shoulder fillets, 226, 227
    stress concentrations, see Stress
    structural steel flat bar, 226, 227
    temperature changes, 181
    thermal effects, see Thermal effects
    torsional loading, 180, 181, 216
    two-segmented post, 222
Axial member
    application, 183
    cross sections, 184

member remains straight, 184
nonuniform section
    corresponding strain, 185
    cylindrical punching tool, 190–193
    end forces, 184–185
    and external load, 184
    internal force, 185
    length dx, small element, 184, 185
    normal stress, 185
    solid conical rod (frustum), 186–187
    water storage tank, 187–190
normal stresses and strains, 184
uniform cross section
    axial force diagram, shaft, 194–195
    composite member, 193
    deflection, 193
    and thrust bearings, 193–194

B

Beams
    bending in
        Castigliano's first theorem, 433
        clamped girder, end load, 343
        composite beams, see Composite beams
        concentrated and distributed loading, 339
        dimensions, cross-sectional, 344
        distributed force, see Distributed forces
        elastic energy of bending, 434
        end force, cantilever, 342, 345
        fabricated beam, cross section, 342
        flexure formula, see Flexure formula
        geometry, cross-sectional, 345
        graphical method, shear and moment diagram, 338
        hinge and smooth roller, support, 347
        lateral deflection, determination, 348
        maximum bending stress, determination, 339–341
        maximum shear force and location, determination, 344
        Maxwell's theorem of reciprocity, 433
        Mohr's theorems, 433
        reaction loading, 337
        shear and moment diagrams, 337–338; see also Shear and moment diagrams
        shear flow, determination, 346
        statically indeterminate beams, 332–334
        stresses and strains, 279
        transverse deflection, determination, 346
        transverse shear, see Transverse shear
        uniform cantilever beam, 344–345
    composite, see Composite beams
    deflection
        analogies, 321
        beam elastic curve, 319–320
        bending moment–deflection relation, 319–320
        boundary conditions, 321–322
        formulas, 336
        free-body diagram, 322–323
        by integration, 322
        lateral deflection, 325–327
        slope relation, 320–321, 415–418
        by superposition, 330–332
        two-segmented beam, elastic curve equation, 327–330
        uniformly distributed force, 322–323
    elastic curve, 319–320
    statically indeterminate, 332–334
Bearing stress
    bearing surface, 68, 88
    contact force, bolt and bar, 68–69
    elemental annular area, bearing surface, 69–70
    loaded post resting, 69, 70
    stress distribution, bearing surface, 69, 70
Bending loading, 8, 9
Bending moment–deflection relation, 319–320
Bending strain and bending stress, 298
Bending strain formula, 335
Bridge circuit
    active strain gauge, 117
    strain-gauge measurement, 116–117
Bridge constant
    description, 118, 120
    mounting configuration, 118
    strain gauges, bridge circuit, 118
Brittle material
    maximum normal stress failure theory, 403, 408
    Mohr's failure criterion, 403–404, 408
Brittle materials, 165
Buckling, 419

C

Calibration constant, 119
Cartesian sign convention, 350, 450
Circular plates, axisymmetric loading
    boundary conditions, 439
    equilibrium equations, 437–439
    moments, 437
    strains, 436–437
    stresses, 437
Circular shafts, torsion
    analysis, 235–236
    pump and stress, 235
    torque transmitting, 234–235
Columns, boundary conditions, 422
Composite beams
    bending stress profile, 307–308
    composite beam section, transformation, 307
    cross-sectional transformation, 307
    reinforced concrete beam, 310
    safety, overall factor, 309–311
    transformed section method, 306–307
Composite circular shaft
    angle of twist per unit length, 268, 269
    bimetallic shaft, 266
    homogeneous and isotropic materials, 266–267
    polar moments of area, 267–268
    shear stress variation, 268
Compressive strain, 100
Constitutive relations, 9
Coordinate-reversal method, 292–295
Coordinate system, 350, 450
Coordinate transformation, 292
Creep, 165
Critical load, pin-jointed column
    compressive force, 420

Index

deflected shape, expression, 421
equilibrium, 420
Cylinders; *see also* Rotating members
   axially restrained ends, 431–432
   rotating cylinder with free ends, 432–433
   thick pressure vessel, 432
   thin pressure vessel, 432
Cylindrical pressure vessels
   infinitesimal element, 374
   internal gauge pressure, 372, 374, 406
   normal stresses, 374
   wall element, 373–374

D

DC bridge (Wheatstone bridge), 120
Displacement and deformation, 119
Distributed forces
   beam segment, FBD, 36
   bending in beams
      cantilever, 339
      and concentrated force, 337
      and concentrated moment, 347
   center of pressure, location, 37–39
   distributed vertical load, baggage trolley, 39–44
   equilibrium configuration, beam, 36
   equivalent point load, 34
   FBD, 40
   hydrostatic force, magnitude, 37–39
   inclination angle determination, 35–37
   internal load, 35–37
   load, 34
   magnitude, 34
Distributed load, 46, 52, 53, 55
Ductile material
   maximum distortion energy failure theory, 401–403, 408
   maximum shear stress failure theory, 401, 408
Ductility, 164

E

Elastic curve, 336
Elasticity
   compatibility equations, 426
   constitutive equations, 425
   equilibrium equations, 426
   strain components, 425
Elastic limit, 164
Endurance limit, 165
Engineering stress, 136
Equilibrium equations, 46
Euler load, *see* Critical load, pin-jointed column

F

Factor of safety (FOS), 144–146
Failure theories
   brittle material, 403–404
   description, 400–401
   ductile material, 401–403
Fatigue failure, 158
FBD, *see* Free-body diagram (FBD)
Flexure analysis, 296–300

Flexure formula
   application, 300
   area addition method, 305
   area removal method, 300–301, 305–306
   assumptions, 296
   bending strain and bending stress, 298
   deformation, 296–297
   description, 336
   flexure analysis, 296–300
   linear variation, bending stress, 299
   maximum bending stress and location, 303–304
   neutral surface, 295–296
   parallel axis theorem, 300
   robotic hand finger, 301–303
FOS, *see* Factor of safety (FOS)
Fracture stress, 142, 164
Free-body diagram (FBD)
   analysis of trusses, 30
   barrel, 44, 45
   bar segment, 82
   beams, 36, 62–63, 141
   bolt, 75, 76
   column segment, 64–65
   definition, 7, 46
   deflection, beams, 322–323
   distributed forces, 40
   drill-bit, 252, 253
   engine post, 75
   internal tube segment, 73–74
   L-shaped component, 23
   materials mechanics, 7
   modified external loading, truss, 32
   mount base, 75, 76
   objectives, 22
   pulley, 79
      shafts, 247, 248, 255, 256
   shear and moment diagrams, 283–287
   spherical pressure vessels, 379, 380
   support reactions, 18–19
   transverse shear, 315
   trolley, 39, 40
   truss, 29, 30
   two beam segments, support reactions, 19–21
   wrench stem, 245–246

G

Gear ratio, 269

H

Hardness, 164
Hooke's law
   average shear stress, strain and modulus, 148–149
   bottom apex, peg determination, 146–147
   constitutive relation, 163–164
   FOS, 144–146
   load and deflection, 143
   physical law/constitutive equation, 133
   rigidity/shear modulus, 144
   shear modulus and needed bolts, 150–151
   shear strain, modulus and deformed shape, 147–148
   stress *vs.* strain, 143–144
Hoop stress, 374

I

Internal bending moment, 335
Internal shear force, 335

L

Landing gear, 26–28
Linear elastic systems, 216
Load *vs.* deflection diagram, 135
Local normal strain
    deformation element, 101–102
    description, 120
    uniaxial case, 100–101
Local shear strain
    definition, 108
    right-angled corner, 109
Longitudinal stress, 374–375

M

Materials mechanics
    aircraft
        loading, 2
        mid-air structural damage, 3
    bending moment, 6
    book organization, 10–12
    definition, 1–3
    engineering fields, 4
    force, 5
    history, 7–8
    homogeneous, 7
    isotropic, 7
    normal force, 5
    problem solution, 8–10
    scenarios, 8, 9
    shear force, 5
    strain, 7
    stress, 6
    subject application, 4
    torque, 6
Maximum in-plane shear strain
    corner segment, 391
    directions of principal strains, 391
    normal strain, 390–391
    and orientation, 391, 405
    $\pi/4$ and $\pi/2$, 390
Maximum in-plane shear stress
    angle, plane, 358–359
    beam cross section, 359–360
    compressive principal stress, 360, 361
    minimum, 358
    Mohr's circle
        analytical results, 364
        corresponding state of stress, 366, 379
        infinitesimal rectangular element, 364, 365
        normal stress, 364
        state of principal stress, 364–366
        torque and axial force, 364, 365
    normal stress, 360, 361
    principal stresses, 355, 359, 360, 361
    third orthogonal direction, 359
Maximum/minimum in-plane shear stress, 405

Maximum shear stress
    hollow shaft, 269
    lug wrench, 246
    solid shaft, 244
Mechanical properties, materials
    aluminum alloy, 177–178
    brittle materials, 157, 165
    carriage arm, machine, 173
    creep, 157, 165
    curve, stress *vs.* strain, 167–169
    cylindrical test specimen, tensile test, 169–171
    ductile material, 155–156, 164
    ductility measures, 156
    fatigue, 158, 159
    frame structure, steel members, 175–176
    hanging cable, parameters, 166–167
    hardness, 157
    Hooke's law, 143–151
    idealized stress *vs.* strain curve, 167
    interrelations, 135
    load–deflection data, 173–175
    load–deformation condition, 169
    metal rod, tensile load, 165–166
    Poisson's ratio, 151–155
    rectangular wooden pole, 171–172
    rubber band, 166
    slabs, polyurethane, 176–177
    strain energy, 158–165
    stress–strain behavior, *see* Stress–strain behavior
    tensile test, 169
    two-member truss, 176
Modulus of resilience, 165
Modulus of toughness, 165
Mohr's circle
    plane strain
        center, 393, 407
        2-D and 3-D case, 392
        description, 392, 393
        direct computation, 392
        maximum in-plane shear strain, 393–395
        radius, 393, 407
    plane stress
        direct calculation, 361–362
        maximum in-plane shear stress, 364–366
        normal stress and shear stress, 405–406
        principal stresses, 362, 363
        transformation result, 362–363
Moment diagram, 335

N

Necking, 136, 142, 164
Neutral axis, 295–296
Neutral surface, 295–296
Normal strain
    average, 102–108
    deformation per unit length, 119
    local, 100–102
Normal stress (s), 87

O

Open-coiled helical springs, 434–436

Index

P

Parallel axis theorem, 300, 336
Percent elongation, 164
Planar shapes, geometric properties, 413
Plane strain problem
    constitutive equations, 426–427
    equilibrium equations, 427
Plane-strain transformation
    definition, 382
    deformation
        element, 383, 384
        new element, 387, 388
        original element, 387
    direct computation, 386
    infinitesimal unstrained rectangular element, 386–387
    normal strains e_x and e_y, 382–383, 407
    orthogonal components, deformation, 384–386
    shear strain, 381–382, 383–384
    triangular element, 384
Plane stress problem
    constitutive equations, 427–428
    equilibrium equations, 428
    polar coordinates
        constitutive equations, 428
        equilibrium equations, 428
        strain components, 428
Plane-stress transformation
    description, 353
    distributed force, 354, 355
    equilibrium equations, 353–354
    infinitesimal element, 353, 354, 355
    local stress, 352–353
    rotated element, 355
    single plane, 352
Point load effect, 282–283
Poisson's ratio
    definition, 151
    linear elastic case, 152
    sleeve and rigid shaft, 154–155
    Young's and shear modulus determination, 152–153
Polar moment of area, 243, 269
PoS, *see* Principle of superposition (PoS)
Principal strains
    coordinate frame, 389
    deformation, original element, 388–389
    direction angles, 388
    in 3-D state, 396
    exaggerated, 389
    infinitesimal unstrained rectangular element, 388, 389
    Mohr's circle, *see* Mohr's circle
    orientation angles, 388, 407
    right-angled corner, 388
Principal stresses
    circular shaft in pure torsion, 356, 357
    compressive, 357, 358
    2-D and 3-D cases, 356
    definition, 355
    in 3-D state, 367
    measurement, 356
    Mohr's circle, *see* Mohr's circle
    and orientation, 405
    planes, 356
    state, plane stress, 357–358
    tensile, 357, 358, 408
Principle of superposition (PoS)
    applied force, 198–199
    applied to thermal problems
        application, 208–209
        axial force application, 210
        cause-effect relationship is linear, 208
        combined force and temperature rise, 210–211
        loading and temperature change, 209–210
    combined load–deflection condition, 197
    description, 195, 216
    force–deflection relationship, 198
    linear elastic systems, 195
    load–deflection condition, 196–197
    load–deformation relationship, 199–200
    load reversal, 196
    tensile force, 197–198
Proportional limit, 164

R

Residual stresses, 215, 216
Rotating members
    discs, 429
    thick cylinders
        equilibrium equations, 430–431
        result, 431
        strain equations, 430
        stress–strain (constitutive) relations, 430
        thermal strains and stresses, 431

S

Saint-Venant's principle
    average stress, 182
    deformed grid shape, 182, 183
    original axial member, 182, 183
    statically equivalent loading, concept, 182, 184, 216
    stress and strain conditions, 181
    stress profiles, 182, 183, 184
Shear and moment diagrams
    area method (graphical method), 288–290
    coordinate-reversal method, 292–295
    coordinate transformation, 292
    derivation steps, 280
    distributed and concentrated loading, two-segmented beam, 285–286
    distributed force and concentrated force, 293–295
    engineering application, 280
    free body diagram, 283–287
    governing relations, 281–282
    internal pin, two-segmented beam, 290–291
    moment curve, 291–292
    point load effect, 282–283
    reversal of direction of traverse, 292
    shear curve, 291
    sign convention, 281
Shear diagram, 335
Shear flow, 269, 316–319
Shear formula, 312–316
Shear loading, 8, 9

Shear strain
  average
    definition, 109
    deformation, flexible element, 110–111
    engine mount, 110
    hung triangular plate, 111–112
    load, plate deformation, 111–112
    plate deformation geometry, 112, 113
  definition, 108
  local, 108–109, 120
  torsion in shafts, 269
Shear stress
  average
    cantilever X-section, 71–72
    complementarity property, plane stress, 80–82, 88
    glued joint, tube, 73–74
    handle and key, FBD, 78
    handle connection, 78
    motor shaft and pulley, 79–80
    single and double shear connectors, 77
    tennis racquet, 72–73
    vibration mount, 75–76
  brittle material, 71
  ductile material, 71
  in object, 88
Shear thermal strains, 120
Sign convention
  bending moment, 281, 335
  shear strain, 239–240
  torque and shear stress in torsional member, 238–239, 269
Single and double shear, 88
Solid noncircular shafts
  material volume, 262
  rectangular section, parameter values, 261
  torsional properties, 261
  zero shear stress, 260–261
Spherical pressure vessels
  free-body segment, 379, 380
  inner radius and wall thickness, 379
  internal gauge pressure, 379
  Mohr's circle, 379, 380
  radial stress, 379
  and wall element, 379, 380
Statically indeterminate structures, axial loading
  compatibility method, 203
  description, 200
  structural integrity, 200
  superposition method, 203–204
  tightening, metal sleeve, 200–202
  two-segmented shaft, 202–203
Statically indeterminate systems, 46
Statically indeterminate torsional members
  circular shaft, steel and brass segments, 257–258
  compatibility conditions, 254, 269
  complete internal torque curve, 258, 260
  FBD, see Free-body diagram (FBD)
  gear ratio, 269
  internal torque diagram, 255–256
  preliminary internal torque curve, 258–260
  shafts, geared pair, 254–256
  step-down gear ratio, 255, 257
Statics
  analysis of trusses, see Analysis of trusses
  cantilever beam, peg support and end load, 53
  distributed forces, see Distributed forces
  distributed load, 52, 53, 55, 56
  equilibrium equation, 16–17, 46
  importance, 15
  light cantilever, 55, 56
  loading of truss, forces and reaction, 49, 50
  nonuniform arm, 52
  overhung traffic light, tension determination, 47–48
  pipe gripper, 49, 50
  reaction forces, determination, 47–48
  reaction load, 52
  statically indeterminate structures, 44–45
  steel plate, reaction and tension, 51–52
  support reactions, see Support reactions, statics
  tension determination, 47
  torquing arm, 54
  vertical force, determination, 48, 49
  water loading, hydrostatic force, 51
Strain
  belt drive, 121
  bronze and steel, 126, 127
  deformation, 99, 119
  energy
    cuboidal element, material, 158, 160
    description, 165
    incremental strain energy density, 158, 160
    infinitesimal change in strain, 158
    modulus of resilience, 161
    modulus of toughness, 162
    shear, 161
  fiberglass plate, 124, 125
  formulation
    angular movement, 241
    cross sections, plane, 236, 237
    element, length dz, 241
    homogeneous and isotropic material, 236
    linear variation, 242
    no longitudinal deformations, 236–238
    radii
      no change in length, 236, 238
      straight, 236, 237
    shaft in torsion, 236
    sign convention, 238–240
    stress state, 238, 239
    torsional deformation geometry, 240–241
  frame structure, 122, 123
  horizontal cable, 121
  indicator, temperature change, 131
  inflated leather soccer ball, 120–121
  materials mechanics, 99
  measurement
    bridge circuit, 116–117
    bridge constant, 118–119
    calibration constant, 119
    foil-type strain gauge, 115, 116
    properties, strain gauge materials, 116
    strain-gauge elements, 115, 116
    strain-gauge rosettes, 398, 407
    strain transformation in plane strain/stress, 398–399
    three-gauge rosette, 399–400
  multispan guideway, expansion slots, 126, 127
  nonuniform rod, variable strain, 123, 124
  normal, see Normal strain
  planar deformation, rectangular plate, 129, 130

Index

plastic plate sheared, bolt and nut, 124, 125
polymer plate, edge reinforcement, 124, 126
rigid concrete platform, 129–131
rigid platform, 121, 122
rubber slab and shear force, 124
shear, *see* Shear strain
slender shaft, threaded segment, 128, 129
steel plate hung, 131
tensioning rod, 127, 128
thermal, *see* Thermal strain
three-dimensional (3-D) state
    absolute maximum shear strain in plane strain, 396–398
    Cartesian coordinate frame, 395
    description, 395
    positive normal strains, 395
    principal strains, 396
    strain transformation, 395–396
transformation
    definition, 350, 380
    3-D state, 395–398
    general state, 381
    maximum in-plane shear strain, 390–391
    measurement, 398–400
    Mohr's circle, *see* Mohr's circle
    plane-strain transformation, *see* Plane-strain transformation
    plane-stress *vs.* plane-strain problems, 381
    principal strains, 388–389
    sign convention, 380–381
    state of plane strain, 411–412
    strain-gauge rosette, 412
two-bar truss, 128, 129
types, 100
uniform axial, rigid ceiling, 129, 130
uniform bar, rigid ceiling, 123
Strain-gauge rosette equations, 407
Strain hardening (work hardening)
    description, 164
    FDB, 141, 142
    rigid beam, hanger rod, 141
    ultimate stress (ultimate strength), 140
Strain sensing equation, 120
Strain transformation, 405
Strength of materials, 1
Stress
    aircraft, 57–58
    under axial loading, *see* Axial loading
    bearing, 68–71
    bolt-secured end plate, anchoring cable, 92, 93
    bonded joint, strips of wood, 94, 95
    concentrations
        brittle material, 211
        cast iron bar, 213–215
        curves and tables, 213
        definition, 212–213, 216
        deformation and stress distribution, 212
        description, 211, 216
        ductile material, 211
        factors, *see* Stress
        fatigue failure, 211
        flat bar, central circular hole, 213, 214
        homogeneous and isotropic material, uniform bar, 211, 212
        material properties, 213
        residual stresses, 215
        two-segmented flat bar, shoulder fillets, 213
        uniform axial loading, 211, 212
    definition, 58–59, 87
    distribution, 93–94
    formula, *see* Flexure formula
    formulation
        internal torque, 243–244
        linear elastic case, 242
        polar moment of area, 243
        stress–strain relationship, 242
    glued joint, 94
    grip pliers, 96, 97
    handle, sluice gate, 97, 98
    horizontal plate
        3-D view, 89, 90
        side sectional view, 89, 90
    magnitude and direction, 59
    materials mechanics, 57
    pipe pliers, 91
    planar truss, 89
    punching press, steel washers, 96, 97
    rectangular rod, 90–91
    riveted joint, two rectangular strips, 91, 92
    shear, *see* Shear stress
    single-bolted wood joint, 97, 98
    steel block, 96
    and strain, 57
    three-dimensional (3-D) state
        absolute maximum shear stress in plane stress, 368–372
        principal stresses, 367
        stress transformation, 367
        x–y–z Cartesian coordinate frame, 366–367
    transformation
        axially loaded bar, 82
        bar segment, FBD, 82
        beam in bending, 410
        billboard structure, 409
        cantilevered beam, 408–409
        cylindrical pressure vessel, 410–411
        definition, 349, 405
        3-D state, 366–372
        equations, 405
        failure, glued joint, 84–85
        general state, 352, 405
        maximum in-plane shear stress, 358–361, 405
        Mohr's circle, *see* Mohr's circle
        plane-stress problem, 352, 405
        plane-stress transformation, 352–355
        principal stresses, 355–358, 405
        shaft, torque and axial force, 408
        sign convention, 351–352
        specification, 351
        tensile member, 408
        thick beam and element, 410, 411
        thin-walled pressure vessels, 372–380
        truss member, 85–87
        variation, 83
U-jaw, second shaft and pin, 91, 92
uniform brass nail, 95
uniform shaft, axial loading, 89
vibration mount, 92, 93

Stress–strain behavior
  constitutive relation, 163–164
  ductile material, 137, 138
  elastic region, 137
  engineering stress *vs.* strain, 136
  logic, offset method, 139
  necking, 142
  nominal strain, 136
  offset method, yield strength, 139
  proportional limit, 137–138
  strain hardening (work hardening), 140–142
  tensile test, 134–136
  test specimen, 136, 137
  true stress–strain diagram, 142–143
  yielding, 138–139
Stress transformation, 88, 405
Support reactions, statics
  body forces, 18
  FBD, 18–19
  internal loading, 21–22, 48, 49
  observations, 20
  surface forces, 18
  transmissibility principle, 22–24
  two beam segments, FBD, 19–21

T

Tensile strain, 100
Thermal effects
  beam, 206–208
  bolt and average determination, 205–206
  compatibility conditions, 205, 216
  expansions (+ve) and contractions (–ve), 204–205, 216
  PoS, *see* Principle of superposition (PoS)
  thermal strain, 205, 216
Thermal strain, 216
  coefficient, thermal expansion (a), 114–115
  description, 114
  frame structure, 126, 128
  shear, 120
Thin-walled pressure vessels
  absolute maximum shear stress
    action plane, 375, 376
    broken-line Mohr's circle, 375
    cylindrical pressure vessel, 377, 378
    design parameters, 377
    gas shock absorber, 376–377
    maximum in-plane shear stress, 375
    Mohr's circle, 375–376
    physical domain, 375
    rivet in double shear, 378–379
  cylindrical pressure vessels, 372–374
  hoop stress, 374
  longitudinal stress, 374–375
  shell structures, 372, 373
  spherical pressure vessels, 379–380
  stress analysis, 372
Thin-walled tubes
  circular and square *X*-sections, 265
  description, 263
  shear stress relation
    axial element, 263, 264
    cross section, 263
    enclosed area element, 263, 264
    equilibrium equation, 264
    shear center, 264
    shear flow, 264
Torque, 269
Torsional loading, 8, 9
Torsion in shafts
  angle of twist, 244–254
  circular hollow shaft, 270
  circular shafts, 234–236
  composite shafts, 266–269
  description, 233
  direct-drive robot arm, torque sensor, 271–272
  drilling
    concrete block, 273, 274
    hard material, 273
  fixed ends and distributed torque, 276
  frustum-shaped hollow shaft, end torques, 274
  gear unit, 275
  gear wheel pair, 276–277
  load, belt-drive and chain-drive, 272–273
  machine tool, drive motor, 271
  motor through step-down gear, 275
  shaft, solid segment and hollow segment, 270
  solid circular shaft, 270
  solid conical member, 277
  solid noncircular shafts, 260–263
  statically indeterminate torsional members, 254–260
  stepped shaft, external torques, 272
  strain formulation, 236–242
  stress formulation, 242–244
  thin-walled tubes, 263–265
  torque, 233–234, 269
  tube, external torque, 276
Transmissibility principle, 46
Transverse shear
  clamped uniform beam, 315
  determination, 314–316
  free-body diagram, 315
  shear flow, 316–319
  shear formula, 312–316
  shear stress and complementary shear stress, 312
Two-force member, 46
Two-force members, 25–26

U

Ultimate stress, 140

W

Water tank
  column segment, FBD, 64, 65
  overhead tank and support column, 63–64
Work hardening, 140

Y

Yielding and buckling
  regions, 423
  resistance design, 422
Yield point, 164
Yield stress (yield strength), 164